AF411220

Sustainable Circular Bioeconomy

Editors

Manoj Kumar Nallapaneni
Md Ariful Haque
Sarif Patwary

Basel • Beijing • Wuhan • Barcelona • Belgrade • Novi Sad • Cluj • Manchester

Editors

Manoj Kumar Nallapaneni
School of Energy and
Environment
City University of Hong Kong
Kowloon Tong
Hong Kong

Md Ariful Haque
Department of Fermentation
Science
Middle Tennessee State
University
Murfreesboro
United States

Sarif Patwary
Human Development and
Consumer Sciences
University of Houston
Houston
United States

Editorial Office
MDPI
St. Alban-Anlage 66
4052 Basel, Switzerland

This is a reprint of articles from the Special Issue published online in the open access journal *Sustainability* (ISSN 2071-1050) (available at: www.mdpi.com/journal/sustainability/special_issues/ circular_bioeconomy_sust).

For citation purposes, cite each article independently as indicated on the article page online and as indicated below:

Lastname, A.A.; Lastname, B.B. Article Title. *Journal Name* **Year**, *Volume Number*, Page Range.

ISBN 978-3-0365-8679-3 (Hbk)
ISBN 978-3-0365-8678-6 (PDF)
doi.org/10.3390/books978-3-0365-8678-6

Contents

About the Editors

Manoj Kumar Nallapaneni

Dr. Nallapaneni is a transdisciplinary energy and sustainability engineer with a PhD in Digital Circular Economy and Circular Power System from the School of Energy and Environment, City University of Hong Kong. He has obtained two Masters degreees, one in Renewable Energy Technologies from Karunya University, India, and the other in Environmental Economics from Annamalai University, India. He holds a Bachelor's degree in Electrical and Electronics Engineering from GITAM University. Before joining the CityU, He worked as a Research Fellow at Universiti Malaysia Pahang, Malaysia, on a project which focused on using solar photovoltaics as urban and rural infrastructure. Earlier to this, he worked as an Assistant Professor in the Department of Electrical and Electronics Engineering at the Bharat Institute of Engineering and Technology, Hyderabad, India, and Energy Engineer at Atiode Solar Systems Limited, Benin City, Nigeria.

Dr. Nallapaneni's research works focus on the topics of simulation, experimental, real-time empirical and location, or climate-specific studies mainly focused on building sustainable and resilient systems (decentralized, networked, and centralized) across critical infrastructure sectors by adopting nexus thinking and systems innovation with an emphasis on circular business models and digitalization. He worked on key sustainability challenges that include design, performance modeling, and analysis of a wide range of clean energy and environmental systems, food–energy–water–waste nexus, industrial symbiosis, waste valorization and material passports, carbon accounting, and pricing. He possess an interdisciplinary skill set that includes performance analytics, techno-economics (spreadsheet and tools), life cycle assessment (emission factor/embodied energy approach and LCA tools), resilience assessment (simple and systemic), leveraging digital innovation (blockchain, IoT, smart contracts, and AI), business model innovation, and nexus systems design with better conceptualization skills.

Md Ariful Haque

Md Ariful Haque (Ph.D.) is an academician, researcher, entrepreneur, and author, currently working as a adjunct faculty and researcher at Middle Tennessee State University. With a diverse educational background, Dr. Ariful earned his bachelor's degree in Agricultural Science from Bangladesh Agricultural University, followed by a master's degree in Industrial Biotechnology from Walailak University in Thailand. He furthered his expertise with a doctoral degree in energy and environment from the City University of Hong Kong.

Dr. Ariful's professional pursuits center around his fervent interest in fermentation, probiotics, and biotransformation. Through his teaching and research, he explores the intricate processes of converting various feedstocks into valuable biopolymers and platform chemicals. This work holds far-reaching applications, spanning industries such as textiles, medical, packaging, and food production. Dr. Ariful is dedicated to translating his innovative ideas into tangible solutions that can revolutionize industries and improve lives. Multifaceted approach coupled with commitment to sustainable practices, Dr. Ariful plays a positive role in both scholarly and practical spheres. In essence, Dr. Md Ariful Haque's journey exemplifies the synergy between academia, research, and entrepreneurship. His academic endeavors and interdisciplinary focus highlight his dedication to creating a more sustainable and innovative world, making him a notable figure in shaping the future of scientific advancement.

Sarif Patwary

With a PhD in fashion sustainability and postdoctoral research experience, Dr. Sarif Patwary has a deep understanding of the environmental and social impacts of fashion production and a track record of developing strategies to reduce these impacts. Currently, he works as a Sr. Sustainability Analyst at Kontoor Brands, Inc., and is dedicated to driving sustainability initiatives within two of its iconic consumer brands: Wrangler®and Lee®.

His expertise lies in conducting sustainability assessments, analyzing data, and developing evidence-based recommendations for improving sustainability performance. He has a proven ability to communicate complex sustainability concepts to diverse audiences, including senior leadership, employees, and stakeholders. He is a proactive and solution-oriented team player, with the ability to work effectively with cross-functional teams to drive change. He is excited to contribute his expertise and passion for the development of a more sustainable fashion industry to this project.

Preface

Welcome to "Sustainable Circular Bioeconomy", a comprehensive exploration of one of the most pressing topics of our time. As we stand at the forefront of the 21st century, it has become increasingly evident that we must shift our focus towards sustainable practices that harmonize with our environment and promote long-term wellbeing.

The concept of a circular bioeconomy encapsulates this urgent need for change. It represents a paradigm shift where we harness the power of nature and leverage the potential of biologically derived resources to create a regenerative and sustainable economic system. This reprint delves deep into the principles, practices, and possibilities of the sustainable circular bioeconomy. Within these pages, you will find a wealth of knowledge and insights contributed by experts, researchers, and practitioners who have dedicated their lives to advancing the borders of sustainability. Each chapter of this reprint is carefully crafted to provide a holistic view of the subject, covering a wide range of topics, including renewable energy, waste management, food–energy–water nexus, digital innovation in sustainability, aquaculture, agriculture, biotechnology, and more.

Our aim in creating this reprint was to inspire and educate readers from diverse backgrounds—academics, policymakers, industry professionals, and concerned citizens—about the immense potential and transformative power of the sustainable circular bioeconomy. Through the integration of biological and technological innovations, we can create a resilient and regenerative future that not only meets our present needs but also ensures the wellbeing of future generations. For the effective realization of this future, we also advocate for focusing on the circularity of bioeconomy from a sustainability and resilience point of view.

We express our deepest gratitude to the esteemed contributors who have shared their expertise and insights to this reprint through the call for papers that we announced in December 2021. Their dedication and passion for sustainability have enriched its content and made this endeavor possible. Our special thanks are expressed to Nichole Wang, Section Managing Editor of *Sustainability*. We also extend our gratitude to the readers who embark on this journey with us, for it is through your engagement and commitment to change that we can collectively shape a better world. As you immerse yourself in the pages ahead, we invite you to open your mind to new possibilities, challenge conventional thinking, and embrace the transformative potential of the sustainable circular bioeconomy by applying the *RePLiCATE Approach* that is suggested within the Editorial of this reprint. Together, let us embark on a path towards a more harmonious, regenerative, and circular future that is sustainable and resilient with boundless hope and determination.

Manoj Kumar Nallapaneni, Md Ariful Haque, and Sarif Patwary
Editors

 *sustainability*

Editorial

It Is Time to Synergize the Circularity of Circular Bioeconomy with Sustainability and Resiliency Principles

Manoj Kumar Nallapaneni [1,2,3,*], Md Ariful Haque [1,4] and Sarif Patwary [5,6]

1   School of Energy and Environment, City University of Hong Kong, Kowloon, Hong Kong
2   Center for Circular Supplies, HICCER—Hariterde International Council of Circular Economy Research, Palakkad 678631, Kerala, India
3   Swiss School of Business and Management Geneva, Avenue des Morgines 12, 1213 Geneva, Switzerland
4   Department of Fermentation Science, Middle Tennessee State University, 1301 E Main Street, Murfreesboro, TN 37132, USA; mhaque@mtsu.edu
5   Human Development and Consumer Sciences, University of Houston, Houston, TX 77204, USA; patwary@ksu.edu
6   Kontoor Brands, Inc., 1250 Revolution Mills Dr Ste 300, Greensboro, NC 27405, USA
*   Correspondence: mnallapan2-c@my.cityu.edu.hk

Citation: Nallapaneni, M.K.; Haque, M.A.; Patwary, S. It Is Time to Synergize the Circularity of Circular Bioeconomy with Sustainability and Resiliency Principles. *Sustainability* **2023**, *15*, 12239. https://doi.org/10.3390/su151612239

Received: 3 August 2023
Revised: 9 August 2023
Accepted: 9 August 2023
Published: 10 August 2023

1. Bioeconomy and Its Circularity

Bioeconomy mainly refers to an economic system based on the sustainable production, conversion, and utilization of biological resources, such as crops, forests, fish, and microorganisms, to produce food, feed, energy, and other products. Following this production, conversion, and utilization principle, the bioeconomy can start replacing most finite and non-renewable resources with renewable and biologically derived resources [1]. Biofuels, bioplastics, bio-based chemicals, bio-based textiles, bio-based fertilizers, organic waste bioremediation, and others are some notable examples of bioeconomy. With recent developments and significant advancements in technologies and processes that aid bioeconomy, bioeconomy is expected to be increasingly important, especially in addressing global challenges such as waste management, climate change, food security, and sustainable development [2]. Additionally, the recent progress in bioeconomy concepts, especially from a sustainable development standpoint of view and the application of a lifecycle perspective, suggest that bioeconomy could promote the transition of the current linear economy model to a more sustainable one called circular economy, leading to the emergence of *circular bioeconomy*, under which circularity is more focused [3].

Bioeconomy's circularity refers to keeping resources in use for as long as possible, extracting their maximum value, and then recovering and regenerating them at the end of their useful life [3]. This includes practices such as recycling, reusing, and remanufacturing products and materials, as well as the restoration of degraded ecosystems following the principles of various *circular economy business models* [4]. One such example of a circular bioeconomy considering the flow of biomass is shown in Figure 1 [5]. Overall, *circular bioeconomy* aims to create a more circular and equitable society combining economic, social, and environmental goals by recognizing the interconnectedness of human well-being, the natural environment, and the economy and seeking solutions that benefit all three.

Now to understand how research solutions are being proposed to benefit human well-being, the natural environment, and the economy, we opened a Special Issue titled *"Sustainable Circular Bioeconomy"* calling for contributions (https://www.mdpi.com/journal/sustainability/special_issues/circular_bioeconomy_sust, accessed on 3 August 2023). The response was positive with a wide range of contributions, based on which we (the editors) carried out a discussion (Section 2.1), paving the way for a fresh research agenda in the field of circular bioeconomy, i.e., *"synergizing the bioeconomy's circularity with sustainability and resiliency"* (Section 2.2). Recommendations on how to synergize and ways to perform this are provided in Section 3.

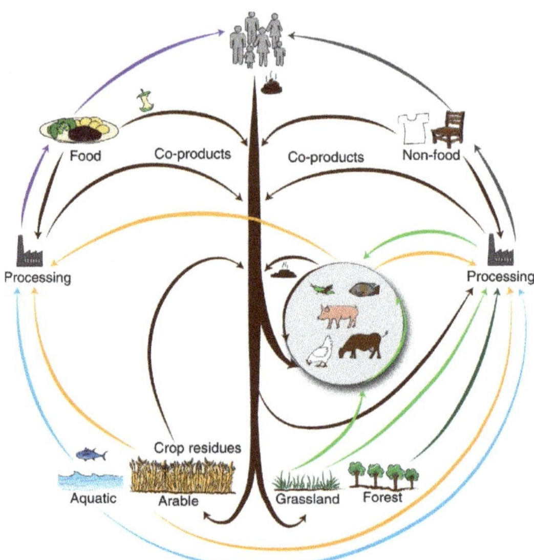

Figure 1. Circular bioeconomy illustration with biomass flows. Adopted from [5] and reprinted with permission from Springer Nature.

2. "Sustainable Circular Bioeconomy" Special Issue in MDPI Sustainability

2.1. Discussion on the Special Issue Contributions

The contributions published in the *"Sustainable Circular Bioeconomy"* Special Issue are varied in their subject fields but broadly fall under the bigger umbrella of circular bioeconomy [6]. The first published article (https://www.mdpi.com/2071-1050/14/1/466, accessed on 26 July 2023) highlights the use of technology to integrate the planning and stakeholder phases with the social, economic, technological, and environmental phases. The focus of key technologies are the Internet of Things (IoT), smart energy grids, GPS tracking systems, and blockchain. The authors have shown how these technologies promote a transition to sustainable progress in the bioeconomy field. The second published article (https://www.mdpi.com/2071-1050/14/2/994, accessed on 26 July 2023) showed the application of agro-lignocellulosic waste as a substrate for producing oyster mushrooms, where the authors just focused on the production process and testing of the mushrooms for their quality. The third published article (https://www.mdpi.com/2071-1050/14/3/1161, accessed on 26 July 2023) is about distiller-dried grains with a soluble diet as a substitute for standard corn-soybean for swine production in the United States of America, where the authors formulated the diet and modeled the life cycle assessment to assess the sustainability of the diet. In the fourth published article (https://www.mdpi.com/2071-1050/14/3/1897, accessed on 26 July 2023), the authors study biomass self-sufficiency status for European member states to meet the European Green Deals by 2050. It is mentioned that most European member states are biomass self-sufficient, but the resilience of such sufficiency relies on the ecological boundary. In the fifth published article (https://www.mdpi.com/2071-1050/14/4/2044, accessed on 26 July 2023), the authors focused on the use of technology to identify situations like aqua farmers involved in constructing illegal fishponds by taking Kolleru Lake in Andhra Pradesh, India as a case study. In the sixth published article (https://www.mdpi.com/2071-1050/14/4/2281, accessed on 26 July 2023), the authors depict the NTFP-based bioeconomic prospect in Kashmir, India. The authors identified that a lack of proper information on the extraction, consumption, and traded quantities of NTFPs in Kashmir, India were significant drawbacks in quantifying the NTFP's contribution to the bioeconomy, suggesting the need for a better decision support system, infrastructure, and regulation to aid bioeconomic prospects. In the seventh published article (https://www.mdpi.com/2071-1050/14/5/3126, accessed on 26 July 2023), the authors detail the environmentally friendly extraction and precipitation process of phenolics and a waste valorization technique of Cocoa Bean Shells to promote bioeconomy. However, the authors did not mention much about the scalability, further process optimization approach, and the process's lifecycle and techno-economic feasibility assessment. In the eighth published article (https://www.mdpi.com/2071-1050/14/6/3369, accessed on 26 July 2023), the authors present the impact of biofuel crop expansion on other crops in the GDP of Thailand. The authors used a computable general equilibrium (CGE) model combined with a life cycle impact assessment. As per the authors, although biofuel promotion could promote Green GDP, policymakers should emphasize the prevention of and the transformation of forests to agricultural land. Without technological advancements, expanding biofuel crops for alternative energy would not ensure efficient resource utilization and prevent environmental degradation. The ninth published article (https://www.mdpi.com/2071-1050/14/15/9678, accessed on 26 July 2023) focused on the accountability of sustainability, where the authors have suggested a model for achieving sustainability in agro-industrial companies. Their model combines the principles and goals of the water-energy-food nexus with existing business excellence models. This model can assist companies in making decisions and managing tradeoffs and synergies as they strive to become more sustainable. In the tenth published article (https://www.mdpi.com/2071-1050/14/16/10299, accessed on 26 July 2023), authors pursued sustainability initiatives after realizing how sustainable progress could emerge in the Brazilian Amazon. This study found a range of new seeds of change; however, more needs to be conducted to support transformation toward sustainable and equitable development in the region. In the eleventh published article (https://www.mdpi.com/2071-1050/14/20/13686, accessed on 26 July 2023), authors from CAZRI in Jodhpur, India, HICCER in Palakkad, India, propose a solar photovoltaic winnower cum-dryer for drying Phoenix dactylifera L. fruits by keeping the socio-economic status of farmers. They performed a techno-economic assessment, showing a high internal rate of return and a shorter payback period. They also showed that their design could serve multiple functions apart from drying, for instance, the effective winnower operation even without natural wind. In the twelfth published article (https://www.mdpi.com/2071-1050/15/1/656, accessed on 26 July 2023), the authors believe that having a scientific understanding of the apparel life cycle among apparel consumers is very important. So with that hypothesis, they investigated three research questions: what is the current norm of clothing acquisition, maintenance, and disposal behavior? What is apparel consumer clothing acquisition, maintenance, and disposal behavior circular-driven? What is a sustainable way of clothing acquisition, maintenance, and disposal? They provided a circular economy lens framework that could serve as new guidelines for consumers to exercise mindful clothing consumption. In the thirteenth published article (https://www.mdpi.com/2071-1050/15/2/1634, accessed on 26 July 2023), the authors suggest using a mixed method approach to assess the implementation and priority level of internationally defined bioeconomy objectives in Latvian policy planning documents. This study found that these objectives were highly prioritized, especially in higher-level policy planning records.

2.2. Need for Synergizing Circularity of Bioeconomy with Sustainability and Resiliency

Considering the case of biomass flows in a circular bioeconomy, as shown in Figure 1 [5], the effectiveness of a circular bioeconomy can only be defined when such flows are aligned with sustainability and resiliency principles. Only then can our societies and ecosystems thrive over the long term in a sustainable development path. Now the question is, what are sustainability and resiliency? How do they matter in this context?

Sustainability refers to the ability to meet the needs of the present without compromising the ability of future generations to meet their own needs. This requires the responsible use and management of natural resources, the efficient use of technology and processes that reduce or limit environmental impacts, and the promotion of social and economic equity. Based on Figure 1 [5], the circular bioeconomy of biomass flows might be sustainable only when the most energy-efficient and less material-intensive processes are used (materials that have lower environmental impacts); at the same time, all waste coming from bioproducts or co-products are managed effectively through appropriate waste management technology.

Resiliency, on the other hand, refers to the ability of systems to bounce back from disturbances and adapt to changing conditions. In the face of ongoing environmental and social challenges such as climate change, resource depletion, and social inequality, it is critical to prioritize sustainability and resiliency in order to build a more equitable and sustainable future for all; however, building a circular bioeconomy aligned with sustainability and resiliency principles might be the best option. Based on Figure 1 [5], the circular bioeconomy of biomass flows may be resilient only when all stakeholders are prepared for disturbances and take appropriate actions to recover from them and ensure the learning of adaptation mechanisms.

Overall, to ensure that circular bioeconomy initiatives are truly sustainable and resilient, they need to be integrated with a wide range of principles from a systems innovation approach [6]. This requires a holistic approach that takes into account the interrelated nature of these factors and involves collaboration and coordination across sectors and stakeholders. However, whether these are actually happening is the question. Suppose we see the discussed circular bioeconomy concepts from the Special Issue in Section 2.1 and other articles published elsewhere; we can observe that almost all the studies lack this systems approach. More or less, most studies are limited to one set of analyses, which may or may not provide sufficient insights into whether the circular bioeconomy initiative is sustainable and has the potential to become a resilient solution. Therefore, it is a needed approach that we look into all three aspects (circularity, sustainability, and resiliency) when we propose a circular bioeconomy initiative. By synergizing circularity with sustainability and resiliency, the bioeconomy can become a powerful tool for achieving long-term, equitable, and environmentally sound development.

3. How to Realize the Circularity of Circular Bioeconomy with Sustainability and Resiliency

Realizing the circularity of circular bioeconomy with sustainability and resiliency requires a comprehensive, integrated approach that considers the interrelated nature of performance (that is otherwise called technical considering resilience and reliability), social, economic, and environmental factors, for instance, the *RePLiCATE (Resilience Performance, Life Cycle Analysis and Techno-Economics) approach* as proposed in Ref. [7], see Figure 2. While pursuing the above said *RePLiCATE* for designing circular bioeconomy concepts, a few other aspects can also be followed which may further provide deeper insights on circular bioeconomy implementation; see Box 1.

By adopting these points, we can realize a circularity of circular bioeconomy with sustainability and resiliency and build a more equitable and sustainable future for all. By embracing circularity, we can minimize waste and pollution and promote the more sustainable use of resources. By promoting sustainability and resiliency, we can create a system that is not only environmentally sustainable but also resilient to external shocks, ensuring that we can meet our needs both now and in the future, which is in line with the developmental goals of sustainability [11]. Thus, synergizing sustainability and resiliency with circularity is a critical need of the hour that needs to be given the utmost importance while developing circular bioeconomy concepts and business initiatives.

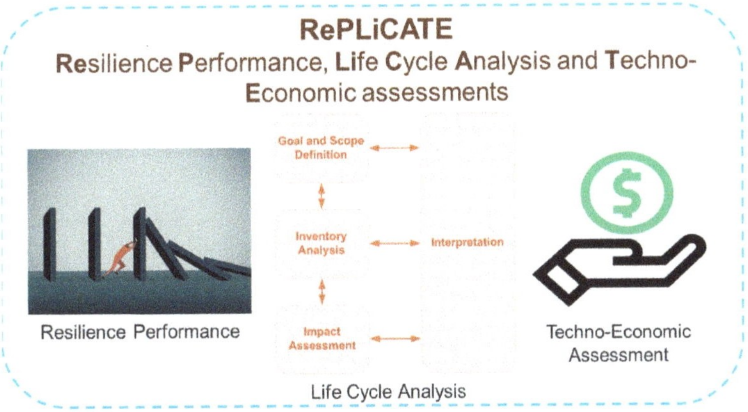

Figure 2. RePLiCATE approach to realize the circularity of circular bioeconomy in line with sustainability and resiliency principles. Adopted from authors own sources [4,7].

Box 1. Points that can help to achieve circular bioeconomy's circularity aligned with sustainability and resiliency principles.

Adopt a Circular Design in a Circular Bioeconomy: This involves designing products and processes to maximize the use of resources and minimize waste. This is only possible if circular design concepts are followed while designing [8].
Design that follows the pre-conditions of RePLiCATE framework: This involves analyzing the products and processes following the *RePLiCATE* framework (Figure 2) pre-conditions such as what should be considered to make a process safe to fail, product a tamper proof, sustainability data quality, recycling technologies and others that are applicable as per circular economy principles and energy efficiency standards, detailed economic parameters considering both the present and future markets and so on to assess circularity, sustainability, resilience, and economic viability [7].
Promote resource efficiency: This involves reducing the use of non-renewable resources and minimizing waste generation by adopting sustainable production and consumption practices [4,9].
Digitalization: This involves the digitalization of circular bioeconomy value chain stakeholders; upon doing so, there is a high possibility for tracing material flows and their characteristics, allowing us to evaluate sustainability and resiliency issues with near accuracy [8–10].
Foster social equity: This involves ensuring that circular bioeconomy initiatives are inclusive and benefit all members of society. This can be achieved through stakeholder engagement, participatory decision-making, and the promotion of fair labor practices [4,9].
Protect biodiversity: This involves minimizing the impact of circular bioeconomy initiatives on ecosystems and promoting the conservation of biodiversity [1,2,8].
Build resilience: This involves designing circular bioeconomy initiatives that are adaptable to changing environmental and social conditions and can withstand disturbances, such as natural disasters and economic shocks [4,7,9].
Foster collaboration: This involves working with multiple stakeholders across sectors and disciplines to ensure that circular bioeconomy initiatives are designed and implemented in a way that maximizes benefits across social, economic, and environmental dimensions [4,9].

Author Contributions: Conceptualization, M.K.N.; writing—original draft preparation, M.K.N. and M.A.H.; writing—review and editing, M.K.N., M.A.H. and S.P.; visualization, M.K.N.; supervision, M.K.N. All authors have read and agreed to the published version of the manuscript.

Acknowledgments: The Special Issue Editors would like to take the opportunity to thank the authors who responded to the call and the MDPI *Sustainability* Journal for approving the topic and allowing us to be in the guest editor roles. We are also deeply indebted to the reviewers whose input was indispensable in selecting the published papers.

Conflicts of Interest: The authors declare no conflict of interest.

References

1. Skydan, O.V.; Yaremova, M.I.; Tarasovych, L.V.; Dankevych, V.Y.; Kutsmus, N.M. Possibilities of Developing Sustainable World by Introducing Bioeconomy: Global Perspective. *Probl. Ekorozwoju* **2022**, *17*, 162–170. [CrossRef]
2. Hetemäki, L.; Kangas, J. Forest Bioeconomy, Climate Change and Managing the Change. In *Forest Bioeconomy and Climate Change*; Springer International Publishing: Cham, Switzerland, 2022; pp. 1–17.
3. Tan, E.C.; Lamers, P. Circular bioeconomy concepts—A perspective. *Front. Sustain.* **2021**, *2*, 701509. [CrossRef]
4. Nallapaneni, M.K. Leveraging Blockchain and Smart Contract Technology for Sustainability and Resilience of Circular Economy Business Models. Ph.D. Thesis, City University of Hong Kong, Hong Kong, China, 2022. Available online: https://scholars.cityu.edu.hk/en/theses/theses(aaaeaed0-d69b-42cb-bb6c-e8f83889815e).html (accessed on 3 August 2023).
5. Muscat, A.; de Olde, E.M.; Ripoll-Bosch, R.; Van Zanten, H.H.; Metze, T.A.; Termeer, C.J.; van Ittersum, M.K.; de Boer, I.J. Principles, drivers and opportunities of a circular bioeconomy. *Nat. Food* **2021**, *2*, 561–566. [CrossRef]
6. D'Adamo, I.; Gastaldi, M.; Morone, P.; Rosa, P.; Sassanelli, C.; Settembre-Blundo, D.; Shen, Y. Bioeconomy of Sustainability: Drivers, Opportunities and Policy Implications. *Sustainability* **2022**, *14*, 200. [CrossRef]
7. Nallapaneni, M.K. RePLiCATE: A framework for sustainable and resilient planning of advanced clean energy systems. In *HKICE Young Scientist Seminar Series Hong Kong Institute for Clean Energy*; City University of Hong Kong: Hong Kong, China, 2022; HKICE_ YS20221125; p. 1.
8. D'Amico, G.; Szopik-Depczyńska, K.; Beltramo, R.; D'Adamo, I.; Ioppolo, G. Smart and Sustainable Bioeconomy Platform: A New Approach towards Sustainability. *Sustainability* **2022**, *14*, 466. [CrossRef]
9. Nallapaneni, M.K.; Chopra, S.S. Sustainability and resilience of circular economy business models based on digital ledger technologies. In Proceedings of the Waste Management and Valorisation for a Sustainable Future, Seoul, Republic of Korea, 26–28 October 2021.
10. Nallapaneni, M.K.; Chopra, S.S. Blockchain-based Artificial Intelligence of Things Nutrient-Rich Food Waste Selection Framework for Food Waste-derived Medical Textiles. In Proceedings of the International Conference on Solid Waste 2023: Waste Management in Circular Economy and Climate Resilience, Institute of Bioresource and Agriculture (IBA) of the Hong Kong Baptist University Hong Kong Convention and Exhibition Centre, Hong Kong, China, 31 May–3 June 2023.
11. Ali, S.M.; Appolloni, A.; Cavallaro, F.; D'Adamo, I.; Di Vaio, A.; Ferella, F.; Gastaldi, M.; Ikram, M.; Kumar, N.M.; Martin, M.A.; et al. Development Goals towards Sustainability. *Sustainability* **2023**, *15*, 9443. [CrossRef]

 *sustainability*

Article

Smart and Sustainable Bioeconomy Platform: A New Approach towards Sustainability

Gaspare D'Amico [1,*], Katarzyna Szopik-Depczyńska [2], Riccardo Beltramo [3], Idiano D'Adamo [4] and Giuseppe Ioppolo [1]

1 Department of Economics, University of Messina, Via dei Verdi 75, 98122 Messina, Italy; giuseppe.ioppolo@unime.it
2 Department of Corporate Management, Institute of Management, University of Szczecin, Cukrowa 8, 71004 Szczecin, Poland; katarzyna.szopik-depczynska@usz.edu.pl
3 Department of Management, University of Turin, Corso Unione Sovietica 218 bis, 10134 Turin, Italy; riccardo.beltramo@unito.it
4 Department of Computer, Control and Management Engineering, Sapienza University of Rome, Via Ariosto 25, 00185 Rome, Italy; idiano.dadamo@uniroma1.it
* Correspondence: gaspare.damico@unime.it

Abstract: The smart and sustainable bioeconomy represents a comprehensive perspective, in which economic, social, environmental, and technological dimensions are considered simultaneously in the planning, monitoring, evaluating, and redefining of processes and operations. In this context of profound transformation driven by rapid urbanization and digitalization, participatory and interactive strategies and practices have become fundamental to support policymakers, entrepreneurs, and citizens in the transition towards a smart and sustainable bioeconomy. This approach is applied by numerous countries around the world in order to redefine their strategy of sustainable and technology-assisted development. Specifically, real-time monitoring stations, sensors, Internet of Things (IoT), smart grids, GPS tracking systems, and Blockchain aim to develop and strengthen the quality and efficiency of the circularity of economic, social, and environmental resources. In this sense, this study proposes a systematic review of the literature of smart and sustainable bioeconomy strategies and practices implemented worldwide in order to develop a platform capable of integrating holistically the following phases: (1) planning and stakeholder management; (2) identification of social, economic, environmental, and technological dimensions; and (3) goals. The results of this analysis emphasise an innovative and under-treated perspective, further stimulating knowledge in the theoretical and managerial debate on the smart and sustainable aspects of the bioeconomy, which mainly concern the following: (a) the proactive involvement of stakeholders in planning; (b) the improvement of efficiency and quality of economic, social, environmental, and technological flows; and (c) the reinforcement of the integration between smartness and sustainability.

Keywords: bioeconomy; digitalization; platform-based bioeconomy; smart and sustainable development; governance

Citation: D'Amico, G.; Szopik-Depczyńska, K.; Beltramo, R.; D'Adamo, I.; Ioppolo, G. Smart and Sustainable Bioeconomy Platform: A New Approach towards Sustainability. *Sustainability* 2022, 14, 466. https://doi.org/10.3390/su14010466

Academic Editor: Julio Berbel

Received: 2 December 2021
Accepted: 30 December 2021
Published: 2 January 2022

Publisher's Note: MDPI stays neutral with regard to jurisdictional claims in published maps and institutional affiliations.

1. Introduction

Current rates of urbanization and industrialization generate a wide range of issues that affect bioeconomy, such as waste recycling [1,2], energy conservation [3], water dissipation [4], traffic congestion [5], social disparities [6], healthcare emergencies [7], loss of biodiversity [8], land utilization difficulties [9], atmospheric and acoustic pollution [10], infrastructure and facilities obsolescence [11,12], food valorisation [13], forest management [14], safety and cyber-security [15,16], sustainable economic development [17], and so on. Consequently, the social, economic, and environmental challenges that emerge from urbanization and industrialization and the opportunities to address these issues more adequately through technology place the bioeconomy at the centre of the academic and managerial debate, as it plays a crucial role in supporting a smart and sustainable transition [18–22].

According to the summit [23], the term smart and sustainable bioeconomy used in this article refers to a centre of "production, utilization, conservation and transformation of biological resources which—through digital technologies—aim to provide real time and continuous data and information that contribute to improve the circularity and efficiency of waste, water, energy, agriculture, health, education, mobility, telecommunications, and governance." In fact, bioeconomy strategies and practices are at the centre of several international frameworks, such as the report on "challenges, visions and ways forward of the cities of the future" implemented by [24], the study on "new perspectives on urbanization of cities in the world" [25], and the 2030 UN Agenda for Sustainable Development [26]. In [27,28], authors identified a wide range of bioeconomy strategies and practices in line with diverse goals and targets of the UN 2030 Agenda on Sustainable Development, such as food security (Goals 1 and 2), water quality (Goal 6), energy efficiency (Goal 7), inclusive economic development (Goal 8), waste prevention and reuse (Goal 12), and prevention of life below water and on land (Goals 14 and 15). Likewise, the report [29] recommends policymakers, urban planners, and managers to strengthen "the sustainable management of resources, facilitating ecosystem conservation, regeneration, restoration and resilience in the face of new and emerging challenges".

Therefore, policymakers, entrepreneurs, and citizens are required to rethink the bioeconomy paradigm through the implementation of smart and sustainable initiatives in order to optimize the social, economic, and environmental processes and operations [30–33]. To this end, it is essential to enhance the understanding and awareness of bioeconomic flows and reorient their circulation in order to support a forward-looking and dynamic vision of the bioeconomy as the engine of smart and sustainable solutions [34–37].

The transition towards a smart and sustainable bioeconomy strives to address through a data-based approach the ever-increasing quantity of renewable biological resources, such as plant resources, agri-food production, forests, marine and livestock resources, microorganisms, algae, as well as waste, by-products and wastewater of agro-industrial origin, and the consequent congestion, in order to reduce the anthropogenic pressure on built and natural settlements [38–41]. Specifically, this new form of smart and sustainable bioeconomy, through the utilization of digital platforms and dashboards [18,42,43], holistically combines a wide range of information and communication technologies (ICTs), such as sensors [44,45], real-time monitoring stations [46], cameras [47], GPS tracking systems [48], big-data analysis techniques [49], artificial intelligence [50], augmented reality [51], blockchain [52], Internet of Things (IoT) [53], cloud computing [54], smart grids [55], satellites [56], nanotechnologies [57], advanced biotechnologies [58], and drones [59].

The information and communication technologies (ICTs) listed above ensure a real time and fully transparent authentication, traceability, treatment, analysis and evaluation of data, and information on bioeconomic resources from source to customer while providing agility, security, and efficiency along the production and distribution processes [60,61]. Consequently, the pervasive and intensive dissemination of fixed and mobile digital devices is revolutionizing the circularity of raw materials and secondary raw materials, by-products, chemicals, biofuels, bioplastics, urban and industrial waste, and wastewater, generating a wide and diversified range of data and information useful for policymakers, planners, managers, agricultural entrepreneurs, scientists, growers, logistics companies, biorefinery workers, chemical and technological companies, etc., able to optimize the use of natural and non-natural resources and improve the quality of their interactions [62,63]. In this regard, information and communication technologies (ICTs) allow a proactive and holistic approach capable of improving the mechanization and commercialization of practices along the bioeconomy chain through innovations, such as precision farming [64], precision livestock [65], sustainable packaging [66], and industry 4.0 [67].

Conversely, the lack of detailed and real-time data and information determines a wide range of uncertainties related, for example, to the timing of procurement, production, distribution and transformation, quality, location, and consumption, which leads to an under-optimization of bioeconomic flows [68]. Hence, the connectivity, variety, proximity, flexibility, coordination capacity, diversity, foresight, interdependence, collaboration, adaptability, creativity, efficiency, agility, self-organization, robustness, and resourcefulness of bioeconomy data and information provided by fixed and mobile digital equipment are therefore essential not only to minimize economic, social, and environmental costs but also as tools to refurbish other dimensions, such as mobility, telecommunications, health, education, and safety.

According to the "Future transitions for the Bioeconomy into Sustainable Development and a Climate-Neutral Economy report," elaborated by the European Commission's Knowledge Centre for Bioeconomy, the bioeconomy employs around 17.5 million people (almost 9% of its workforce), generating 614 billion euros of added value (about 5% of its GDP). Furthermore, if we include the tertiary bioeconomy sector based on digital services, which amounts to 872 billion euros, we reach an overall European extent of the bioeconomy of 1.5 trillion euros (almost 10% of its GDP) [69].

This growing importance of bioeconomy has provided a wide range of strategies and practices at the global level [23]. In this sense, the development of bioeconomy policies has become increasingly complex and varied [70]. In general, the strategies and practices of the bioeconomy tend to differ on the basis of factors related for example to technological advances [71], the availability of natural resources [72], cultural and institutional progress [73], and the development of the economic system [74]. In Germany, for example, the bioeconomy is clearly recognized as an inter-sectoral concept and refers not only to biological resources but also embraces social aspects, such as multi-level governance, stakeholders' management, and people empowerment [75,76]. Otherwise, Japan prioritizes the "biotechnological" vision, emphasizing the role of digital technologies such as Big Data, Artificial Intelligence, and Internet of Things [77]. On the other hand, the United States focuses more on the safety and security aspects, such as the cyber protection from biological threats, the development of biotechnologies for military use, and the preservation of sensitive infrastructure and biological data [78]. Given their significant and varied availability of biological resources, Costa Rica [79] and South Africa [80] embrace sustainable bioprospecting practices for scientific and commercial purposes, while Thailand [81], Italy [82], and Nordic Council of Ministers [83] are committed to the conservation of biodiversity and natural ecosystems for tourism activities.

Therefore, the current theoretical and managerial discussion is increasingly focused on the role of biotechnology, nanotechnology, and information and communication technologies (ICTs) in the bioeconomy [84]. In this regard, Costa Rica [79] coined the term "advanced bioeconomy" in order to highlight the importance of digitalization in improving the circularity of natural and non-natural resources. At the same time, the intensive and pervasive use of digital technologies in the bioeconomy highlights a wide range of challenges and issues within the social, economic, and environmental spheres [85].

On the basis of reports, master plans, and documents elaborated by governments, ministries, departments, agencies, and research centres, the following article provides a detailed overview of the bioeconomy strategies and practices implemented globally in order to develop a multidimensional platform able to holistically integrate the phases that characterize the smart and sustainable bioeconomy decision-making process. In summary, the proposed overview aims to explain the different approaches developed at a global level and to strengthen the understanding of (a) planning and stakeholder management; (b) identification of the social, economic, environmental, and technological dimensions; and (c) setting of the goals to be pursued. To do this, an in-depth analysis was conducted on a wide range of countries, such as South Africa, Costa Rica, USA, Japan, Malaysia, Thailand, Austria, Finland, France, Germany, Ireland, Italy, and so on, which have developed a plethora of smart and sustainable bioeconomy initiatives in order to improve the planning, collection, monitoring, and analysis of economic, social, environmental, and technological flows. In this sense, by identifying and analysing a wide range of smart and sustainable strategies and practices, the study fills a gap in the theoretical and managerial literature of the bioeconomy, providing a further piece in the debate between policymakers, entrepreneurs, scientists, planners, and citizens.

Hence, the study is structured in this manner: Section 2 outlines the methodology. Section 3 proposes the smart and sustainable bioeconomy platform, identifying and analysing the phases that characterize the smart and sustainable bioeconomy at a global level. Section 4 offers considerations on the role of smart and sustainable bioeconomy in future challenges. Finally, Section 5 provides the conclusions.

2. Materials and Methods

This paper provides a systematic review of the literature of strategies and practices implemented worldwide in order to develop a multidimensional platform capable of analysing and integrating the phases that characterize the smart and sustainable bioeconomy decision-making process. To do this, a wide range of globally implemented bioeconomy strategies and practices was investigated. The research took place in three different phases: identification, operational, and results (see Figure 1).

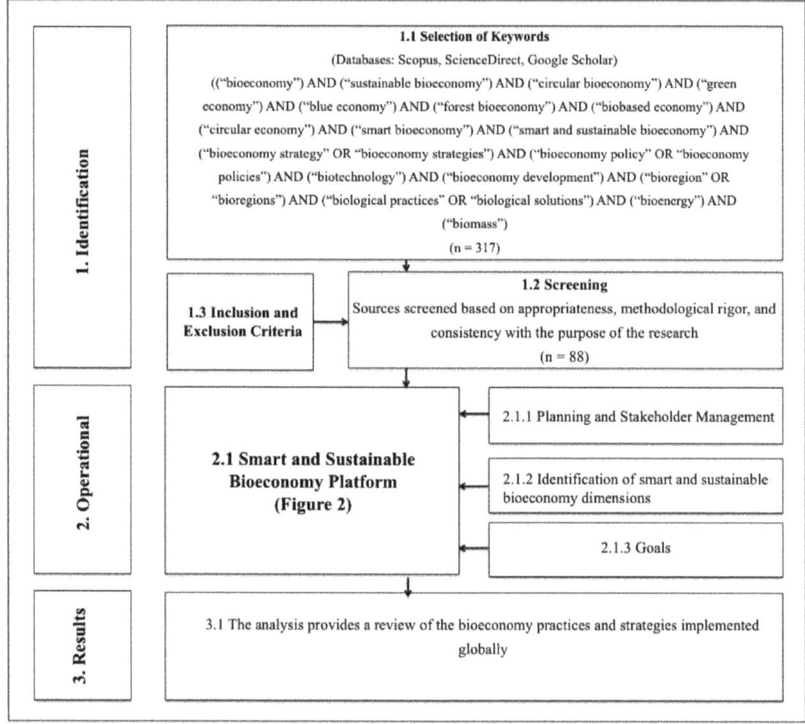

Figure 1. Methodology. Source: Authors.

In Phase 1—Identification, the search question, the keywords, a series of inclusion and exclusion criteria, and the search databases are outlined. With regard to the aim and research question of this paper, the study aims to provide a clear and exhaustive analysis, developing a platform capable of integrating in a systemic and holistic way the aspects of each operational phase of the smart and sustainable bioeconomy decision-making process. Therefore, the study focuses on the following research question: with which strategies and practices is the global context facing the transition towards the smart and sustainable bioeconomy? In this regard, the study of the smart and sustainable bioeconomy offers a multidisciplinary perspective of circular economy. Specifically, the scientific areas embrace agricultural, forest and marine economics, logistics, industrial organization, strategic management, technology and innovation management, data science, information and communication technologies, environmental and IT engineering, energy management, sustainable development, public policy analysis, geography, urban governance, territorial planning, etc.

In terms of sources, the investigation was based on the exploration and integration of a wide range of sources, such as urban and industrial reports and master plans, government documents, non-academic research, official websites, publications of ministries, departments, divisions, agencies, committees, research institutes, universities, etc. In parallel, scientific literature, such as peer-reviewed journal articles, book chapters, conference proceedings, and any other source in line with the development of the smart and sustainable bioeconomy, was used to verify and integrate basic information. As for the databases, ScienceDirect, Google Scholar, Scopus, and institutional websites were used. To do this, the following keywords were used: ("bioeconomy") AND ("sustainable bioeconomy") AND ("circular bioeconomy") AND ("circular bioeconomy") AND ("green economy") AND ("blue economy") AND ("forest bioeconomy") AND ("biobased economy") AND ("circular economy") AND ("smart bioeconomy") AND ("smart and sustainable bioeconomy") AND ("bioeconomy strategy" OR "bioeconomy strategies") AND ("bioeconomy policy" OR "bioeconomy policies") AND ("biotechnology") AND ("bioeconomy development") AND ("bioregion" OR "bioregions") AND ("biological practices" OR "biological solutions") AND ("bioenergy") AND ("biomass").

In order to further refine the search, all the selected sources ($n = 317$) were screened following different inclusion and exclusion criteria, in line with the objective and the research question. Specifically, the inclusion criteria include (1) appropriateness with the purpose of the study; (2) theoretical and managerial robustness; (3) scientific rigor; and (4) consistency with the long-term and novel perspective of the study. As exclusion criteria, sources with partial information and inconsistent with the research topic were not included in the search. The screening generated a detailed and complete overview ($n = 88$) that encompasses the strategies and policies of smart and sustainable bioeconomy implemented by Ethiopia, Ghana, Kenya, Malawi, Mali, Mauritius, Mozambique, Namibia, Nigeria, Rwanda, Senegal, South Africa, Tanzania, Uganda, Argentina, Brazil, Canada, Colombia, Costa Rica, Ecuador, Mexico, Paraguay, Uruguay, the United States of America, Australia, China, India, Indonesia, Japan, Malaysia, New Zealand, Russia, South Korea, Sri Lanka, Thailand, Austria, Belgium, Croatia, Czech Republic, Denmark, Finland, France, Germany, Ireland, Italy, Latvia, Lithuania, the Netherlands, Norway, Portugal, Slovenia, Spain, Sweden, and the UK (see Table A1 in Appendix A). The large number and varied typology of countries taken into consideration undoubtedly constitutes a strength of the study, as it represents almost all the countries involved in the transition towards the smart and sustainable bioeconomy. In this regard, the plurality of bioeconomy strategies and practices highlights the difference, the gap, and the prevailing focus between countries with diverse social, economic, environmental, and technological infrastructures. In this sense, a large number of countries around the world are not equipped to collect, monitor, analyse, and evaluate bioeconomic flows through real-time monitoring stations, sensors, digital tracking systems, artificial intelligence, blockchain, cloud computing, etc.

In Phase 2—Operational. Based on the analysis of the previous phase, a platform was created, capable of providing a theoretical and managerial approach to the development of the smart and sustainable bioeconomy. Specifically, the platform for the smart and sustainable bioeconomy represented in Figure 2 is characterized by the three following phases: (1) planning and stakeholder management; (2) identification of smart and sustainable bioeconomy dimensions; and (3) goals. Furthermore, each phase provided a holistic perspective, emphasizing aspects related to smartness and sustainability.

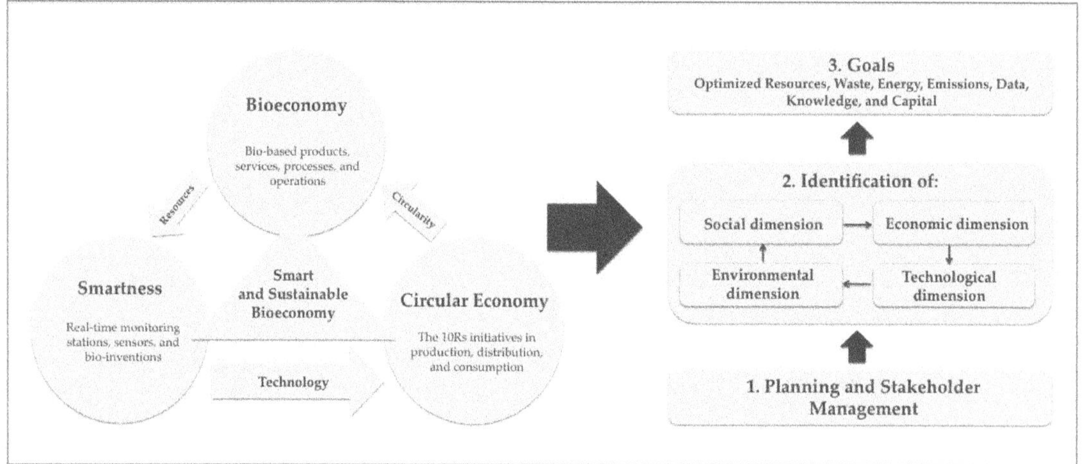

Figure 2. Smart and sustainable bioeconomy platform. Source: Authors.

In Phase 3—Results. The grouping of the initiatives developed by the countries taken into consideration in the investigation around the phases that characterize the smart and sustainable bioeconomy platform provides a framework capable of providing a review of the bioeconomy practices and strategies implemented globally.

3. Results

Based on the literature, the strategies and practices of the bioeconomy implemented globally were considered, analysed, and grouped. The result is a multidimensional platform illustrated in Figure 2 able to describe the smart and sustainable development in the bioeconomy in a holistic perspective. In detail, the phases that characterize the proposed smart and sustainable bioeconomy platform are divided into: (1) planning and stakeholder management; (2) identification of smart and sustainable bioeconomy dimensions; and (3) goals (Figure 2, vertical reading).

3.1. Planning and Stakeholder Management

Regarding the first phase (as represented in Figure 2), most of the bioeconomy strategies and practices investigated involve a wide range of stakeholders coordinated by government institutions. In general, these co-design activities take various forms, such as inter-ministerial committees, working groups, expert and public consultations, inter-ministerial collaborations, and partnerships. At the same time, bioeconomy policies planning are delegated to representatives of ministries, departments, agencies, committees, executive offices, councils, cabinets, associations, research centres, and steering groups. For example, the Bio-Circular-Green Economy (BCG) strategy in Thailand is characterized by an expert consultation process coordinated by the Thai minister of science and technology [81]. Similarly, South Africa [80], Costa Rica [79], Japan [86], Malaysia [87], Austria [88], and Latvia [89] involve in the strategy the ministries and departments of science, innovation, and technology. In the USA, the President of the United States has coordinated the Office of Science and Technology Policy of the White House in the elaboration of the Bioeconomy Blueprint [78]. Otherwise, the Council of Science, Technology, and Innovation of Japan has decentralized to the Japan Science and Technology Agency (JST), the New Energy and Industrial Technology Development Organization (NEDO), the National Institute of Technology and Evaluation (NITE), and the Japan Agency for Medical Research and Development (AMED) [86]. Likewise, the Finnish Technical Research Centre VTT, which operates under the mandate from the ministry of economic affairs and labour and the

Finnish innovation fund SITRA, were involved in the planning activities [90]. In Malaysia, the bioeconomy is entrusted to the Bioeconomy Corporation owned by the Malaysian ministry of finance, administered by the National Bioeconomy Council (NBC), supported by the Bioeconomy International Advisory Panel, and chaired by the Malaysian Prime Minister [87]. Differently, the Austrian inter-ministerial collaboration involves the federal ministry of transport, innovation, and technology and the federal ministry of sustainability and tourism [88,91]. In general, the ministry of economy coordinates bioeconomy strategies and practices in Finland, France, Italy, Latvia, Spain, and the UK. Industry and trade ministries have been involved in South Africa, Costa Rica, Japan, Malaysia, and Norway. Specifically, Japan embraces the Japan External Trade Organization (JETRO) and the Japan International Cooperation Agency (JICA).

The ministries and departments of agriculture, forestry, and fisheries are active in most of the countries considered. South Africa also integrates the department of environmental affairs [80]. Costa Rica includes the ministry of agriculture and livestock and the ministry of environment and energy [79]. Japan involves the National Agriculture and Food Research Organization (NARO) [86]. In addition to the ministry of agriculture, agrifood, and forests, France includes the ministry of ecology, sustainable development, and energy [92]. The Presidency of the Italian Council of Ministers has delegated the minister of the environment, land, and sea; the committee of the Italian regions; the Territorial Cohesion Agency; and the Italian Technology Clusters for Green Chemistry, Agrifood, and Blue Growth, the former drafting of the BIT I strategy [93]. Similarly, the strategy adopted by Latvia is characterized by an inter-ministerial group formed by the ministry of agriculture, economy, environmental protection, and regional development [89]. Otherwise, the Austrian bioeconomy strategy was implemented through a public consultation supervised by the Vienna University of Natural Resources and Life Sciences [91]. Furthermore, South Africa, Japan, Finland, France, Germany, Italy, and Latvia involve ministries from health, education, research, and welfare. In this regard, Norway has established a wide range of partnerships between the ministries of education, research, local government, modernization, and foreign affairs [94,95].

3.2. Identification of Smart and Sustainable Bioeconomy Dimensions

The investigation of the dimensions of the smart and sustainable bioeconomy involves a wide range of sectors, such as energy, waste, water, education, governance, and health, which mainly depend on environmental, social, economic, and technological characteristics of the context in which they circulate. However, the analysis of the countries taken into consideration and engaged in the transition towards the smart and sustainable bioeconomy shows a predominance of the fields of automated agriculture, industrial biotechnology and nanotechnology, smart grids for the optimized circulation of biomass, genetics, genomics, chemistry, medicine, marine and terrestrial biodiversity, and biorefinery.

The strategies and practices of smart and sustainable bioeconomy shared among the countries analysed emphasise the importance of industrial districts and knowledge-sharing centres in the fields of biotechnology, nanotechnology, genomics, genetics, and precision automation. In this regard, the UK is characterised as a thriving environment for innovation, entrepreneurship, and scientific research. In recent years, through the Synbio for Growth program, start-ups related to biology have received nearly 500 million pounds of funding in order to develop increasingly innovative bioeconomic products and processes [96,97]. At the same time, the economic initiatives adopted in Spain and Malaysia include digital technologies as centralised refrigeration systems, temperature tracking sensors, and prediction systems able to ensure greater nutritional quality and to reduce waste during the processing, packaging, storage, and distribution phases of the cold chain. Furthermore, in the economic dimension, it is interesting to specify the role of sustainable and virtual tourism within the naturalistic areas of Costa Rica, Finland, and Thailand.

Regarding the agricultural dimension, South Africa ranks first among African countries in agricultural biotechnology, producing more than 85% of genetically modified corn and soybeans. In this sense, by strengthening native crops (e.g., fortified sorghum, rooibos, and shrub honey), the bioeconomy strategy aims to satisfy the market demand for niche natural products [80]. At the same time, agriculture 4.0 initiatives, such as real-time monitoring of fertilizer and water use, precision automation, innovative plant selection methods to cope with drought, flood and insect resistance, and systems of vertical and modular agriculture, are present in Costa Rica, Thailand, Austria, France, Germany, Ireland, Italy, Latvia, Spain, and the UK. Differently, Malaysia has implemented a wide range of initiatives covering the development of animal vaccines, biological fertilizers and pesticides, plant micropropagation, and livestock farming through tracking systems [87]. Austria and Finland focus on how to improve forest resource management. As nearly 80% of Finland's total area is characterised by forests, the Finnish forestry industry is a leader in wood processing, implementing a multitude of low-water consumption processes [90]. Likewise, Austria and Japan promote forestry in the sustainable construction sector in order to minimise environmental impacts, using bio-based chemicals, bioplastics, and compostable and biodegradable materials. Regarding the energy dimension, the strategic axes of the bioeconomy plans of Austria, Costa Rica, and France focus on the production of bioenergy derived from the residual biomass from urban and industrial processes and operations in order to replace fossil fuels with high environmental impact for powering public transport, heating homes, biofertilizers, animal feed, etc. In this sense, Japan aims to use biomaterials with high performance in terms of weight, durability, and safety [86]. In Malaysia, the National Biomass Strategy focuses on the reuse of palm oil to generate bioenergy, biofuels, and bio-based organic products. On the other hand, Thailand has created a capillary system of power plants connected through blockchain-enabled smart grids with the aim of producing clear energy from renewable resources. At the same time, the strategy aims to convert biomass and agricultural by-products into bioplastics, fibres and pharmaceutical products. In general, most countries that include the energy component in the bioeconomy strategy integrate digital technologies such as smart grids, weather forecasting and monitoring systems, and so on. Within the energy dimension, the biorefinery initiatives are adopted by a multitude of countries globally. Indeed, Ireland, Latvia, Norway, Spain, and the UK dedicate great attention to the development of biorefineries in order to ensure a sustainable conversion of residual biomass (e.g., biolubricants, bioplastics, food additives, cosmetics, solvents, chemicals, etc.). Specifically, in Ireland, we highlight the AgriChemWhey project led by Glandia integrated with the dairy processing industry; the BioMarine Ingredients marine biorefinery, which converts raw materials into proteins, oil, and calcium; and the Biorefinery Glas project, which optimises the circularity of glass. In the UK, other examples include the alliance of several biorefineries as BioPilotsUK and the regional innovation cluster, BioVale, in Yorkshire and the Humber, which focus on bio-waste reuse and advanced biorefining. In order to address water scarcity in numerous areas of the country, South Africa is promoting improvements in wastewater treatment through computerized management of water flows. In Europe, the Finnish forestry industry is already leading in this sector by developing technologies for water recycling in its processes. Likewise, the Italian government has launched several projects, such as the PRIMA and BLUEMED initiatives, in order to promote sustainable management of water in the Mediterranean region [82]. The Spanish bioeconomic strategy summarizes the importance of the efficient use of water resources, promoting adequate water management and its reuse in other dimensions, such as construction, logistics, and health. Regarding waste management, Costa Rica intends develops the sustainable management and valorisation of residual solid waste, interurban biological corridors, and urban design approaches inspired by biological principles, processes, and systems. Given the increase in global marine plastic pollution, the Japanese strategy focuses on organic waste and wastewater, converting waste into high-value substances. Similarly, in Ireland, particular emphasis is placed on management and the valorisation of marine waste. The bioeconomy strategy in Germany further focuses on waste streams (e.g., organic waste, urban and industrial wastewater, carbon dioxide and synthesis gas). Furthermore, the strategy highlights the need for innovative methods and processes for the efficient processing and recycling of challenging resources, such as metals or phosphorus.

In the health dimension, South Africa focuses on supporting research and development initiatives of bio-based chemicals and industrial biotechnology in order to better tackle infant mortality, HIV, tuberculosis, and malaria infections. At the same time, bioprospection plays a crucial role in the development of new drugs, vaccines, diagnostics, and medical devices. On the other hand, in European countries, such as Austria, Italy, Germany, Latvia, and the UK, the health dimension mainly encompasses healthy diets and eating habits and psycho-physical well-being. The USA, through the Bioeconomy Blueprint, underlines the positive impacts of genetic engineering, synthetic biology, and bioinformatics on public health. To this end, U.S. federal agencies are incentivized to prioritise bio-based and sustainable materials in public procurement and their implementation and dissemination through technology transfer and easier market access. France, Japan, and Malaysia mainly focus on biopharmaceuticals, regenerative and precision medicine, omics technologies, nutrition, sport, and digital healthcare. Specifically, the digital health strategies and practices aim to generate personalized and categorised nutrition plans through a detailed research of consumer behaviours and preferences.

3.3. Goals

From an economic perspective, the smart and sustainable bioeconomy strategies and practices adopted aim to increase the competitiveness of the agricultural and industrial sectors in national and international markets. Specifically, the increase of the employment rate is the goal set by South Africa, the USA, Malaysia, France, Germany, Ireland, and Latvia. At the same time, the USA has mainly focused on the elaboration of training programs and career updating. Differently, Germany aims to increase employment rate in rural areas. The Latvian strategy intends to address the structural changes in agriculture such as the reduction of small and medium-sized enterprises and the decrease in the workforce due to the progressive digitalisation of processes. Therefore, the development of the smart and sustainable bioeconomy aims to decarbonise the production and consumption processes. In this sense, Costa Rican and Italian strategies have coined the term "circular bioeconomy" to emphasise the circularity of biological resources. According to Costa Rica, the circular bioeconomy contributes to reducing the carbon footprint of production processes and generates new market niches for consumers interested in minimizing their impacts on the environment. Similarly, Japan integrates circularity into the bioeconomy strategy in order to meet diversified needs. Furthermore, a key aspect recalled in a multitude of strategies is the development of public-private partnerships. Specifically, the USA, Japan, Austria, and the UK underline the need for a collaborative environment where industry and government interact dynamically in the implementation of regulatory processes that favour investment in research and development and commercialization of bio-inventions. Japan, for example, encourages the evolvement of international hubs capable of attracting the best start-ups in the field of biotechnology. Conversely, Austria focuses on how to mobilise private capital and the financial systems in the development of smart and sustainable bioeconomy initiatives.

From an environmental point of view, the smart and sustainable bioeconomy strategies and practices investigated aim to address climate change and environmental conservation through a plethora of initiatives related to waste and water management, renewable energy, and land-use optimization. In this sense, the reports of Austria and France refer to the achievement of the targets of the Paris Agreement on the climate. According to the Austrian guidelines, the smart and sustainable bioeconomy will significantly contribute to the reduction of greenhouse gas emissions by 2030. At the same time, Japan, Latvia, and Thailand expressly recall the Sustainable Development Goals (SDGs) of the United Nations 2030 Agenda. Differently, South Africa, Norway, and Spain do not explicitly indicate the Sustainable Development Goals among their objectives, but various initiatives lead to reducing greenhouse gas emissions and contributing to a more sustainable use of biological resources. Japan includes CO_2 reduction, land-use improvement, water management optimization, and food security. In this sense, the Irish bioeconomic strategy is based on the principles of sustainability, cascade, precaution, and food first. Likewise, Italy envisages the three following macro-areas: (1) certifications and quality standards; (2) agri-food, forestry, and marine pilot initiatives at local level; and (3) safeguarding biodiversity and ecosystem services.

Finally, the social dimension includes a wide range of ethical and legal issues. For example, Costa Rica and Japan prioritise social inclusion and equity aspects. In addition, Costa Rica takes into account the creation of opportunities for country's youth and indigenous communities. Conversely, Malaysia focuses primarily on people's health and well-being through reduced health care costs, early disease detection, and cheaper and more accessible medicines. Italy, on the other hand, promotes various initiatives in order to increase awareness, updating of skills, education, attitude, training, and entrepreneurship throughout the bioeconomy. Finally, Germany considers the importance of systems thinking and holistic approaches capable of creating synergies, identifying such conflicts, and minimizing them on the basis of scientific knowledge.

4. Discussion

The strategies and practices of smart and sustainable bioeconomy developed globally and investigated in this study confirm the advances in the scientific literature on the role of technology in the circularity of social, environmental, and economic flows of industrial and urban processes and operations [98]. At the same time, the initiatives identified and analysed support the observations of [99] on the smart and sustainable bioeconomy, where they emphasise the contribution of biotechnology, omics technologies, nanotechnology, precision mechanics, blockchain, and smart grids. Therefore, the theoretical and managerial debate demonstrates the significance assigned to the smart and sustainable bioeconomy [100]. In this sense, the smart and sustainable bioeconomy platform certifies that the planning and stakeholders management; the identification of social economic, environmental, and technological dimensions; and the definition of the goals is a challenging and complex issue that requires a multidimensional and holistic approach, in which a wide range of aspects must be taken into account simultaneously.

However, the stakeholders involved; the economic, social, environmental, and technological dimensions; and the goals of the smart and sustainable bioeconomy argue that the context in which the countries perform is much more hybrid and multi-layered with respect to the reductive conception that the economic, social, environmental, and technological pillars are important and to be pursued. Therefore, we do not claim that these issues have not been previously described and emphasized in the literature, but we declare that the proposed platform holistically highlights the actors engaged, the activated smart and sustainable dimensions of bioeconomy, and the goals to be achieved of a wide range of countries involved globally in this transition.

Firstly, the scientific literature on smart and sustainable bioeconomy underlines the crucial role of proactive, participatory and multi-level governance [101–103]. The authors [33,104] affirm that a distributed governance of the bioeconomy characterized by scalable coordination, consensus transmission protocols, and flexibility in decision-making processes is necessary for greater adaptability and efficiency of processes and operations. In this regard, Section 3.1 'Planning and Stakeholder Management', confirms the necessary bottom-up approach, specifying the role of a plethora of actors, such as ministries, departments and councils of science, innovation and technology, ministries of economy and trade, agriculture, fisheries and forests, foreign trade organizations, international cooperation agencies, research centres, universities, and biotechnology and nanotechnology companies.

In accordance with Section 3.2 'Identifications of the dimensions of the smart and sustainable bioeconomy', digital technologies, such as real-time monitoring stations, smart grids, weather forecasting systems, automatic irrigation systems, precision machinery, etc., improve the fluidity and timeless of flows that circulate in agriculture, fisheries, forests, logistics, health, education, waste, and water. In this sense, the theoretical literature on smart and sustainable bioeconomy emphasises the pre-eminent role of technology in the extraction, tracking, and evaluation phases [105,106]. However, the development of digital technologies in the bioeconomy can encounter criticalities in underdeveloped or developing countries not equipped with adequate economic, technological, social, and environmental structures [107].

Section 3.3 'Goals' highlights a multitude of purposes that confirm and support the current theoretical literature. In summary, from an economic point of view, smart and sustainable bioeconomy strategies and practices embrace the following: (a) the planning and development of agricultural processes and operations with low environmental impacts; (b) the design of industrial parks, international hubs, and start-up clusters in order to share knowledge; (c) the competitiveness of industrial sectors in national and international markets; (d) the creation and enhancement of highly skilled employment in the fields of biotechnology, genetics, and nanotechnology; and (e) the ability to attract and mobilise private capital and funding for the development of digital technologies and bio-inventions. However, the significant investments in planning, installation, integration, maintenance, and redefinition of digital technologies for the smart and sustainable bioeconomy undoubtedly represent a barrier to entry for underdeveloped countries [108]. In this regard, the search for national and international funding is crucial in the transition towards the smart and sustainable bioeconomy [109]. At European level, the European Structural and Investment Funds (ESIF) [110], which embrace regional funds (ERDF) [111] and agriculture and rural development funds (EAFRD) [112], and Smart Specialization Strategies (RIS3) [113,114] aim to facilitate the modernization of the bioeconomy throughout the Europe.

Regarding the environmental perspective, the identified smart and sustainable bioeconomy strategies and practices confirm relevant issues, such as waste, water and wastewater management, energy efficiency, and land use [115,116]. In this sense, Austria and France based their goals with the Paris Agreement on Climate, while Japan, Latvia, and Thailand recall the environmental goals of the United Nations 2030 Agenda on Sustainable Development.

Finally, the social dimension of the smart and sustainable bioeconomy includes several aspects, such as social inclusion, ethics, and legality [117,118]. Regarding the social sphere, the balance of governance between the various ministries, departments, cabinets, agencies, committees, municipal, regional or state owned utilities, divisions, universities, and research centres involved in planning, monitoring, and evaluating smart and sustainable bioeconomy policies emphasizes the need for multidimensional and participatory decision-making processes [101,119–121]. In this regard, the term "orchestration" is coined as a fundamental aspect for understanding the evolution of complex systems towards inclusive and participatory models [122–125]. At the same time, motivational, behavioural, and cognitive issues persist, such as the lack of (a) awareness of the benefits of proactive co-participation of the stakeholders involved; (b) knowledge of technological devices functioning; (c) citizens', entrepreneurs', and businesses' understanding of the practices of production, distribution, and consumption of sustainable bioeconomic products and services; and (d) trust in safeguarding the privacy and security of sensitive data. In this regard, the smart and sustainable bioeconomy highlights technical challenges relating to the quality and robustness of the data and information collected, their degree of security, and their ability to be converted into useful feedback [126]. Furthermore, various countries investigated embrace the health and psycho-physical well-being of people, focusing on reducing healthcare costs through personalized medicine, prevention, nutrition, and more accessible medicines.

Therefore, the proposed platform indicates that the planning and stakeholders management, the identification of the smart and sustainable dimensions of the bioeconomy, and the definition of the goals to be pursued must be carried out taking into consideration the social, economic, environmental, and technological factors holistically. This bottom-up and multidimensional approach is confirmed and emphasised by the theoretical literature on bioeconomy and our investigation of strategies and practices adopted globally.

5. Conclusions

The era of growing urbanization and datafication pushes us to rethink how to tackle sustainable development. In this context, smart and sustainable bioeconomy offers a renewed perspective towards resilient and intelligent future. In recent years, smart and sustainable bioeconomy initiatives are gaining increasing importance in the technical-spatial context in order to collect, monitor, process, and evaluate a large amount of data and to improve the quality and efficiency of industrial and urban processes and operations and therefore the functioning of our countries. In this regard, the importance of smart and sustainable bioeconomy is demonstrated by the numerous strategies and practices implemented by countries, such as Austria, Costa Rica, Finland, Germany, Ireland, Latvia, Malaysia, South Africa, Thailand, and so on. Therefore, the bioeconomy enabled by sensors, real-time monitoring stations, tracking systems, Internet of Things, smart grids, precision mechanics, automation, etc., have the potential to improve the circularity of dimensions, such as waste, water and wastewater, energy, land, biodiversity, economy, health, safety, education, and agriculture. The aim of this study is to provide a clear and comprehensive overview of the concept of smart and sustainable bioeconomy, developing a platform capable of integrating a wide range of bio-initiatives implemented at a global level. Specifically, the smart and sustainable bioeconomy platform illustrated in Figure 2 describes the phases of planning and stakeholder management; identification of economic, social, environmental, and technological dimensions; and the definition of the goals that characterise the smart and sustainable bioeconomy decision-making process. In this sense, the proposed platform improves the understanding of the functioning of smart and sustainable bioeconomy. At the same time, the exploration of the smart and sustainable bioeconomy requires not only a qualitative perspective of the strategies and practices as proposed in this study but also further quantitative research to assess and interpret their social, economic, environmental, and technological impacts. However, the difficulties encountered in obtaining quantitative data on the initiatives investigated undoubtedly represent a limitation to our research. Therefore, the future perspective of this paper is to enrich it with a quantitative approach in order to provide a complete and exhaustive point of view for future analyses.

Hence, in order to summarise the results of this study, we underline and list the following highlights: (a) the effective and efficient implementation of the smart and sustainable bioeconomy requires continuous planning, monitoring, and analysis of the social, economic, technological, and environmental dimensions; (b) the smart and sustainable bioeconomy can improve the participation, accountability, and comprehension of citizens, local authorities, and companies; and (c) the smart and sustainable bioeconomy generates a multitude of social, economic, and environmental challenges still under observation by the scientific and managerial community today.

Author Contributions: Conceptualization, G.D.; supervision, G.I.; visualization, K.S.-D., R.B. and I.D.; writing—original draft, G.D., K.S.-D., R.B., I.D. and G.I.; writing—review and editing, G.D., K.S.-D., R.B., I.D. and G.I. All authors have read and agreed to the published version of the manuscript.

Funding: This research received no external funding.

Data Availability Statement: Not applicable.

Acknowledgments: This research is carried out by the Commodity Science Team of the "Lean & Quality Solutions Lab," Department of Economics, University of Messina. The authors thank the editor and anonymous referees for their valuable observations.

Conflicts of Interest: The authors declare no conflict of interest.

Appendix A

Table A1. Summary of sources grouped in the geographical areas. Source: Authors.

Africa													
	Ethiopia	Ghana	Kenya	Malawi	Mali	Mauritius	Mozambique	Namibia	Nigeria	Rwanda	Senegal	South Africa	Tanzania	Uganda
Source	[127–129]	[130]	[131–134]	[135]	[136,137]	[138]	[139]	[140]	[141]	[142]	[143]	[80]	[144]	[145]

Americas									
	Argentina	Brazil	Canada	Colombia	Costa Rica	Ecuador	Mexico	Paraguay	Uruguay	USA
Source	[146]	[147,148]	[149,150]	[151–154]	[79]	[155]	[156]	[157]	[158–160]	[78]

Asia/Pacific										
	Australia	China	India	Indonesia	Japan	Malaysia	New Zealand	Russia	South Korea	Sri Lanka	Thailand
Source	[161–163]	[164–167]	[168–170]	[171,172]	[86]	[87]	[173,174]	[175,176]	[177,178]	[179, 180]	[81]

Europe												
	Austria	Belgium	Croatia	Czech Republic	Denmark	Finland	France	Germany	Ireland	Italy	Latvia	Lithuania	Netherlands
Source	[88,91]	[181]	[182]	[183]	[184–186]	[90]	[92]	[75,187]	[188]	[82,93]	[89]	[189]	[190]
	Norway	Portugal	Slovenia	Spain	Sweden	UK							
Source	[94,95]	[191,192]	[193]	[194,195]	[196]	[96,97]							

References

1. Mak, T.M.W.; Xiong, X.; Tsang, D.C.W.; Yu, I.K.M.; Poon, C.S. Sustainable food waste management towards circular bioeconomy: Policy review, limitations and opportunities. *Bioresour. Technol.* **2020**, *297*, 122497. [CrossRef] [PubMed]
2. Aldieri, L.; Ioppolo, G.; Vinci, C.P.; Yigitcanlar, T. Waste recycling patents and environmental innovations: An economic analysis of policy instruments in the USA, Japan and Europe. *Waste Manag.* **2019**, *95*, 612–619. [CrossRef]
3. Muscat, A.; de Olde, E.M.; Kovacic, Z.; de Boer, I.J.M.; Ripoll-Bosch, R. Food, energy or biomaterials? Policy coherence across agro-food and bioeconomy policy domains in the EU. *Environ. Sci. Policy* **2021**, *123*, 21–30. [CrossRef]
4. Nagarajan, D.; Lee, D.-J.; Chen, C.-Y.; Chang, J.-S. Resource recovery from wastewaters using microalgae-based approaches: A circular bioeconomy perspective. *Bioresour. Technol.* **2020**, *302*, 122817. [CrossRef]
5. Andersson, I.; Grundel, I. Regional policy mobilities: Shaping and reshaping bioeconomy policies in Värmland and Västerbotten, Sweden. *Geoforum* **2021**, *121*, 142–151. [CrossRef]
6. Sanz-Hernández, A.; Sanagustín-Fons, M.V.; López-Rodríguez, M.E. A transition to an innovative and inclusive bioeconomy in Aragon, Spain. *Environ. Innov. Soc. Transit.* **2019**, *33*, 301–316. [CrossRef]
7. Haines, A. Health in the bioeconomy. *Lancet Planet. Health* **2021**, *5*, e4–e5. [CrossRef]
8. Primmer, E.; Varumo, L.; Krause, T.; Orsi, F.; Geneletti, D.; Brogaard, S.; Aukes, E.; Ciolli, M.; Grossmann, C.; Hernández-Morcillo, M.; et al. Mapping Europe's institutional landscape for forest ecosystem service provision, innovations and governance. *Ecosyst. Serv.* **2021**, *47*, 101225. [CrossRef]
9. Liobikiene, G.; Chen, X.; Streimikiene, D.; Balezentis, T. The trends in bioeconomy development in the European Union: Exploiting capacity and productivity measures based on the land footprint approach. *Land Use Policy* **2020**, *91*, 104375. [CrossRef]
10. Musonda, F.; Millinger, M.; Thrän, D. Optimal biomass allocation to the German bioeconomy based on conflicting economic and environmental objectives. *J. Clean. Prod.* **2021**, *309*, 127465. [CrossRef]
11. Salvador, R.; Puglieri, F.N.; Halog, A.; de Andrade, F.G.; Piekarski, C.M.; De Francisco, A.C. Key aspects for designing business models for a circular bioeconomy. *J. Clean. Prod.* **2021**, *278*, 124341. [CrossRef]
12. Souza, T.M.L.; Morel, C.M. The COVID-19 pandemics and the relevance of biosafety facilities for metagenomics surveillance, structured disease prevention and control. *Biosaf. Health* **2021**, *3*, 1–3. [CrossRef]
13. Sharma, P.; Gaur, V.K.; Sirohi, R.; Varjani, S.; Kim, S.H.; Wong, J.W.C. Sustainable processing of food waste for production of bio-based products for circular bioeconomy. *Bioresour. Technol.* **2021**, *325*, 124684. [CrossRef]
14. Sanz-Hernández, A. Privately owned forests and woodlands in Spain: Changing resilience strategies towards a forest-based bioeconomy. *Land Use Policy* **2021**, *100*, 104922. [CrossRef]
15. Kearns, P.W.E.; Kleter, G.A.; Bergmans, H.E.N.; Kuiper, H.A. Biotechnology and Biosafety Policy at OECD: Future Trends. *Trends Biotechnol.* **2021**, *39*, 965–969. [CrossRef]
16. Sun, D.; Wu, L.; Fan, G. Laboratory information management system for biosafety laboratory: Safety and efficiency. *J. Biosaf. Biosecurity* **2021**, *3*, 28–34. [CrossRef]
17. Refsgaard, K.; Kull, M.; Slätmo, E.; Meijer, M.W. Bioeconomy—A driver for regional development in the Nordic countries. *New Biotechnol.* **2021**, *60*, 130–137. [CrossRef] [PubMed]
18. Watanabe, C.; Naveed, N.; Neittaanmäki, P. Digital solutions transform the forest-based bioeconomy into a digital platform industry—A suggestion for a disruptive business model in the digital economy. *Technol. Soc.* **2018**, *54*, 168–188. [CrossRef]
19. Watanabe, C.; Naveed, N.; Neittaanmäki, P. Digitalized bioeconomy: Planned obsolescence-driven circular economy enabled by Co-Evolutionary coupling. *Technol. Soc.* **2019**, *56*, 8–30. [CrossRef]
20. D'Adamo, I.; Falcone, P.M.; Morone, P. A new socio-economic indicator to measure the performance of bioeconomy sectors in Europe. *Ecol. Econ.* **2020**, *176*, 106724. [CrossRef]
21. Nallapaneni, M.K.; Chopra, S.S. Sustainability and resilience of circular economy business models based on digital ledger technologies. In Proceedings of the Waste Management and Valorisation for a Sustainable Future, Seoul, Korea, 26–28 October 2021.
22. Kershaw, E.H.; Hartley, S.; McLeod, C.; Polson, P. The sustainable path to a circular bioeconomy. *Trends Biotechnol.* **2021**, *39*, 542–545. [CrossRef]
23. Global Bioeconomy Summit. *Global Bioeconomy Policy Report (IV): A Decade of Bioeconomy Policy Development around the World*; Secretariat of the Global Bioeconomy Summit: Berlin, Germany, 2020. Available online: https://gbs2020.net/wp-content/uploads/2020/11/GBS-2020_Global-Bioeconomy-Policy-Report_IV_web.pdf (accessed on 2 December 2021).
24. European Commission. *Cities of Tomorrow—Challenges, Visions, Ways Forward*; Publications Office of the European Union: Luxembourg, 2011. Available online: https://ec.europa.eu/regional_policy/sources/docgener/studies/pdf/citiesoftomorrow/citiesoftomorrow_final.pdf (accessed on 2 December 2021).
25. OECD. Cities in the world: A new perspective on urbanisation. In *OECD Urban Studies*; OECD Publishing: Paris, France, 2020. [CrossRef]
26. United Nations (UN). *Transforming Our World: The 2030 Agenda for Sustainable Development (A/RES/70/1)*; General Assembly; United Nations: New York, NY, USA, 2015. Available online: https://www.un.org/ga/search/view_doc.asp?symbol=A/RES/70/1&Lang=E (accessed on 1 December 2021).
27. Calicioglu, Ö.; Bogdanski, A. Linking the bioeconomy to the 2030 sustainable development agenda: Can SDG indicators be used to monitor progress towards a sustainable bioeconomy? *New Biotechnol.* **2021**, *61*, 40–49. [CrossRef] [PubMed]

28. Ronzon, T.; Sanjuán, A.I. Friends or foes? A compatibility assessment of bioeconomy-related sustainable development goals for European policy coherence. *J. Clean. Prod.* **2020**, *254*, 119832. [CrossRef] [PubMed]
29. New Urban Agenda. United Nations Conference on Housing and Sustainable Urban Development (Habitat III) in Quito, Ecuador; United Nations General Assembly (A/RES/71/256). 2017. Available online: https://uploads.habitat3.org/hb3/NUA-English.pdf (accessed on 1 December 2021).
30. Van Lancker, J.; Wauters, E.; Van Huylenbroeck, G. Managing innovation in the bioeconomy: An open innovation perspective. *Biomass Bioenergy* **2016**, *90*, 60–69. [CrossRef]
31. Kircher, M.; Breves, R.; Taden, A.; Herzberg, D. How to capture the bioeconomy's industrial and regional potential through professional cluster management. *New Biotechnol.* **2018**, *40*, 119–128. [CrossRef]
32. Stegmann, P.; Londo, M.; Junginger, M. The circular bioeconomy: Its elements and role in European bioeconomy clusters. *Resour. Conserv. Recycl. X* **2020**, *6*, 100029. [CrossRef]
33. Dieken, S.; Dallendörfer, M.; Henseleit, M.; Siekmann, F.; Venghaus, S. The multitudes of bioeconomies: A systematic review of stakeholders' bioeconomy perceptions. *Sustain. Prod. Consum.* **2021**, *27*, 1703–1717. [CrossRef]
34. Näyhä, A. Transition in the Finnish forest-based sector: Company perspectives on the bioeconomy, circular economy and sustainability. *J. Clean. Prod.* **2019**, *209*, 1294–1306. [CrossRef]
35. Bröring, S.; Laibach, N.; Wustmans, M. Innovation types in the bioeconomy. *J. Clean. Prod.* **2020**, *266*, 121939. [CrossRef]
36. Kuckertz, A.; Berger, E.S.C.; Brändle, L. Entrepreneurship and the sustainable bioeconomy transformation. *Environ. Innov. Soc. Transit.* **2020**, *37*, 332–344. [CrossRef]
37. Fava, F.; Gardossi, L.; Brigidi, P.; Morone, P.; Carosi, D.A.R.; Lenzi, A. The bioeconomy in Italy and the new national strategy for a more competitive and sustainable country. *New Biotechnol.* **2021**, *61*, 124–136. [CrossRef]
38. D'Adamo, I.; Gastaldi, M.; Morone, P.; Rosa, P.; Sassanelli, C.; Settembre-Blundo, D.; Shen, Y. Bioeconomy of sustainability: Drivers, opportunities and policy implications. *Sustainability* **2021**, *14*, 200. [CrossRef]
39. Wydra, S. Measuring innovation in the bioeconomy—Conceptual discussion and empirical experiences. *Technol. Soc.* **2020**, *61*, 101242. [CrossRef]
40. Gonçalves, M.; Freire, F.; Garcia, R. Material flow analysis of forest biomass in Portugal to support a circular bioeconomy. *Resour. Conserv. Recycl.* **2021**, *169*, 105507. [CrossRef]
41. Ioppolo, G.; Heijungs, R.; Cucurachi, S.; Salomone, R.; Kleijn, R. Urban metabolism: Many open questions for future. In *Pathways to Environmental Sustainability: Methodologies and Experiencesi*; Springer: Cham, Switzerland, 2014; pp. 23–32.
42. Naveed, N.; Watanabe, C.; Neittaanmäki, P. Co-evolutionary coupling leads a way to a novel concept of R&D—Lessons from digitalized bioeconomy. *Technol. Soc.* **2020**, *60*, 101220.
43. Pelli, P.; Lähtinen, K. Servitization and bioeconomy transitions: Insights on prefabricated wooden elements supply networks. *J. Clean. Prod.* **2020**, *244*, 118711. [CrossRef]
44. Aiello, G.; Giovino, I.; Vallone, M.; Catania, P.; Argento, A. A decision support system based on multisensor data fusion for sustainable greenhouse management. *J. Clean. Prod.* **2018**, *172*, 4057–4065. [CrossRef]
45. D'Amico, G.; L'Abbate, P.; Liao, W.; Yigitcanlar, T.; Ioppolo, G. Understanding sensor cities: Insights from technology giant company driven smart urbanism practices. *Sensors* **2020**, *20*, 4391. [CrossRef]
46. Jedermann, R.; Praeger, U.; Lang, W. Challenges and opportunities in remote monitoring of perishable products. *Food Packag. Shelf Life* **2017**, *14*, 18–25. [CrossRef]
47. Tavakoli, H.; Gebbers, R. Assessing nitrogen and water status of winter wheat using a digital camera. *Comput. Electron. Agric.* **2019**, *157*, 558–567. [CrossRef]
48. Shamshiri, R.R. Exploring GPS data for operational analysis of farm machinery. *Res. J. Appl. Sci. Eng. Technol.* **2013**, *5*, 3281–3286. [CrossRef]
49. Rizzi, J.; Nystuen, I.; Debella-Gilo, M.; Søvde, N.E.; Solbjørg, E. Big data and machine learning in the bioeconomy sector: Preliminary results from the Norwegian case. In *Geophysical Research Abstracts*; EGU2019-16256-1; EGU General Assembly 2019; EGU: Munich, Germany, 2019; Volume 21.
50. OECD. *The Digitalisation of Science, Technology and Innovation: Key Developments and Policies*; OECD Publishing: Paris, France, 2020. [CrossRef]
51. European Commission. *100 Radical Innovation Breakthroughs for the Future*; Directorate-General for Research and Innovation: Brussels, Belgium, 2019. Available online: https://ec.europa.eu/info/sites/default/files/research_and_innovation/knowledge_publications_tools_and_data/documents/ec_rtd_radical-innovation-breakthrough_052019.pdf (accessed on 1 December 2021).
52. Klerkx, L.; Rose, D. Dealing with the game-changing technologies of Agriculture 4.0: How do we manage diversity and responsibility in food system transition pathways? *Global Food Secur.* **2020**, *24*, 100347. [CrossRef]
53. Nallapaneni, M.K.; Dash, A.; Singh, N.K. Internet of things (IoT): An opportunity for energy-food-water nexus. In Proceedings of the 2018 International Conference on Power Energy, Environment and Intelligent Control (PEEIC), Greater Noida, India, 13–14 April 2018; pp. 68–72.
54. European Commission. *The Junction of Health, Environment and the Bioeconomy: Foresight and Implications for European Research & Innovation Policies*; Directorate-General for Research and Innovation: Brussels, Belgium, 2015. Available online: https://www.kowi.de/Portaldata/2/Resources/horizon2020/coop/Foresight-Junction-Health-ENV-Bioeconomy-.pdf (accessed on 1 December 2021).

55. Electricity Advisory Committee. Smart Grid: Enabler of the New Energy Economy. 2008. Available online: https://www.ctc-n.or g/sites/www.ctc-n.org/files/resources/final-smart-grid-report.pdf (accessed on 2 December 2021).

56. Food and Agriculture Organization of the United Nations (FAO). *Country Fact Sheet on Food and Agriculture Policy Trends*; FAO: Rome, Italy, 2018. Available online: http://www.fao.org/3/i7696e/i7696e.pdf (accessed on 1 December 2021).

57. Parisi, C.; Vigani, M.; Rodríguez-Cerezo, E. Agricultural Nanotechnologies: What are the current possibilities? *Nanotoday* **2015**, *10*, 124–127. [CrossRef]

58. Arujanan, M.; Singaram, M. The biotechnology and bioeconomy landscape in Malaysia. *New Biotechnol.* **2018**, *40*, 52–59. [CrossRef] [PubMed]

59. Frisvold, G.B.; Moss, S.M.; Hodgson, A.; Maxon, M.E. Understanding the U.S. bioeconomy: A new definition and landscape. *Sustainability* **2021**, *13*, 1627. [CrossRef]

60. Miehe, R.; Bauernhansl, T.; Beckett, M.; Brecher, C.; Demmer, A.; Drossel, W.-G.; Elfert, P.; Full, J.; Hellmich, A.; Hinxlage, J.; et al. The biological transformation of industrial manufacturing—Technologies, status and scenarios for a sustainable future of the German manufacturing industry. *J. Manuf. Syst.* **2020**, *54*, 50–61. [CrossRef]

61. Galanakis, C.; Rizou, M.; Aldawoud, T.M.S.; Ucak, I.; Rowan, N.J. Innovations and technology disruptions in the food sector within the COVID-19 pandemic and post-lockdown era. *Trends Food Sci. Technol.* **2021**, *110*, 193–200. [CrossRef]

62. Rantala, S.; Swallow, B.; Paloniemi, R.; Raitanen, E. Governance of forests and governance of forest information: Interlinkages in the age of open and digital data. *For. Policy Econ.* **2020**, *113*, 102123. [CrossRef]

63. Wydra, S.; Hüsing, B.; Köhler, J.; Schwarz, A.; Schirrmeister, E.; Voglhuber-Slavinsky, A. Transition to the bioeconomy—Analysis and scenarios for selected niches. *J. Clean. Prod.* **2021**, *294*, 126092. [CrossRef]

64. Vaintrub, M.O.; Levit, H.; Chincarini, M.; Fusaro, I.; Giammarco, M.; Vignola, G. Review: Precision livestock farming, automats and new technologies: Possible applications in extensive dairy sheep farming. *Animal* **2021**, *15*, 100143. [CrossRef]

65. Zhang, M.; Wang, X.; Feng, H.; Huang, Q.; Xiao, X.; Zhang, X. Wearable Internet of Things enabled precision livestock farming in smart farms: A review of technical solutions for precise perception, biocompatibility, and sustainability monitoring. *J. Clean. Prod.* **2021**, *312*, 127712. [CrossRef]

66. Escursell, S.; Llorach-Massana, P.; Roncero, M.B. Sustainability in e-commerce packaging: A review. *J. Clean. Prod.* **2021**, *280*, 124314. [CrossRef]

67. D'Amico, G.; Szopik-Depczyńska, K.; Dembińska, I.; Ioppolo, G. Smart and sustainable logistics of Port cities: A framework for comprehending enabling factors, domains and goals. *Sustain. Cities Soc.* **2021**, *69*, 102801. [CrossRef]

68. Mertens, A.; Van Lancker, J.; Buysse, J.; Lauwers, L.; Van Meensel, J. Overcoming non-technical challenges in bioeconomy value-chain development: Learning from practice. *J. Clean. Prod.* **2019**, *231*, 10–20. [CrossRef]

69. Fritsche, U.; Brunori, G.; Chiaramonti, D.; Galanakis, C.; Hellweg, S.; Matthews, R.; Panoutsou, C. *Future Transitions for the Bioeconomy towards Sustainable Development and a Climate-Neutral Economy—Knowledge Synthesis Final Report*; Publications Office of the European Union: Luxembourg, 2020. Available online: https://op.europa.eu/en/publication-detail/-/publication/54a1e67 9-f634-11ea-991b-01aa75ed71a1/language-en (accessed on 1 December 2021).

70. Woźniak, E.; Tyczewska, A.; Twardowski, T. Bioeconomy development factors in the European Union and Poland. *New Biotechnol.* **2021**, *60*, 2–8. [CrossRef] [PubMed]

71. Wohlgemuth, R.; Twardowski, T.; Aguilar, A. Bioeconomy moving forward step by step—A global journey. *New Biotechnol.* **2021**, *61*, 22–28. [CrossRef]

72. Borgström, S. Reviewing natural resources law in the light of bioeconomy: Finnish forest regulations as a case study. *For. Policy Econ.* **2018**, *88*, 11–23. [CrossRef]

73. Sanz-Hernández, A.; Esteban, E.; Garrido, P. Transition to a bioeconomy: Perspectives from social sciences. *J. Clean. Prod.* **2019**, *224*, 107–119. [CrossRef]

74. Ferreira, V.; Pié, L.; Terceño, A. Economic impact of the bioeconomy in Spain: Multiplier effects with a bio social accounting matrix. *J. Clean. Prod.* **2021**, *298*, 126752. [CrossRef]

75. The Federal Government. *National Bioeconomy Strategy*; Federal Ministry of Education and Research, Division "Sustainable Economy; Bio-Economy": Berlin, Germany, 2020. Available online: https://www.bmbf.de/upload_filestore/pub/BMBF_Nation ale_Biooekonomiestrategie_Langfassung_eng.pdf (accessed on 1 December 2021).

76. Purkus, A.; Lüdtke, J. A systemic evaluation framework for a multi-actor, forest-based bioeconomy governance process: The German Charter for Wood 2.0 as a case study. *For. Policy Econ.* **2020**, *113*, 102113. [CrossRef]

77. The Cabinet Office. *Decision of the Council for Integrated Innovation Strategy*; German Embassy: Tokyo, Japan, 2020. Available online: https://www.dwih-tokyo.org/files/2020/10/bio2020_honbun_en_rev-1.pdf (accessed on 2 December 2021).

78. The White House. *National Bioeconomy Blueprint*; The White House: Washington, DC, USA, 2012. Available online: http://www.ascension-publishing.com/BIZ/Bioeconomy-Blueprint.pdf (accessed on 29 November 2021).

79. Costa Rica. National Bioeconomy Strategy. 2020. Available online: https://gbs2020.net/wp-content/uploads/2020/09/PolicyBri ef-Bioeconomy-Strategy-Costa-Rica.pdf (accessed on 1 December 2021).

80. *The Bio-Economy Strategy*; Department of Science and Technology: Pretoria, South Africa, 2013. Available online: https://www.go v.za/sites/default/files/gcis_document/201409/bioeconomy-strategya.pdf (accessed on 27 November 2021).

81. Thailand. BCG in Action. 2021. Available online: https://www.nxpo.or.th/th/en/bcg-in-action/ (accessed on 26 November 2021).

82. *Implementation Action Plan (2020–2025) for the Italian Bioeconomy Strategy BIT II*; National Bioeconomy Coordination Group of the Presidency of Council of Ministers: Rome, Italy, 2020. Available online: http://cnbbsv.palazzochigi.it/media/1962/implementation-action-plan_bioeconomy_28_-7-2020.pdf (accessed on 25 November 2021).
83. Nordic Council of Ministers. *Nordic Bioeconomy Programme. 15 Action Points for Sustainable Change*; Nordic Council of Ministers: Copenhagen, Denmark, 2018. Available online: https://norden.diva-portal.org/smash/get/diva2:1222743/FULLTEXT01.pdf (accessed on 1 December 2021).
84. Orejuela-Escobar, L.; Landázuri, A.C.; Goodell, B. Second generation biorefining in Ecuador: Circular bioeconomy, zero waste technology, environmental and sustainable development: The nexus. *J. Bioresour. Bioprod.* **2021**, *6*, 83–107. [CrossRef]
85. Vorgias, C. Bioeconomy challenges and open issues. *J. Biotechnol.* **2018**, *280*, S10. [CrossRef]
86. Japan Association of Bioindustries Executives. *Bioeconomy Vision of Japan for 2030*; JABEX: Tokyo, Japan, 2016. Available online: https://www.jba.or.jp/jabex/pdf/2016/JABEX_vision_digest(english160420).pdf (accessed on 1 December 2021).
87. Federal Government, Malaysia. *Bioeconomy Transformation Programme*; Ministry of Science, Technology and Innovation: Putrajaya, Malaysia, 2015. Available online: http://www.bioeconomycorporation.my/wp-content/uploads/2011/11/publications/BTP_AR_2015.pdf (accessed on 29 November 2021).
88. *Bioeconomy—RTI Strategy*; Federal Ministry of Education, Science and Research: Wien, Austria, 2018. Available online: https://nachhaltigwirtschaften.at/resources/nw_pdf/biooekonomie-fti-strategie-ag2-2018.pdf (accessed on 29 November 2021).
89. *Latvian Bioeconomy Strategy 2030*; Ministry of Agriculture: Riga, Latvia, 2018. Available online: https://www.zm.gov.lv/public/files/CMS_Static_Page_Doc/00/00/01/46/58/E2758-LatvianBioeconomyStrategy2030.pdf (accessed on 23 November 2021).
90. *The Finnish Bioeconomy Strategy*; Ministry of the Environment and Ministry of Employment and the Economy: Helsinki, Finland, 2014. Available online: https://www.biotalous.fi/wp-content/uploads/2014/08/The_Finnish_Bioeconomy_Strategy_110620141.pdf (accessed on 24 November 2021).
91. *Bioeconomy—A Strategy for Austria*; Council of Ministers: Wien, Austria, 2019. Available online: https://www.bmbwf.gv.at/en/Topics/Research/Research-in-Austria/Strategic-focus-and-advisory-bodies/Strategies/Bioeconomy-Strategy.html (accessed on 29 November 2021).
92. *A Bioeconomy Strategy for France. 2018–2020 Action Plan*; Ministry of Agriculture and Food: Paris, France, 2018. Available online: https://agriculture.gouv.fr/bioeconomy-strategy-france-2018-2020-action-plan (accessed on 24 November 2021).
93. Braia, L. The Smart Specialisation Strategy and the Bioeconomy Cluster in the Basilicata Region. Brussels. 2016. Available online: https://www.slideshare.net/LucaBraia1/luca-braia-the-smart-specialisation-strategy-and-the-bioeconomy-cluster-in-the-basilicata-region (accessed on 23 November 2021).
94. Bardalen, A. The Norwegian Bioeconomy Strategy—Structural Changes and Green Shift in the Economy. Riga. 2016. Available online: https://www.norden.lv/Uploads/2016/08/26/1472194554_.pdf (accessed on 26 November 2021).
95. Capasso, M.; Klitkou, A. *Socioeconomic Indicators to Monitor Norway's Bioeconomy in Transition*; Nordic Institute for Studies in Innovation, Research and Education: Oslo, Norway, 2020. Available online: https://www.forskningsradet.no/contentassets/a5128e4978d74df090fd7858d767b010/nifureport2020-5-1.pdf (accessed on 28 November 2021).
96. *Growing the Bioeconomy*; HM Government: London, UK, 2018. Available online: https://assets.publishing.service.gov.uk/government/uploads/system/uploads/attachment_data/file/761856/181205_BEIS_Growing_the_Bioeconomy__Web_SP_.pdf (accessed on 29 November 2021).
97. National Plan for Industrial Biotechnology. *Driving Progress to 2025*; Scottish Enterprise: Glasgow, UK, 2019. Available online: https://www.sdi.co.uk/media/1673/national-plan-for-ib-2019-pdf.pdf (accessed on 29 November 2021).
98. Dupont-Inglis, J.; Borg, A. Destination bioeconomy—The path towards a smarter, more sustainable future. *New Biotechnol.* **2018**, *40*, 140–143. [CrossRef]
99. Rahman, M.; Morsaline Billah Md Hack-Polay, D.; Alam, A. The use of biotechnologies in textile processing and environmental sustainability: An emerging market context. *Technol. Forecast. Soc. Chang.* **2020**, *159*, 120204. [CrossRef]
100. Krüger, A.; Schäfers, C.; Busch, P.; Antranikian, G. Digitalization in microbiology—Paving the path to sustainable circular bioeconomy. *New Biotechnol.* **2020**, *59*, 88–96. [CrossRef]
101. Egea, F.J.; Torrente, R.G.; Aguilar, A. An efficient agro-industrial complex in Almería (Spain): Towards an integrated and sustainable bioeconomy model. *New Biotechnol.* **2018**, *40*, 103–112. [CrossRef] [PubMed]
102. Ioppolo, G.; Cucurachi, S.; Salomone, R.; Shi, L.; Yigitcanlar, T. Integrating strategic environmental assessment and material flow accounting: A novel approach for moving towards sustainable urban futures. *Int. J. Life Cycle Assess.* **2019**, *24*, 1269–1284. [CrossRef]
103. Näyhä, A. Finnish forest-based companies in transition to the circular bioeconomy—Drivers, organizational resources and innovations. *For. Policy Econ.* **2020**, *110*, 101936. [CrossRef]
104. D'Amato, D.; Korhonen, J. Integrating the green economy, circular economy and bioeconomy in a strategic sustainability framework. *Ecol. Econ.* **2021**, *188*, 107143. [CrossRef]
105. Tylecote, A. Biotechnology as a new techno-economic paradigm that will help drive the world economy and mitigate climate change. *Res. Policy* **2019**, *48*, 858–868. [CrossRef]
106. Wang, S.; Li, Y.; Chen, Q.; Feng, X.; Shi, X.; Huang, Y.; Mei, L.; Li, W.; Liu, H.; Qi, X.; et al. Integration of biosafety surveillance through Biosafety Surveillance Conceptual Data Model. *Biosaf. Health* **2019**, *1*, 98–104. [CrossRef]

107. Solarte-Toro, J.C.; Alzate, C.A.C. Biorefineries as the base for accomplishing the sustainable development goals (SDGs) and the transition to bioeconomy: Technical aspects, challenges and perspectives. *Bioresour. Technol.* **2021**, *340*, 125626. [CrossRef] [PubMed]
108. D'Amico, G.; Arbolino, R.; Shi, L.; Yigitcanlar, T.; Ioppolo, G. Digital technologies for urban metabolism efficiency: Lessons from urban agenda partnership on circular economy. *Sustainability* **2021**, *13*, 6043. [CrossRef]
109. Honjo, Y.; Nagaoka, S. Initial public offering and financing of biotechnology start-ups: Evidence from Japan. *Res. Policy* **2018**, *47*, 180–193. [CrossRef]
110. Guillen, J.; Asche, F.; Carvalho, N.; Fernández Polanco, J.M.; Llorente, I.; Nielsen, R.; Nielsen, M.; Villasante, S. Aquaculture subsidies in the European Union: Evolution, impact and future potential for growth. *Mar. Policy* **2019**, *104*, 19–28. [CrossRef]
111. Agovino, M.; Casaccia, M.; Crociata, A.; Sacco, P.L. European Regional Development Fund and pro-environmental behaviour. The case of Italian separate waste collection. *Socio-Econ. Plan. Sci.* **2019**, *65*, 36–50. [CrossRef]
112. Sin, A.; Nowak, C. Comparative analysis of EAFRD's measure 121 ("Modernization of agricultural holdings") implementation in Romania and Poland. *Procedia Econ. Financ.* **2014**, *8*, 678–682. [CrossRef]
113. Polido, A.; Pires, S.M.; Rodrigues, C.; Teles, F. Sustainable development discourse in Smart Specialization Strategies. *J. Clean. Prod.* **2019**, *240*, 118224. [CrossRef]
114. Deakin, M.; Mora, L.; Reid, A. The research and innovation of Smart Specialisation Strategies: The transition from the Triple to Quadruple Helix. In Proceedings of the 27th International Scientific Conference on Economic and Social Development, Rome, Italy, 1–2 March 2018; pp. 94–103.
115. O'Brien, M.; Wechsler, D.; Bringezu, S.; Schaldach, R. Toward a systemic monitoring of the European bioeconomy: Gaps, needs and the integration of sustainability indicators and targets for global land use. *Land Use Policy* **2017**, *66*, 162–171. [CrossRef]
116. Angouria-Tsorochidou, E.; Teigiserova, D.A.; Thomsen, M. Limits to circular bioeconomy in the transition towards decentralized biowaste management systems. *Resour. Conserv. Recycl.* **2021**, *164*, 105207. [CrossRef]
117. Pätäri, S.; Arminen, H.; Albareda, L.; Puumalainen, K.; Toppinen, A. Student values and perceptions of corporate social responsibility in the forest industry on the road to a bioeconomy. *For. Policy Econ.* **2017**, *85*, 201–215. [CrossRef]
118. Lazaro-Mojica, J.; Fernandez, R. Review paper on the future of the food sector through education, capacity building, knowledge translation and open innovation. *Curr. Opin. Food Sci.* **2021**, *38*, 162–167. [CrossRef]
119. Giurca, A.; Metz, T. A social network analysis of Germany's wood-based bioeconomy: Social capital and shared beliefs. *Environ. Innov. Soc. Transit.* **2018**, *26*, 1–14. [CrossRef]
120. Böcher, M.; Töller, A.E.; Perbandt, D.; Beer, K.; Vogelpohl, T. Research trends: Bioeconomy politics and governance. *For. Policy Econ.* **2020**, *118*, 102219. [CrossRef]
121. Guerrero, J.E.; Hansen, E. Company-level cross-sector collaborations in transition to the bioeconomy: A multi-case study. *For. Policy Econ.* **2021**, *123*, 102355. [CrossRef]
122. Berthet, E.T.; Hickey, G.M.; Klerkx, L. Opening design and innovation processes in agriculture: Insights from design and management sciences and future directions. *Agric. Syst.* **2018**, *165*, 111–115. [CrossRef]
123. Wallin, I.; Pülzl, H.; Secco, L.; Sergent, A.; Kleinschmit, D. Research trends: Orchestrating forest policy-making: Involvement of scientists and stakeholders in political processes. *For. Policy Econ.* **2018**, *89*, 1–3. [CrossRef]
124. Cordella, A.; Paletti, A. Government as a platform, orchestration, and public value creation: The Italian case. *Gov. Inf. Q.* **2019**, *36*, 101409. [CrossRef]
125. Otero-Muras, I.; Carbonell, P. Automated engineering of synthetic metabolic pathways for efficient biomanufacturing. *Metab. Eng.* **2021**, *63*, 61–80. [CrossRef]
126. Pelai, R.; Hagerman, S.M.; Kozak, R. Biotechnologies in agriculture and forestry: Governance insights from a comparative systematic review of barriers and recommendations. *For. Policy Econ.* **2020**, *117*, 102191. [CrossRef]
127. Ethiopia. Ethiopia's Climate-Resilient Green Economy. 2011. Available online: https://www.preventionweb.net/files/61504_eth iopiacrge.pdf (accessed on 26 November 2021).
128. *National Strategy and Plan of Action for Pharmaceutical Manufacturing Development in Ethiopia (2015–2025)*; Ministry of Health and Ministry of Industry: Addis Ababa, Ethiopia, 2015. Available online: https://www.who.int/phi/publications/Ethiopia_strategy _local_poduction.pdf (accessed on 26 November 2021).
129. International Council of Biotechnology Associations. *Biotechnology: Driving Solutions for Sustainable Development*; ICBA: Washington, DC, USA, 2019. Available online: https://www.bio.org/sites/default/files/2019-11/ICBA%202019_SDG%20Brochure_Final. pdf (accessed on 27 November 2021).
130. *Draft Bioenergy Policy for Ghana*; Energy Commission: Accra, Ghana, 2010. Available online: https://www.cleancookingalliance.o rg/binary-data/RESOURCE/file/000/000/69-1.pdf (accessed on 28 November 2021).
131. Kenya. A National Biotechnology Development Policy. 2006. Available online: https://www.embrapa.br/documents/1355145/148 54343/National+biotechnology+Awareness+Strategy.pdf/d424fec3-7dfa-470f-9aee-923b9af24d04 (accessed on 26 November 2021).
132. *Strategy for Developing the Bio-Diesel Industry in Kenya (2008–2012)*; Ministry of Energy, Renewable Energy Department: Nairobi, Kenya, 2008. Available online: https://kerea.org/wp-content/uploads/2012/12/National-Biodiesel-Strategy-2008-2012.pdf (accessed on 1 December 2021).

133. *Draft Strategy and Action Plan for Bioenergy and LPG Development in Kenya*; Ministry of Energy and Petroleum: Nairobi, Kenya, 2015. Available online: https://kepsa.or.ke/download/draft-strategy-and-action-plan-for-bioenergy-and-lpg-development-in-kenya/ (accessed on 24 November 2021).

134. *Green Economy Strategy and Implementation Plan 2016–2030*; Government of Kenya, Ministry of Environment, Natural Resources: Nairobi, Kenya, 2016. Available online: http://www.environment.go.ke/wp-content/uploads/2018/08/GESIP_Final23032017.pdf (accessed on 24 November 2021).

135. *Malawi National Export Strategy*; Government of Malawi: Lilongwe, Malawi, 2013. Available online: https://kulimamalawi.org/wp-content/uploads/2019/12/Malawi-National-Export-Strategy-NES-Main-Volume.pdf (accessed on 26 November 2021).

136. *Stratégie Nationale de Development des Biocarburants au Mali*; Agence Nationale de Développment des Biocarburants: Bamako, Mali, 2009. Available online: http://www.bio-step.eu/fileadmin/BioSTEP/Bio_strategies/Mali_Biofuel_strategie_2009.pdf (accessed on 27 November 2021).

137. Strategie de Developpment de la Maitrise de l'energie au Mali. *Groupe de la Banque Africaine de Developpment*; Department Regionale Ouest II ORWB: Bamako, Mali, 2010. Available online: https://www.afdb.org/fileadmin/uploads/afdb/Documents/Project-and-Operations/%C2%B2Mali%20-%20Strat%C3%A9gie%20de%20d%C3%A9veloppement%20de%20m%C3%AEtrise%20de%20l\T1\textquoteright%C3%A9nergie_02.pdf (accessed on 28 November 2021).

138. Cervigni, R.; Scandizzo, P.L. *The Ocean Economy in Mauritius: Making It Happen, Making It Last*; World Bank Group: Washington, DC, USA, 2017. Available online: http://hdl.handle.net/10986/28562 (accessed on 29 November 2017).

139. *Green Growth Mozambique—Policy Review and Recommendations for Action*; African Development Bank, Energy, Environment and Climate Change Department (ONEC): Mozambique, 2015. Available online: https://www.afdb.org/fileadmin/uploads/afdb/Documents/Generic-Documents/Transition_Towards_Green_Growth_in_Mozambique_-_Policy_Review_and_Recommendations_for_Action.pdf (accessed on 29 November 2021).

140. Namibia. The National Programme on Research, Science, Technology and Innovation. 2014. Available online: https://www.ncrst.na/files/downloads/ee9_NCRST_NPRSTI%202014%20to%202017.pdf (accessed on 26 November 2021).

141. Abuja, Nigeria. Official Gazette of the Nigerian Bio-Fuel Policy and Incentives. 2007. Available online: https://storage.googleapis.com/cclow-staging/o3f5giwh2pmpdx3ncoasbht74v8g?GoogleAccessId=laws-and-pathways-staging%40soy-truth-247515.iam.gserviceaccount.com&Expires=1622990071&Signature=n6hQcVEqz08201WbnRzW16uj0%2BtkdANdbCi962CxeamEWyilq y3DQMeqNO0TlwtJIL7WEpcWyFHPqnMcfpCyU2mGxLgh9cbPWrbzs0K5ZOkIf6NtOeO8jVkk88VMRRDrwmT2Shr9%2Fh8je EjxBAGSAqlSJ5LXK0ggdPNJrlDNmK%2BRGwtxRnoiPMknKD6EhyRjwcNQY2GfD1XNaOQgFOeiFj%2FJIjSCSTspPN4L71F %2Bc8GNGE4Tr8H5RWd6rlCPr8uyosv2iuiS4s3Jap7FEakxt0Vcq37WmUBVVrqRBrYLUvmqtqPA8C6FY9O3GXCzsbzgsCdSw jwCoa9xcHIApTe3Qg%3D%3D&response-content-disposition=inline%3B+filename%3D%22f%22%3B+filename%2A%3DUTF-8%27%27f&response-content-type=application%2Fpdf (accessed on 26 November 2021).

142. Rwanda. Green Growth and Climate Resilience. *National Strategy for Climate Change and Low Carbon Development*. 2011. Available online: https://storage.googleapis.com/cclow-staging/3bz89qjih4msz1lzm4hix4o0n1ar?GoogleAccessId=laws-and-pathways-staging%40soy-truth-247515.iam.gserviceaccount.com&Expires=1622990647&Signature=c2ozAjQWfCDpBb6Tq2okElgUE7yin 5qcz4x4c7MfK572kFvNIRJvynqx%2FPJZbjI18Own45RAoyapjcqSueBQULf%2FiNGkgG1VxBEX3N3iMkc8mbqrcNTXkGt2%2B REG6DdGAc8S0vjOQJAJfHKYxyAdMVbzXQFnr2oVTT%2BTpQ5WqHTtINcx1T7hUhea%2FAtY65f4TW%2B6NTR%2Buln9i kuokw%2B2BRKnmyLH4 Okz FN1VMMmrWCXyRXXwVdyWhLHehKHYYKQpFwzA2vBXEo%2BY5F5LP1XrmjYUGkAqz dSFMuPbz7W8rYAt8E1FBrVaXdIAo0K4E2bhj0VMbX3%2FQjOtqs2IjzaW1Q%3D%3D&response-content-disposition=inline%3B+filename%3D%22f%22%3B+filename%2A%3DUTF-8%27%27f&response-content-type=application%2Fpdf (accessed on 26 November 2021).

143. Senegal Economic Update. *Learning from the Past for a Better Future*; World Bank Group: Senegal, 2014. Available online: https://openknowledge.worldbank.org/bitstream/handle/10986/21504/942580WP0Box380alEconomicUpdate0010.pdf?sequence=1&isAllowed=y (accessed on 26 November 2021).

144. *National Biotechnology Policy*; Ministry of Communication, Science and Technology: Dodoma, Tanzania, 2010. Available online: http://www.tzonline.org/pdf/Biotecchnology_Policy_WEBB1.pdf (accessed on 26 November 2021).

145. The Potential of Bio-Fuel in Uganda. *An Assessment of Land Resources for Bio-fuel Feedstock Suitability*; National Environment Management Authority: Kampala, Uganda, 2010. Available online: https://wedocs.unep.org/bitstream/handle/20.500.11822/9199/-The%20Potential%20of%20Bio-fuel%20in%20Uganda%3A%20An%20assessment%20of%20land%20resources%20for%20bio-fuel%20feedstock%20suitability-2010Bio-fuels%20in%20Uganda.pdf?sequence=3&isAllowed=y (accessed on 26 November 2021).

146. Argentina. La Bioeconomía como Estrategia para el Desarrollo Argentino. 2019. Available online: https://fibamdp.files.wordpress.com/2020/06/la-bioeconomicc81a-como-estrategia-para-el-desarrollo-argentino.pdf (accessed on 24 November 2021).

147. *National Strategy for Science, Technology and Innovation*; Ministry of Science, Technology, Innovations and Communications: Brasilia, Brazil, 2016. Available online: https://portal.insa.gov.br/images/documentos-oficiais/ENCTI-MCTIC-2016-2022.pdf (accessed on 1 December 2021).

148. *Plano de Ação em Ciência, Tecnologia e inovação em Bioeconomia*; Ministério da Ciência, Tecnologia, Inovações e Comunicações (MCTIC): Brasilia, Brazil, 2018. Available online: https://antigo.mctic.gov.br/mctic/export/sites/institucional/ciencia/SEPED/Arquivos/PlanosDeAcao/PACTI_BIOECONOMIA_web.pdf (accessed on 2 December 2021).

149. A Forest Bioeconomy Framework for Canada. Canadian Council of Forest Ministers. 2017. Available online: https://www.ccfm.org/wp-content/uploads/2017/08/10a-Document-Forest-Bioeconomy-Framework-for-Canada-E.pdf (accessed on 26 November 2021).

150. Canada's Bioeconomy Strategy. Leveraging our Strengths for a Sustainable Future. *Bioindustrial Innovation Canada*. 2019. Available online: https://www.fpac.ca/wp-content/uploads/b22338_1906a509c5c44870a6391f4bde54a7b1.pdf (accessed on 26 November 2021).

151. *Política para el Desarrollo Commercial de la Biotecnología a Partir del uso Sostenible de la Biodiversidad*; Departamento Nacional de Planeación, República de Colombia: Bogotá, Colombia, 2011. Available online: http://repositorio.colciencias.gov.co/bitstream/handle/11146/231/1541-Desarrollo%20Ccial%20Biotecnolog%c3%ada%20CONPES%203697_2011_%201.pdf?sequence=1&isAllowed=y (accessed on 26 November 2021).

152. *Programa Nacional de Biocomercio Sostenible (2014–2024)*; Ministerio de Ambiente y Desarrollo Sostenible: Bogotá, Colombia, 2014. Available online: https://www.minambiente.gov.co/images/NegociosVerdesysostenible/pdf/biocomercio_/PROGRAMA_NACIONAL_DE_BIOCOMERCIO_SOSTENIBLE.pdf (accessed on 26 November 2021).

153. Colombia Bio. 2016. Available online: https://minciencias.gov.co/sites/default/files/upload/paginas/colombiabio-program-2016.pdf (accessed on 26 November 2021).

154. *Política de Crecimiento Verde*; Consejo Nacional de Política Económica Y Social, Departamento Nacional de Planeación, República de Colombia: Bogotá, Colombia, 2018. Available online: http://www.andi.com.co/Uploads/CONPES%20CV%203934.pdf (accessed on 24 November 2021).

155. Ecuador. *The Bioeconomy: A Catalyst for the Sustainable Development of Agriculture and Rural Territories in LAC*; Food and Agriculture Organization of the United Nations: San Jose, Costa Rica, 2019. Available online: http://www.iica-ecuador.org/sisbio/doc_informacion/IICA_Cap4_Eng_V4.pdf (accessed on 27 November 2021).

156. Mexico. *La Bioeconomía. Nuevo Marco para el Crecimiento Sostenible en América Latina*; Editorial Pontificia Universidad Javeriana: Bogotà, Colombia, 2019. Available online: https://agritrop.cirad.fr/592946/7/ID592946.pdf (accessed on 27 November 2021).

157. Paraguay. Estrategia Nacional y Plan de Acción para la Conservación de la Biodiversidad del Paraguay 2015–2020. *Programa de las Naciones Unidas para el Desarrollo (PNUD) y Fondo para el Medio Ambiente Mundial (FMAM). Asunción*. 2016, p. 190. Available online: https://www.cbd.int/doc/world/py/py-nbsap-v2-es.pdf (accessed on 27 November 2021).

158. *Plan Sectorial Biotecnología*; Gabinete Productivo: Montevideo, Uruguay, 2012. Available online: https://www.miem.gub.uy/sites/default/files/plan_sectorial_biotecnologia.pdf (accessed on 27 November 2021).

159. Uruguay Agrointeligente. *Los Desafíos para Desarrollo Sostenible*; Ministerio De Ganadería, Agricultura Y Pesca: Montevideo, Uruguay, 2017. Available online: https://www.gub.uy/ministerio-ganaderia-agricultura-pesca/sites/ministerio-ganaderia-agricultura-pesca/files/2019-12/libro%20completo%20con%20hipervinculos.pdf (accessed on 27 November 2021).

160. *Aportes para una Estrategia de Desarrollo 2050*; Dirección de Planificación: Montevideo, Uruguay, 2019. Available online: https://observatorioplanificacion.cepal.org/sites/default/files/plan/files/Estrategia_Desarrollo_2050.pdf (accessed on 27 November 2021).

161. Smith, M. *Developing a Bioeconomy in South Australia. Adelaide Thinker in Residence*; Government of South Australia, Department of the Premier and Cabinet: Adelaide Australia, 2005. Available online: https://www.dunstan.org.au/wp-content/uploads/2018/12/TIR_Reports_2004_Smith.pdf (accessed on 29 November 2021).

162. Bird, J. Opportunities for Primary Industries in the Bioenergy Sector. In *National RD&E Strategies*; Rural Industries Research and Development Corporation: Kingston, Australia, 2011. Available online: https://www.agrifutures.com.au/wp-content/uploads/publications/11-079.pdf (accessed on 29 November 2021).

163. *National Marine Science Plan (2015–2025)*; National Marine Science Committee, Australian Government: Coffs Harbor, Australia, 2015. Available online: https://www.marinescience.net.au/wp-content/uploads/2018/06/National-Marine-Science-Plan.pdf (accessed on 29 November 2021).

164. Beijing, China. The Outline of the 12th Five-Year Program for National Economic and Social Development of the People's Republic of China. 2011. Available online: https://www.asifma.org/wp-content/uploads/2018/05/prc-12th-fyp1.pdf (accessed on 29 November 2021).

165. *Circular of the State Council on Issuing the National 13th Five-Year Plan for the Development of Strategic Emerging Industries*; PRC State Council: Beijing, China, 2016. Available online: https://cset.georgetown.edu/wp-content/uploads/Circular-of-the-State-Council-on-Issuing-the-National-13th-Five-Year-Plan-for-the-Development-of-Strategic-Emerging-Industries.pdf (accessed on 29 November 2021).

166. KPMG. The 13th Five-Year Plan—China's Transformation and Integration with the World Economy. 2016. Available online: https://assets.kpmg/content/dam/kpmg/cn/pdf/en/2016/10/13fyp-opportunities-analysis-for-chinese-and-foreign-businesses.pdf (accessed on 29 November 2021).

167. Linster, M.; Yang, C. *China's Progress towards Green Growth: An International Perspective*; OECD Green Growth Papers, No. 2018/05; OECD Publishing: Paris, France, 2018.

168. *Biotechnology Policy 2015–2020*; Government of Andhra Pradesh, India: Andhra Pradesh, India, 2015. Available online: https://investuttarakhand.com/themes/backend/acts/act_english1575360796.pdf (accessed on 29 November 2021).

169. *National Biotechnology Development Strategy 2015–2020*; Department of Biotechnology, Ministry of Science & Technology, Government of India: New Delhi, India, 2015. Available online: http://dbtindia.gov.in/sites/default/files/DBT_Book-_29-december_2015.pdf (accessed on 29 November 2021).

170. BIRAC. *India Bioeconomy Report 2020*; Biotechnology Industry Research Assistance Council: New Delhi, India, 2020. Available online: https://birac.nic.in/webcontent/1594624763_india_bioeconomy_rep.pdf (accessed on 29 November 2021).

171. Government Regulation on National Energy Policy. National Energy Policy, The President of the Republic of Indonesia. 2014. Available online: https://policy.asiapacificenergy.org/sites/default/files/National%20Energy%20Policy%202014_0.pdf (accessed on 29 November 2021).

172. Food and Agriculture Organization of the United Nations. *Assessing the Contribution of Bioeconomy to Countries' Economy*; FAO: Rome, Italy, 2018. Available online: http://www.fao.org/3/I9580EN/i9580en.pdf (accessed on 1 December 2021).

173. *Primary Sector Science Roadmap*; Ministry for Primary Industries: Wellington, New Zealand, 2017. Available online: https://www.mpi.govt.nz/dmsdocument/18383-primary-sector-science-roadmap-te-ao-turoa-strengthening-new-zeala nds-bioeconomy-for-future-generations (accessed on 29 November 2021).

174. *Agritech Industry Transformation Plan*; Ministry of Business, Innovation & Employment, New Zealand Government: Wellington, New Zealand, 2020. Available online: https://www.mbie.govt.nz/dmsdocument/11572-growing-innovative-industries-in-new- zealand-agritech-industry-transformation-plan-july-2020-pdf (accessed on 29 November 2021).

175. Popov, V.; Prospects of Bioeconomy in the Russian Federation. National Technology Platform, Bioindustry and Bioresources— Biotech2030. 2012. Available online: https://www.oecd.org/sti/emerging-tech/Popov.pdf (accessed on 29 November 2021).

176. *Russian Government Roadmap for Development of Biotechnology*; Government of the Russian Federation: Moscow, Russia, 2013. Available online: https://apps.fas.usda.gov/newgainapi/api/report/downloadreportbyfilename?filename=Russian%20Government %20Roadmap%20for%20Development%20of%20Biotechnology_Moscow_Russian%20Federation_9-20-2013.pdf (accessed on 28 November 2021).

177. *Biotechnology in Korea*; Italian Trade Agency: Rome, Italy, 2014. Available online: https://www.infomercatiesteri.it/public/imag es/paesi/123/files/Korea%20Biotech%20Report.pdf (accessed on 29 November 2021).

178. *Biotechnology in Korea*; Ministry of Education, Science and Technology: Seoul, Korea, 2020. Available online: https://www.kribb. re.kr/eng/file/Biotechnology_in_Korea_0905.pdf (accessed on 29 November 2021).

179. Bioenergy in Sri Lanka: Resources, Applications and Initiatives. Practical Action Consulting. 2010. Available online: https: //assets.publishing.service.gov.uk/media/57a08b12ed915d622c000aab/PISCES_Sri_Lanka_Bioenergy_Working_Paper.pdf (accessed on 29 November 2021).

180. Musafer, N. Biomass energy in Sri Lanka: Retrospective and prospective analysis. In Proceedings of the International Conference on Advanced Materials for Clean Energy and Health applications (AMCEHA 2019), Jaffna, Sri Lanka, 6–8 February 2019. Available online: https://www.bioenergysrilanka.lk/wp-content/uploads/2019/02/Biomass-Energy-in-Sri-Lanka.pdf (accessed on 29 November 2021).

181. *Bioeconomy in Flanders*; Environment, Nature and Energy Department, Flemish Government: Brussels, Belgium, 2019. Available online: https://publicaties.vlaanderen.be/view-file/13902 (accessed on 29 November 2021).

182. *Rural Development Programme of the Republic of Croatia for the Period 2014–2020*; Ministry of Agriculture, Directorate for Management of EU Funds for Rural Development, EU and International Co-operation: Zagreb, Croatia, 2016. Available online: https: //ruralnirazvoj.hr/files/documents/Programme_2014HR06RDNP001_3_1_en.pdf (accessed on 27 November 2021).

183. *Bioeconomy Concept in the Czech Republic from the Perspective of the Ministry of Agriculture*; Ministry of Agriculture: Praha, Czechia, 2019. Available online: http://eagri.cz/public/web/file/630927/Koncepce_biohospodarstvi_v_CR_z_pohledu_MZe_na_leta _2019_24.pdf (accessed on 27 November 2021).

184. *Denmark at Work. Plan for Growth for Water, Bio & Environmental Solutions*; The Danish Government: Copenhagen, Denmark, 2013. Available online: https://eng.em.dk/media/10603/12-03-13-summary-plan-for-growth-for-water-bio-etc.pdf (accessed on 27 November 2021).

185. *Research 2025—Promising Future Research Areas*; Danish Agency for Science and Higher Education: Copenhagen, Denmark, 2018. Available online: https://ufm.dk/en/publications/2018/filer/forsk25_katalog_eng_enkelt.pdf (accessed on 26 November 2021).

186. *Strategy for Circular Economy*; Ministry of Environment and Food and Ministry of industry, Business and Financial Affairs, The Danish Government: Copenhagen, Denmark, 2018. Available online: https://stateofgreen.com/en/uploads/2018/10/Strategy-f or-Circular-Economy-1.pdf (accessed on 25 November 2021).

187. *National Policy Strategy on Bioeconomy*; Federal Ministry of Food and Agriculture: Berlin, Germany, 2014. Available online: https://www.bioways.eu/download.php?f=62&l=en&key=c21c2ea7e095424f3545c66da7b98821 (accessed on 23 November 2021).

188. *National Policy Statement on the Bioeconomy*; Government of Ireland: Dublin, Ireland, 2018. Available online: https://assets.gov.ie/ 2244/241018115730-41d795e366bf4000a6bc0b69a136bda4.pdf (accessed on 23 November 2021).

189. *Integrated Science, Studies and Business Centres (Valleys)*; Ministry of Education, Science and Sport, Republic of Lithuania: Vilnius, Lithuania, 2019. Available online: https://www.smm.lt/web/en/science1/science_1 (accessed on 25 November 2021).

190. *The Position of the Bioeconomy in the Netherlands*; Ministry of Economic Affairs and Climate Policy: The Hague, The Netherlands, 2018. Available online: https://www.government.nl/documents/leaflets/2018/04/01/the-position-of-the-bioeconomy-in-th e-netherlands (accessed on 26 November 2021).

191. Neto, C.P. *Forest-Based Biorefineries: The Pulp and Paper Industry*; RAIZ—Forest and Paper Research Institute: Aveiro, Portugal; The Navigator Company: Setúbal, Portugal, 2019. Available online: https://scar-europe.org/images/CASA/Events/Portugal_20ma y2019/presentations/Carlos_Neto.pdf (accessed on 28 November 2021).

192. Vasconcelos, V.; Moreira-Silva, J.; Moreira, S. *Portugal Blue Bioeconomy Roadmap—BLUEandGREEN*; CIMAR: Matosinhos, Portugal, 2019; p. 68. Available online: https://www2.ciimar.up.pt/pdfs/resources/roadmap_digital_hGBit_.pdf (accessed on 28 November 2018).

193. *Transition to a Green Economy in Slovenia*; Ministry of the Environment and Spatial Planning: Ljubljana, Slovenia, 2016. Available online: https://www.oneplanetnetwork.org/sites/default/files/transition_to_a_green_economy_in_slovenia.pdf (accessed on 29 November 2021).
194. Overbeek, G.; de Bakker, E.; Beekman, V.; Davies, S.; Kiresiewa, Z.; Delbrück, S.; Ribeiro, B.; Stoyanov, M.; Vale, M. Review of Bioeconomy Strategies at Regional and National Levels. 2016. Available online: http://www.bio-step.eu/fileadmin/BioSTEP/Bio_documents/BioSTEP_D2.3_Review_of_strategies.pdf (accessed on 29 November 2021).
195. Escudero, J. Overview of State of Play on Bioeconomy in Spain; 3rd Workshop "Facilitating Development of Bioeconomy Policy—Needs and Gaps"; Brussels, Belgium. 2019. Available online: https://www.scar-swg-sbgb.eu/lw_resource/datapool/_items/item_64/ws3_spain.pdf (accessed on 29 November 2021).
196. *Swedish Research and Innovation Strategy for a Bio-Based Economy*; Swedish Energy Agency VINNOVA, The Swedish Research Council for Environment, Agricultural Sciences and Spatial Planning: Stockholm, Sweden, 2012. Available online: https://www.formas.se/download/18.462d60ec167c69393b91e60f/1549956092919/Strategy_Biobased_Ekonomy_hela.pdf (accessed on 29 November 2021).

sustainability

MDPI

Article

Utilization of Agro-Industrial Wastes for the Production of Quality Oyster Mushrooms

Morzina Akter [1], Riyadh F. Halawani [2], Fahed A. Aloufi [2], Md. Abu Taleb [2], Sharmin Akter [1] and Shreef Mahmood [1,*]

1   Department of Horticulture, Hajee Mohammad Danesh Science and Technology University, Dinajpur 5200, Bangladesh; momehoque16@gmail.com (M.A.); sharminkhandaker1993@gmail.com (S.A.)
2   Department of Environmental Science, Faculty of Meteorology, Environment and Arid Land Agriculture, King Abdulaziz University, Jeddah 21589, Saudi Arabia; rhalawani@kau.edu.sa (R.F.H.); faloufi@kau.edu.sa (F.A.A.); taleb@manarat.ac.bd (M.A.T.)
*   Correspondence: smahmood@hstu.ac.bd; Tel.: +880-1712289013

Abstract: The objective of this study was to utilize agro-lignocellulosic wastes for growing oyster mushroom which become problematic for disposal. *Pleurotus ostreatus* was cultivated on five agro-industrial wastes: rice straw (RS), wheat straw (WS), corncobs (CC), saw dust and rice husk @ 3:1 (SR) and sugarcane bagasse (SB). Approximately 500 g sized polypropylene bags (20.32 × 30.48 cm) were used for each substrate. The SR significantly improved the number of fruiting body (27.80), size of the fruiting body (5.39 g), yield (115.13 g/packet), ash and shortened the days for stimulation to primordial initiation and harvest (9.2 days). The maximum percentage of visual mycelium growth with the least time (15.0 days) to complete the mycelium running was found in SB, whereas the highest biological efficiency value (56.5) was calculated in SR. The topmost value of total sugar (33.20%) and ash (10.87 g/100 g) were recorded in WS, whereas the utmost amount of protein (6.87 mg/100 g) and total polyphenolics (196.88 mg GAE/100 g) were detected from SB and SR, respectively. Overall SR gave the highest amount of the fruiting body with the topmost polyphenols and ash, moderate protein and total sugar, and secured maximum biological efficiency too. The results demonstrate that saw dust with rice husk could be used as an easy alternative substrate for oyster mushroom cultivation.

Keywords: oyster mushroom; agro-industry; wastes; productivity; quality

Citation: Akter, M.; Halawani, R.F.; Aloufi, F.A.; Taleb, M.A.; Akter, S.; Mahmood, S. Utilization of Agro-Industrial Wastes for the Production of Quality Oyster Mushrooms. *Sustainability* 2022, 14, 994. https://doi.org/10.3390/su14020994

Academic Editors: Nallapaneni Manoj Kumar, Md. Ariful Haque and Sarif Patwary

Received: 1 December 2021
Accepted: 13 January 2022
Published: 17 January 2022

Publisher's Note: MDPI stays neutral with regard to jurisdictional claims in published maps and institutional affiliations.

1. Introduction

Developing countries such as Bangladesh suffer much from a food insecurity problem, mainly due to inadequate and imbalanced diet intake. The problem is further compounded by the rapid growth of the population in the country. As a consequence, people, especially children and women, are experiencing chronic malnutrition problems. Mushrooms could substantiate this malnutrition problemto some extent, as edible mushrooms are rich sources of protein, vitamins, minerals and also contain a number of secondary plant metabolites [1–4]. Bangladesh is an agrobased country and various agroindustries generate a large amount of lignocellulosic byproducts annually that are worthy of being transformed. Some of these byproducts are used as feeds for livestock and the compost industry, and some are still treated as waste. These wastes are mainly burned for cooking purposes or disposed into surrounding environments, leading to various environmental problems. However, these agro-industrial wastes can potentially be used in cultivating mushrooms, which, in turn, contribute to minimizing malnutrition problems and could reduce the environmental pollution [5]. In addition, such uses helplandless and marginal farmers to increase their income through intensive indoor farming and create employment opportunities, especially for unemployed youth and women folk. These actions would directly impact Sustainable Development Goals.

Edible mushrooms are saprophytic fungi and have the ability to degrade lignocellulosic materials by their extensive enzymes [6]. Among the edible mushrooms, *Pleurotusostreatus* is ranked first in Bangladesh because of its adaptability in local climatic conditions and ability to grow on a wide range of substrates [5]. Different studies also reported the potential uses of various agro-industrial residues, including cotton waste, wheat straw, sawdust, rice straw, sugarcane bagasse, and corncobs, in mushroom cultivation [7–9]. In Bangladesh, rice straw is usually used as a substrate to cultivate mushrooms; however, its demand is increasing day by day because of the expansion of cattle farming. The availability of sufficient rice straw all year round, in all parts of the country, is also uncertain. Therefore, the potentiality of other agro-industrial wastes, such as wheat straw, rice husk, corn cob, and sugarcane bagasse, etc., needs to be evaluated to identify options that are cost effective, and can provide a better yield and quality of mushroom. Proper use of these agro-industrial wastes as substrates for mushroom cultivation could improve the economic status of the farmers, contribute to alleviating nutritional problems and would reduce environmental pollutions. In this context, the present study has been undertaken to evaluate the productivity and quality of oyster mushrooms using different locally produced agro-industrial wastes.

2. Materials and Methods

2.1. Location and Treatments

The present investigation was carried out both at the Laboratories of the Horticulture, and Food Processing and Preservation, Hajee Mohammad Danesh Science and Technology University (HSTU), Bangladesh. The single factor experiment consisted of five treatments, i.e., different types of substrates: rice straw (RS), wheat straw (WS), corn cobs (CC), mixture of 75% saw dust and 25% rice husk (SR), and sugarcane bagasse (SB).

2.2. Collection and Preparation of Substrates

Rice straw, wheat straw, corn cobs and sugarcane bagasse were collected locally from the Agricultural Farm, Parbatipurupazila, Dinajpur and saw dust from Horticulture Center, Dinajpur. All the substrates were chopped (2 cm length), except the mixture of saw dust and rice husk, and immersed in water for about 24 h to achieve 65–70% moisture. The next day, after removing excess water from the substrates, all those substrates were boiled for 1 h and cooled. The pasteurized substrates were cooled and used for mushroom cultivation.

2.3. Preparation of Spawn Packet

Packets of spawn were prepared separately with polypropylene bags (20.32 × 30.48 cm) with each type of the substrates. Firstly, a layer of the prepared substrates was placed into a polypropylene packet and, afterwards, approximate 125 g of the cultured mother spawn was spread on the outer side of the substrate. Depending on the type of substrates, the weight of the spawn packet was approximately 500 g. The spawning process was repeated again following the same procedure, and the top most layer of the spawn was covered with the minimum amount of substrate. The neck of the packet was covered with a heat resistant plastic neck and plugged with cotton. Afterward, the neck was covered with brown paper by placing a rubber band to hold it in place. All the packets were placed on the floor of the laboratory with the necessary hygienic measures.

2.4. Cultivation of Spawn Packets and Harvest of Fruiting Bodies

When all packets were covered by mycelium, then the cotton plug, brown paper, rubber bands were removed. In the case of saw dust mixed with rice husk packets, the upper position of both sides of the plastic packet were cut into a "D" shape with a sharp knife. However, in the cases of the other substrates, four (4) cuts were made in a rectangular (5 × 1 cm) shape. After removing the plastic sheet, the substrate of the cut surface was scraped to remove the thin whitish mycelium layer. The packets were placed separately on the floor of the culture room and covered with a brown paper. High humidity was maintained in the culture room by spraying water thrice daily. The light in the culture room was totally cutoff, but the ventilation was maintained throughout the culture time. The humidity and temperature of the culture room was recorded at 3h intervals. Harvesting was performed as the fruiting bodies came out from the cut surface of the packet and attained the maximum size.

2.5. Parameters Recorded

2.5.1. Physical Parameters

Parameter	Procedure of Measurement
Percent (%) visual mycelium	This was measured for each substrate before the mycelium surrounded the packet. It was noted at 4, 8, 12 and 15 days after inoculation (DAI) and the percentages were estimated with the observation of the naked eye.
Days required to complete the mycelium running in spawn	This indicates the days required from inoculation to the completion of the running of the mycelium. When the whole spawn packet turned white with the growth of the mycelium, then it was noted as the indication of the completion of the mycelium running of spawn.
Days required from stimulation to the primordia initiation	The days required from cutting the spawn packet to primordia initiation were recorded and measured at both the first and the second flush.
Days required from stimulation to harvest	The time (days) required from stimulation to harvesting was counted as the sum of days required from stimulation to harvesting and was recorded.
Number of effective fruiting bodies per packet (NFBP)	W-developed fruiting bodies were considered as effective fruiting bodies which were counted and expressed in number per packet. However, the tiny fruiting bodies were not counted.
Diameter and length of stalk (cm)	The diameter and length of the stalk of each fruiting body was recorded from the top to the base of the stalk using an electric digital caliper (Model: Guanglu, China).
Diameter and thickness of cap (cm):	The cap diameter and thickness over one gram (wt.) was measured using an electric digital slide caliper (Model: Guanglu, China).
Individual and total weight of fruiting bodies per packet	The individual weight of each fruiting body (IWFB) was measured without removing the lower hard portion. The weight of all fruiting bodies per packet was weighed without the lower hard and dirty portions. The weight was measured using an electric balance (Model: PA 214, USA).
Biological efficiency	The following formula was used to calculate the biological efficiency [10]. $$\text{Biological efficiency } (\%) = \frac{\text{Weight of fresh mushrooms harvested per packet}}{\text{Weight of dry substrate per bag}} \times 100$$
Ash content	For determining ash, 1 g of each fruiting body was taken into a crucible. The crucible was placed in a muffle furnace for 6 h at 600 °C. Then, total ash was calculated using the following equation [11]: $$\text{Ash content } (\text{g}/100 \text{ g}) = \frac{\text{Weight of Ash}}{\text{Weight of Sample taken}} \times 100$$

2.5.2. Total Sugar Content (mg/100 g fw)

The total soluble sugar content of each fruiting body was determined by using the colorimetric method [12]. For this, firstly, 2 mL previously extracted of supernatant was diluted with 1 mL phenol solution (5%). Subsequently, 5 mL of H_2SO_4 (95.5%) was added to the samples. The testtubes were then allowed to stand for 10 min and vortexed for 30 s. The test tubes were kept in a water bath at room temperature for 20 min for color development. Finally, the absorbance was recorded using a UV-VIS spectrophotometer (PG Instrument Ltd., Bristol, UK) at a wavelength of 490 nm. The standard curve for the total soluble sugar determination was constructed by using glucose solutions whose concentrations ranged between 0 to 0.25 mg/mL.

2.5.3. Protein Content (mg/100 g of Fresh wt.)

The protein concentrations were determined using the colorimetric method [13]. Coomassie Brilliant Blue G-250 (0.04 mg/mL) and ortho-phosphoric acid (85%) were used as protein reagent in the assay. One gram of afresh sample was taken for preparing the extraction solution [14]. The fresh sample was extracted in 5 mL of 100 mM Tris-HCl (pH 7.5) using a homogenizer (Model: VELP Scientifica, Usmate Velate, Italy). After vigorously vortexing, the mixture was kept in a refrigerator at 4–5 °C for one hour and afterward centrifuged at 5300 rpm for 15 min at 4 °C. One hundred microliters (100 μL) of the supernatant was mixed with 1400 μL distilled water, to which previously prepared 1.5 mL Bearden solution was added. After vortexing, the absorbance was recorded at 595 nm by using a UV/VIS spectrophotometer (PG Instrument Ltd., Bristol, UK). The content of protein in the sample was calculated using bovine serum albumin (BSA, Sigma-Aldrich, Saint Louis, MO, USA) as the standard.

2.5.4. Total Phenolics (mg GAE/100 g of Fresh wt.)

The total phenolic compounds in the fruiting body were estimated by Folin-Ciocalteu reagent (FC) and the colorimetric method [15]. The extraction was performed using 1 g fresh sample [16]. The mushroom tissue was extracted in 4 mL methanol (80%) containing 2.7% HCl (37%), shaken for 2 h on an orbital shaker (200 rpm) at room temperature and centrifuged at 5300 rpm for 15 min at 4 °C. The extraction procedure was repeated again and the supernatants were combined for the total phenolic assay. Three hundred microliters (300 μL) of the extract was diluted with 2.25 mL of Folin-Ciocalteu reagent and 2.25 mL of sodium carbonate solution (60 g/L), respectively. The samples were vortexed and left for 90 min at room temperature. After incubation, the absorbance was recorded at 765 nm by using a UV/VIS spectrophotometer. Then, the content of total phenolics was quantified from a standard curve of gallic acid.

2.6. Statistical Analyses

The study was designed as a complete randomized design (CRD), with five treatments and each having five replicates. Data were subjected to analysis of variance (ANOVA) with the Statgraphics Plus Version 2.1 statistical program [17]. Comparisons of the treatment means was performed by the Fisher's Least Significant Difference (Lsd) test at 5% level of significance.

3. Results

3.1. Growth and Development of Mycelia and Fruiting Body

In general, the growth of mycelia in various substrates increased with the passage of time and notable variation ($p \leq 0.05$) was found among the substrates in different days after inoculation (DAI), except at 4 DAI (Table 1). At 16 DAI, the maximum growth was recorded in SB (100) and WS (97.0), while the lowest growth was in SR (46.6%). The same substrate (SB) also took the fewest days (15.0) from the day of inoculation to complete the mycelium running, but WS needed the most days (38.2) to complete the mycelium running. It was also observed that SR required significantly fewer days (2.6) from stimulation to primordia initiation in the first flush while in the second flush, the fewest days were required in WS (12.8). SR also required significantly fewer days (6.6 days) from stimulation to harvest in the first flush but in the second flush, fewer days were required (20.6 days) in WS (Table 2). In contrast, the WS, RS, SB and CC substrates took 8.0, 10.8, 13.0 and 13.4 days, respectively, for stimulation to harvest in the first flush, while in the second flush SB took the most days (31.2 days) and no harvest was possible in the CC substrate after the second flush.

Table 1. Percent visual mycelium of oyster mushroom on different agro-industrial waste substrates.

Substrates	Percent (%) Visual Mycelium at Different DAI			
	4	8	12	15
Rice straw	8.80 [NS]	22.80 c	31.40 b	66.00 b
Wheat straw	9.20	21.40 c	31.00 b	97.00 a
Corn cob	10.60	27.20 b	58.60 a	60.00 c
Saw dust with rice husk	9.20	16.20 d	31.80 b	46.60 d
Sugarcane bagasse	10.80	31.60 a	60.00 a	100.00 a
Lsd	2.01	3.72	3.90	3.86
CV (%)	15.67	11.82	6.95	3.96

[NS] Nonsignificant; the means with the same letter(s) in a column do not differ significantly as per Lsd test ($p \leq 0.05$).

Table 2. Growth of mycelium and number of fruiting bodies of oyster mushroom on different agro-industrial waste substrates.

Substrates	Complete Mycelium Running in Spawn	First Flush (Days)		Second Flush (Days)		Number of Fruiting Bodies Per Packet	
		Stimulation to Primordial Initiation	Stimulation to Harvest	Stimulation to Primordial Initiation	Stimulation to Harvest	1st Flush	2nd Flush
Rice straw	19.20 c	6.00 b	10.80 b	20.40 b	25.00 b	12.20 c	12.00 a
Wheat straw	16.60 d	4.60 c	8.00 c	12.80 c	20.60 c	13.60 b	12.20 a
Corn cob	20.60 b	7.80 a	13.40 a	-	-	8.80 d	-
Saw dust with rice husk	38.20 a	2.60 d	6.60 d	18.80 b	26.60 b	21.20 a	6.60 b
Sugarcane bagasse	15.00 e	5.40 bc	13.00 a	25.40 a	31.20 a	12.00 c	5.80 b
Lsd	1.10	0.85	0.79	1.95	2.66	1.31	1.10
CV (%)	3.82	12.27	5.78	7.54	7.69	7.30	9.01

The means with the same letter(s) in a column do not differ significantly as per Lsd test ($p \leq 0.05$).

3.2. Number, Size and Yield of Fruiting Bodies, and Biological Efficiency

The number of fruiting bodies per packet (NFBP) varied from 8.80 to 21.20 among the substrates (Table 2). The highest NFBP was counted in SR (21.20), followed by WS (13.6), RS (12.20) and SB (12.0), while the lowest was in CC (8.80) in the first flush whereas in the second flush, the highest NFBP was recorded in WS (12.20) and RS (12.0). When combining the NFBP of both the first and second flushes, then the highest NFBP was obtained from SR (21.2 + 6.6 = 27.8) and the lowest from CC (8.80). Regarding size, the highest diameter and length of stalk was measured in the substrate SR (1.34 and 0.41 cm) in the first flush but in the second flush, WS produced the longest stalk and maximum diameter (1.37 and 0.29 cm) (Table 3). In all cases, the lowest length and diameter of stalk was measured in SB. Similar to the stalk, the SR substrate also had the highest diameter and thickest cap (2.01 and 0.28 cm), while the lowest was recorded in CC (1.60 and 0.16 cm) in the first flush. It was also noted that the variations in thickness of cap among the substrates were not significant in the second flush.

Table 3. Size of fruiting body on different agro-industrial waste substrates.

Substrates	First Flush (cm)				Second Flush (cm)			
	Size of Stalk		Size of Cap		Size of Stalk		Size of Cap	
	Length	Diameter	Diameter	Thickness	Length	Diameter	Diameter	Thickness
Rice straw	1.04 bc	0.36 b	2.00 a	0.22 b	1.03 d	0.24 bc	1.66 c	0.19 [NS]
Wheat straw	1.13 b	0.39 ab	2.00 a	0.20 b	1.37 a	0.29 a	1.82 b	0.18
Corn cob	0.99 c	0.25 c	1.60 c	0.16 c	-	-	-	-
Saw dust with rice husk	1.34 a	0.41 a	2.01 a	0.28 a	1.21 b	0.26 ab	1.91 a	0.17
Sugarcane bagasse	0.89 d	0.22 c	1.84 b	0.20 b	1.12 c	0.21 c	1.68 c	0.17
Lsd	0.10	0.05	0.09	0.03	0.06	0.04	0.05	0.02
CV (%)	9.26	9.58	3.74	9.52	3.79	12.65	2.53	9.62

[NS] Nonsignificant; the means with the same letter(s) in a column do not differ significantly as per Lsd test ($p \leq 0.05$).

The IWFB and total weight of fruiting bodies per packet ranged from 2.22 to 5.39 g and 27.68 to 115.13 g, respectively; no harvest was possible in CC substrate at second flush (Table 4). In both flushes, the highest IWFB was measured in the substrate SR (142.58 g) followed bythe second highest in WS (127.36 g) and the lowest IWFB was in the CC substrate (27.68 g). In the first flush, the highest number of fruiting bodies per packet was harvested from SR (115.13 g) but in the second flush WS yielded the highest fruiting body (62.29 g). In all cases, a moderate yield was obtained from RS (107.90 g) and SB (64.41 g), and the lowest was from CC (27.68 g). Regarding the biological efficiency of oyster mushroom, it was significantly influenced by the substrates, with high SR (56.5) performing the best followed by WS (48.3), RS (38.8) and SB (30.4) and lowest value (20.5) was obtained from CC.

Table 4. Yield and biological efficiency of oyster mushroom on different agro-industrial waste substrates.

Substrates	First Flush (g)		Second Flush (g)		Biological Efficiency (%)
	Individual Weight of Fruiting Body	Weight of Fruiting Bodies Per Packet	Individual Weight of Fruiting Body	Weight of Fruiting Bodies Per Packet	
Rice straw	4.69 b	60.32 c	3.86 b	47.58 b	38.80 c
Wheat straw	4.73 b	65.07 b	4.59 a	62.29 a	48.30 b
Corn cob	2.75 d	27.68 e	-	-	20.50 e
Saw dust with rice husk	5.39 a	115.13 a	4.02 b	27.45 c	56.50 a
Sugarcane bagasse	3.60 c	47.61 d	2.22 c	16.80 d	30.40 d
Lsd	0.30	4.81	0.29	6.52	7.49
CV (%)	5.28	5.83	6.09	12.62	10.43

The means with the same letter(s) in a column do not differ significantly as per Lsd test ($p \leq 0.05$).

3.3. Quality of Fruiting Body

The quality of mushroom depends on its biochemical constituents. The biochemical constituents studied in this study varied significantly ($p \leq 0.05$) among the substrates (Table 5). The content of ash ranged from 6.94 to 10.87 g/100 g in the fruiting bodies and the highest amount of ash was determined in the fruiting bodies grown on both WS (10.87) and SR (10.05) substrates, and the lowest was from SB (6.94 g/100 g). Regarding total sugar, the maximum sugar was detected in the fruiting bodies grown on WS substrate (22.41), which were statistically identical with RS (21.78). On the contrary, both CC and SB substrates had the minimum total sugar statistically, while mushroom grown on the SR substrate contained moderate total sugar (18.65 mg/100 g). The maximum protein content in the fruiting body was in SB substrate which was statistically similar with RS, SR and CC whereas the minimum from the WS (Table 5). The concentration of total polyphenols ranged from 109.59 to 196.88 mg and varied significantly ($p \leq 0.05$) among the substrates compared. The maximum concentration of polyphenols was extracted in the fruiting body obtained from SR (196.88 mg), whereas the lowest from SB (109.59 mg). The RS, WS and CC substrates gave statistically similar amounts of total polyphenols.

Table 5. Biochemical constituents of oyster mushroom on different agro-industrial waste substrates.

Substrates	Ash (g/100 g)	Total Sugar (mg/100 g)	Protein (mg/100 g)	Polyphenols (mg GAE/100 g)
Rice straw	9.02 b	21.78 a	6.12 b	165.48 b
Wheat straw	10.87 a	22.41 a	5.67 c	152.68 b
Corn cob	8.97 b	16.65 c	6.33 b	156.90 b
Saw dust with rice husk	10.05 a	18.65 b	6.48 b	196.88 a
Sugarcane bagasse	6.94 c	16.23 c	6.87 a	109.59 c
Lsd	0.85	1.48	0.37	16.23
CV (%)	4.05	7.51	3.66	14.88

The means with the same letter(s) in a column do not differ significantly as per Lsd test ($p \leq 0.05$).

4. Discussion

The variation in the growth and development of mycelia and fruiting bodies with different substrates might be due to the composition of different substrates. The proper amount of alpha-cellulose, hemi-cellulose and lignin enhance the growth and development of mycelia whereas the presence of polyphenolic compounds retard the growth and development of mycelia [18]. The higher mycelia growth and spawn running in SB may be due to the availability of a higher level of nutrients at the beginning of inoculation. Although the lowest growth of mycelia was recorded in SR substrate, it took the fewest days from stimulation to harvest. The content of cellulose and lignin in SR might favor the growth of fruiting body. The present findings are in accordance with a previous study where authors [19] reported that sawdust amended with paddy straw provided suitable conditions for spawn running. The slower growth and development of fruiting bodies in CC might be due to the presence of a higher level of nitrogen and/or polyphenols, which inhibit the growth and development of mycelia. In other studies, the rapid growth and development of the mycelia of king oyster mushroom (*Pleurotuseryngii*) on CC and milky mushroom (*Calocybeindica*) on WS have been reported more than other substrates [7,8] which might be due the variation in the chemical composition of substrates and the different species of mushroom used in the study. In this study, oyster mushrooms produced the maximum number of fruiting bodies on the SR substrate, which might be due to the fact that this mixture contains comparatively higher amounts of cellulose, hemicelluloses and lignin, which might favor the growth and development of oyster mushrooms in the present study [20,21]. The favorable conditions of the SR substrate enhanced the growth of fruiting bodies and thereby produced the biggest stalk and cap. Similar findings were also reported earlier [22]; however, some other studies showed variations in the size of stalk and cap of fruiting bodies, which might be due to the variation in the strains of oyster mushrooms, as well as different substrates and growing conditions [23,24].

In the present investigation, SR produced the highest IWFB followed by WS, which might be due to their larger size of stalk and cap. On the contrary, RS and SB substrates gave moderate IWFB; this is logical, as these substrates yielded a medium size of stalk and cap. However, the lowest value of IWFB was obtained from the CC substrate because of the characteristics that contribute to the lowest yield value. From the results of the present experiment, it is evident that SR yielded the highest number of fruiting bodies (first harvest + second harvest) over other substrates. The reason for this may be the physical nature and high cellulose, hemicelluloses and lignin of the SR substrate, which were suitable for the oyster mushroom cultivation. The present result is in close proximity with an earlier study, where the authors opined that maximum yield, biological efficiency and the number of fruiting bodies of oyster mushrooms was obtained from sawdust [20]. The lowest value of all yield contributing parameters in the second flush could be linked with a lower availability of simpler carbon at the first flush while leaving few carbon compounds for the subsequent flushes [25]. Biological efficiency is used to assess the efficiency of substrate bioconversion into fruiting bodies [9]. From an economic point of view, BE value should be over 50% [9]. In this study, only the SR substrate exceeded a 50% level of BE, as this substrate yielded the highest fruiting bodies per packet. Oyster mushrooms grown on a CC substrate have a much lower BE than earlier studies [9,26,27], the reason for this may be that the adverse C:N ratio retarded the growth of the mycelium, thereby influencing the overall yield and BE. However, similar BE for RS, WS and SB substrates have been reported earlier by several authors [28–31].

Mushrooms grown on SR and WS have higher levels of ash, which might be due to the fact that they accumulated minimum moisture in their fruiting bodies and similar values of ash have been reported by several authors [7,10,25,32]. In this study, the amount of total sugar was detected in the range of 16.23 to 22.41 mg/100 g, while protein values ranged from 5.67 to 6.87 mg/100 g. The lower value of total sugars and protein content in the fruiting bodies might be due to the different protocols used for protein estimation and also most of the authors quantified the carbohydrate and protein content on the dry weight basis not on fresh weight basis. The significant differences in total sugar content in mushrooms may possibly be due to the C:N and various chemical composition of the substrates [33].Since the WS and RS substrates are rich in carbohydrate and fiber, as a result their fruiting bodies are also found to be rich in sugars. This is in conformity with several reports [10,25,32], where WS and RS produced carbohydrate rich mushrooms. It was also observed that the fruiting bodies grown on the SB substrate contained the highest amount of protein, which might be due to the availability of higher levels of nitrogen in this substrate. A similarly higher level of protein in the mushroom has also been reported by authors [33]. Polyphenolics are strong antioxidant compounds and, in this study, mushroom grown on SR substrate exhibited the highest amount of total ployphenols than other substrates. However, insufficient literature related to the polyphenol content in mushroom is available to make a conclusive statement on mushroom polyphenol in relation to different substrates.

5. Conclusions

Among the substrates used in this study, sugarcane bagasse exhibited faster mycelia growth and time from inoculation to mycelium running than other substrates; however, this did not correspond with time from stimulation to primordial initiation and stimulation to harvest, size, yield and quality of mushroom. In all cases, rice straw and corncob substrates showed slower growth and also gave poor yield compared to other substrates. In some cases, wheat straw performed better than sawdust with rice husk but, due to moderate yield and slower mycelium running rate, it may not be economical for small scale cultivation. Based on the present results, it is apparent that most of the yield contributing characteristics and biological efficiency were better in sawdust with the rice husk substrate. In addition, the highest concentration of polyphenols and moderate amount of total sugar and protein were detected from the same substrate. Therefore, saw dust in combination with rice husk (3:1) can be used as an alternative source for the small scale cultivation of oyster mushrooms.

Author Contributions: The work was conducted as a collaboration among all the authors. Authors M.A. and S.M. designed the experiment and M.A., S.M. and R.F.H. analyzed the data. M.A. and S.A. prepared the visualization and S.A., F.A.A. and M.A. organized the first draft of the manuscript. Author M.A.T., S.M. and F.A.A. wrote the manuscript and S.A., R.F.H. and F.A.A. edited the manuscript, and M.A.T., R.F.H. and F.A.A. were responsible for fund acquisition. All authors have read and agreed to the published version of the manuscript.

Funding: This research was partially funded by the Ministry of Science and Technology, Bangladesh.

Institutional Review Board Statement: Not applicable.

Informed Consent Statement: Not applicable.

Data Availability Statement: Mother culture of *Pleurotus ostreatus* was used in this study which was kindly provided by the Horticulture Center, Department of Agricultural Extension, Dinajpur 5200, Bangladesh.

Acknowledgments: Authors are expressing their appreciation to the Ministry of Science and Technology, Bangladesh for the partial financial support to complete the research project.

Conflicts of Interest: The authors declare that they have no conflict of interest.

References

1. Hirano, R.; Sasamoto, W.; Matsumoto, A.; Itakura, H.; Igarashi, O.; Kondo, K. Antioxidant ability of various flavonoids against DPPH radicals and LDL oxidation. *J. Nutr. Sci. Vitaminol.* **2001**, *47*, 357–362. [CrossRef]
2. Khan, M.A.; Khan, L.A.; Hossain, M.S.; Tania, M.; Uddin, M.N. Investigation on the nutritional composition of the common edible and medicinal mushrooms cultivated in Bangladesh. *Bangladesh J. Mushroom* **2009**, *3*, 21–28.
3. Kalač, P. A review of chemical composition and nutritional value of wild-growing and cultivated mushrooms. *J. Sci. Food Agric.* **2013**, *93*, 209–218. [CrossRef]
4. Valverde, M.E.; Hernándea-Pérez, T.; Paredes-López, O. Edible mushrooms: Improving human health and promoting quality life. *Int. J. Microbial.* **2015**, *2015*, 376387. [CrossRef]
5. Ferdousi, J.; Riyadh, J.A.; Hossain, M.I.; Saha, S.R.; Zakaria, M. Mushroom production benefits, status, challenges and opportunities in Bangladesh: A review. *Annu. Res. Rev. Biol.* **2019**, *34*, 1–13. [CrossRef]
6. Kumla, J.; Suwannarach, N.; Sujarit, K.; Penkhrue, W.; Kakumyan, P.; Jatuwong, K.; Vadthanarat, S.; Lumyong, S. Cultivation of mushrooms and their lignocellulolytic enzyme production through the utilization of agro-industrial waste. *Molecules* **2020**, *25*, 2811. [CrossRef]
7. Sardara, H.; Alib, M.A.; Anjuma, F.N.; Hussaina, S.; Naza, S.; Karimd, S.M. Agro-industrial residues influence mineral elements accumulation and nutritional composition of king oyster mushroom (*Pleurotus eryngii*). *Sci. Hort.* **2017**, *225*, 327–334. [CrossRef]
8. Sardar, H.; Anjum, M.A.; Nawaz, A.; Naz, S.; Ejaz, S.; Ali, S.; Haider, S.T. Effect of different agro-wastes, casing materials and supplements on the growth, yield and nutrition of milky mushroom (*Calocybe indica*). *Folia Hort.* **2020**, *32*, 115–124. [CrossRef]
9. Sardar, H.; Anjum, M.A.; Nawaz, A.; Ejaz, S.; Ali, M.A.; Khan, N.A.; Nawaz, F.; Raheel, M. Impact of various agro-industrial wastes on yield and quality of Pleurotus sajor-caju. *Pak. J. Phytopathol.* **2016**, *28*, 87–92.
10. Prasad, S.; Rathore, H.; Sharma, S.; Tiwari, G. Yield and proximate composition of *Pleurotus florida* cultivated on wheat straw supplemented with perennial grasses. *Indian J. Agric. Sci.* **2018**, *88*, 91–94.
11. Raghuramulu, N.; Madhavan, N.K.; Kalyanasundaram, S. *A Manual of Laboratory Techniques*. National Institute of Nutrition; Indian Council of Medical Research: Hyderabad, India, 2003; pp. 56–58.
12. Dubois, M.K.A.; Giles, J.K.; Hamilton, P.A.; Smith, F. Colorimetric method for determination of sugars and related substances. *Anal. Chem.* **1956**, *28*, 350–356. [CrossRef]
13. Bearden, J.C. Quantitation of submicrogram quantities of protein by an improved protein-dye binding assay. *Biochim. Biophys. Acta (BBA)-Protein Struct.* **1978**, *533*, 525–529. [CrossRef]
14. McCown, B.H.; Beck, G.E.; Hall, T.C. Plant leaf and stem proteins. Extraction and electrophoretic separation of the basic, water-soluble fraction. *Plant Physiol.* **1968**, *43*, 578–582. [CrossRef] [PubMed]
15. Singleton, V.L.; Rossi, J.A. Colorimetry of total phenolics with phosphomolybdic phosphotungstic acid reagents. *Am. J. Enol. Vitic.* **1965**, *16*, 144–158.
16. Velioglu, Y.S.; Mazza, G.; Gao, L.; Oomah, B.D. Antioxidant activity and total phenolics in selected fruits, vegetables, and grain products. *J. Agric. Food Chem.* **1998**, *46*, 4113–4117. [CrossRef]
17. STSC. *Statgraphics Users Guide*; STSC: Spencer, MA, USA, 1987.
18. Wang, D.; Sakoda, A.; Suziki, M. Biological efficiency and nutritional value of *Pleurotus ostreatus* cultivated on spent beer grain. *Bioresour. Technol.* **2001**, *78*, 293–300. [CrossRef]
19. Khan, A.M.; Khan, S.M. Studies on the cultivation of Oyster mushroom *Pleurotus ostreatus* on different substrates. *Pak. J. Phytopath* **2001**, *13*, 140–143.
20. Shah, Z.A.; Asar, M.; Ishtiaq, M. Comparative study on cultivation and yield performance of oyster mushroom on different substrates (wheat straw, leaves, saw dust). *Pak. J. Nutr.* **2004**, *3*, 159–160.
21. Dos Santos, R.M.; Neto, W.P.F.; Silverio, H.A.; Martins, D.F. Cellulose nano crystals from pineapple leaf, a new approach for the reuse of this agro-waste. *Ind. Crops Prod.* **2013**, *50*, 707–714. [CrossRef]
22. Zhang, R.; LiXiu, J.H.; Fadel, J.G. Oyster mushroom cultivation with rice and wheat straw. *Bioresour. Technol.* **1998**, *82*, 277–284. [CrossRef]
23. Khlood, A.; Ahmad, A. Production of oyster mushroom (*Pleurotus ostreatus*) on olive cake agro-waste. *Dirasat Agric. Sci.* **2004**, *32*, 64–70.
24. Sarker, N.C.; Hossain, M.M.; Sultana, N.; Mian, I.H.; Karim, A.J.M.S.; Amin, S.M. Performance of Different Substrates on the growth and Yield of *Pleurotus ostreatus* (Jacquin ex Fr.) Kummer. *Bangladesh J. Mushroom* **2007**, *1*, 44–49.
25. Adenipekun, C.O.; Omolaso, P.O. Comparative study on cultivation, yield performance and proximate composition of *Pleurotus pulmonarius* Fries. (Quelet) on rice straw and banana leaves. *World J. Agric. Sci.* **2015**, *11*, 151–158.

26. Rofiqah, U.; Kurniawan, A.; Aji, R.W.N. Effect of temperature in ionic liquids pretreatment on structure of lignocellulose from corncob. *J. Phys. Conf. Ser.* **2019**, *1373*, 012018. [CrossRef]

27. Pointner, M.; Kuttner, P.; Obrlik, T.; Jager, A.; Kahr, H. Composition of corncobs as a substrate for fermentation of biofuels. *Agron. Res.* **2014**, *12*, 391–396.

28. El-Tayeb, T.S.; Abdelhafez, A.A.; Ali, S.H.; Ramadan, E.M. Effect of acid hydrolysis and fungal bio treatment on agro-industrial wastes for obtainment of free sugars for bioethanol production. *Braz. J. Microbiol.* **2012**, *43*, 1523–1535. [CrossRef] [PubMed]

29. Limayema, A.; Ricke, S.C. Lignocellulosic biomass for bioethanol production: Current perspectives, potential issues and future prospects. *Prog. Energy Comb. Sci.* **2012**, *38*, 449–467. [CrossRef]

30. Motte, J.C.; Trably, E.; Escudié, R.; Hamelin, J.; Steyer, J.P.; Bernet, N.; Delgenes, J.P.; Dumas, C. Total solids content: A key parameter of metabolic pathways in dry anaerobic digestion. *Biotechnol. Biofuels* **2013**, *6*, 164. [CrossRef]

31. Zainudin, M.H.M.; Rahman, N.A.; Abd-Aziz; S.; Funaoka, M.; Shinano, T.; Shirai, Y. Utilization of glucose recovered by phase separation system from acid-hydrolysed oil palm empty fruit bunch for bioethanol production. *Sci. Pertanika J. Trop. Agric.* **2012**, *35*, 117–126.

32. Pardo-Giménez, A.; Pardo, J.E.; Dias, E.S.; Rinker, D.L.; Caitano, C.E.C.; Zied, D.C. Optimization of cultivation techniques improves the agronomic behavior of *Agaricus subrufescens*. *Sci. Rep.* **2020**, *10*, 8154. [CrossRef]

33. Adedokun, O.M.; Akuma, A.H. Maximizing agricultural residues: Nutritional properties of straw mushroom on maize husk, waste cotton and plantain leaves. *Nat. Res.* **2013**, *4*, 534–537. [CrossRef]

 *sustainability*

Article

Assessing the Environmental Footprint of Distiller-Dried Grains with Soluble Diet as a Substitute for Standard Corn–Soybean for Swine Production in the United States of America

Md Ariful Haque [1,2,*], Zifei Liu [2,*], Akinbile Demilade [2] and Nallapaneni Manoj Kumar [1,*]

1 School of Energy and Environment, City University of Hong Kong, Kowloon, Hong Kong
2 Department of Biological and Agricultural Engineering, Kansas State University, Manhattan, KS 66506, USA; akinbiledemilade@k-state.edu
* Correspondence: mahaque3@cityu.edu.hk (M.A.H.); zifeiliu@k-state.edu (Z.L.); mnallapan2@cityu.edu.hk or nallapanenichow@gmail.com (N.M.K.)

Citation: Haque, M.A.; Liu, Z.; Demilade, A.; Kumar, N.M. Assessing the Environmental Footprint of Distiller-Dried Grains with Soluble Diet as a Substitute for Standard Corn–Soybean for Swine Production in the United States of America. *Sustainability* 2022, 14, 1161. https://doi.org/10.3390/su14031161

Academic Editor: Idiano D'Adamo

Received: 25 November 2021
Accepted: 18 January 2022
Published: 20 January 2022

Publisher's Note: MDPI stays neutral with regard to jurisdictional claims in published maps and institutional affiliations.

Abstract: The swine diet formulation in the United States of America (U.S.A.) is entering a new era of decision making to promote low-carbon pork production systems. As a part of the decision-making process, the precision nutrition approaches to customize diet and alternative feeding options that are economically viable and environmentally sustainable are given priority. Hence, the objective of this study is to identify an alternative diet over a standard corn–soybean meal diet. The byproducts from the supply chain of human food and biofuels, i.e., distiller-dried grain with solubles (DDGS), are chosen as an alternative option to formulate a swine diet. First, two alternative byproduct diets with low and high DDGS inclusion (10.1% and 28.8%, respectively) were formulated using the least-cost technique. Second, a life cycle inventory was created, followed by data collection from the key sources, including DATA SMART-2017, USDA, RIA-GREET 2018, and the relevant literature. Third, in SimaPro 8.5.2.0 (PRé Sustainability: LE Amersfoort, The Netherlands), the ReCiPe 2016, the midpoint method by economic allocation was used to investigate the environmental footprint of the formulated diets to inform sustainability decisions of swine-farm managers. The considered functional unit is the 'lb diet', and the system boundary is the farm gate that considers only the feed production stage. The observed results include global warming potential, land use, water consumption, fossil resources scarcity, and terrestrial ecotoxicity. The comparative results of a 28.8% DDGS diet over the standard corn–soybean meal diet for the displacement ratio of 0.69 show an approximate global warming potential saving of 0.04 kg CO_2 eq. per lb DDGS feed at the feed production stage. Moreover, the DDGS displacement ratio of 0.69 does not significantly impact water consumption and fossil resources; however, it can reduce land use by 26% and terrestrial ecotoxicity by 8% compared to the standard diet. Overall, the quantified environmental footprint results of the byproduct DDGS diets indicate that the footprints of DDGS diets were lower than the standard diet.

Keywords: DDGS; alternative diet for swine; byproduct diet; global warming potential; standard corn–soybean meal

1. Introduction

Byproducts from bioprocesses that are often discarded as waste can be placed under circular bioeconomy practices. Such practices enable increased resource efficiency, which is believed to accelerate Sustainable Development Goals (SDGs) [1]. Resource utilization under circular bioeconomy principles also offers economic developments when implemented through the green technology planning decision model suggested by Ikram et al. [2]. Byproduct resources from various food, animal, vegetable, and sugar industries can be a potential animal food source. In swine diets, the major by-products from biobased industries are used. These include the distiller- and brewer-dried grains from ethanol and brewing industries. Moreover, there are many instances where the leftover stillage from on-farm alcohol fuel production units is used in swine diets [3]. Among all the byproducts, distiller-dried grains are common and mostly occur in feed ingredients in the pork industry in the U.S.A. More recently, the U.S.A. has seen the ethanol industry rising; as a result, the proportion and allocation of grain processing have increased. This resulted in a potential swine feed co-product, i.e., corn distiller-dried grains with solubles (DDGS) [4].

The appropriate amount of ingredients in a swine diet largely depends on many factors. These include protein quality, nutrients, palatability, storage life, cost, and amino acids. Apart from these, the age of the pigs to which the diet is to be fed, the ingredients' production environment, and sometimes the presence of anti-nutritional elements could also be an influence.

DDGS has added advantages of all these attributes and has been a very good alternative feed ingredient in the U.S.A. swine industry. DDGS substitutes soybean meal (SBM), di-calcium phosphate, and corn in swine diets, providing lysine, phosphorus, and energy. In DDGS, lysine is very restrictive to 0.7%, whereas phosphorus is relatively high (0.71%) [3]. In terms of energy, DDGS is approximately equal to corn, and the protein content in DDGS is relatively high at around 27%. As a result, amino acid balance is retained. Furthermore, the amino acids in DDGS appear to be less readily accessible than those in SBM. DDGS, on the other hand, can be used successfully in swine diets when supplemented with synthetic amino acids.

The rate of DDGS inclusion may vary depending on the nutrient quality in it and the growth stage of the swine. Stein (2007) reported that the inclusion of 30% DDGS on grow-finishing swine does not negatively affect swine growth performance; however, low lysine and high fiber content affect digestibility and pig performance [5]. For lactating sows, weanlings, and grow-finishing swine, inclusion of up to 30% can be made while for gestating sows the inclusion can be up to 50% depending on the quality of DDGS fed to the swine [5].

On the other hand, the variation of nutrient contents in the feed ingredients led to difficulties in the feed cost comparison. Therefore, relative values are quite useful for comparison purposes. To ease this complexity, Klashing (2012) defined the term displacement ratio as the amount of a feedstuff that is displaced when one unit of DDGS is added [6]. The unit addition of an alternative feedstuff, namely how much can replace the traditional corn and soybean from the diet, is not only important for cost calculation but also environmental sustainability. Indeed, the paper aimed to evaluate the environmental footprint of an alternative feed ingredient, DDGS, at different inclusion rates using the least-cost technique to formulate the diet. The life cycle assessment (LCA) method, as agreed upon by the ISO 14040 and ISO 14044:2006 standards, is used to evaluate the environmental footprint of a product or process throughout the entire life cycle. This method is rather popular and extensively used in the EU for different production sectors, including agriculture [7,8]. In LCA, the allocation has a significant impact on the environmental footprint. However, when allocation cannot be avoided, the hierarchy of allocation rules as per ISO 14044 was followed. As per ISO 14044, it should preferably be based on the physical relationship between the inputs and outputs [9].

According to the National Pork Board, the pork sector in the U.S.A. has achieved tremendous progress in terms of select environmental impact categories. Approximately 75.9% of the land, 25.1% of the water, and 7.0% of the energy use have been reduced during the last 55 years. However, this improvement is attributed only to the high productivity and efficiency in the pork production system [10]. To attain this high productivity, choosing a diet with lower environmental impacts, always ensures that the % of DDGS addition to the standard diet should not have a negative effect on swine growth. There are few research studies available for the environmental impact evaluation of feed ingredients in North American swine diets [11–14]. This study would thereby help facilitate an estimation of the environmental footprint of ingredients and diets as one of the baseline studies in the swine industry.

The novelty of this study will help the swine producer to choose a cost-effective and environmentally friendly diet. On a national level, the use of low-cost by-product DDGS in the diet will reduce the global warming potential (GWP), land use (LU), water consumption (WC), terrestrial ecotoxicity (TE), and fossil resources (FR).

2. Methods

2.1. Diet Formulation

All the major ingredients, including corn, SBM, and DDGS, used in this study are grown or produced in the US crop production region 3 and are assumed to be representative of the U.S.A. Formulations of the diets were based on surveys from experts and the least-cost formulation principles (Table 1). The nutrient budget of the diets was maintained according to the US national resources council nutrition requirement [15] and PIC nutritional requirement for finishing pigs [16]. Nutritional values were taken from the National Hog Farmer report published in 2020 as the second option [10]. Nutritional values other than those in the National Hog Farmer report were calculated with the help of the Animal Science Department, Kansas State University. An inclusion range of 10.1–28.8% DDGS was applied in the DDGS diets (see Table A1 in Appendix A).

Table 1. Growing finishing swine diets—control/standard and with different DDGS inclusion.

Ingredient Use	Corn-SBM	Corn-SBM-10.1% DDGS	Corn-SBM-28.8% DDGS
	lb/Pig (from 50 to 280 lb Body Weight)		
Corn	520.07	476.6	387.6
Soybean meal	119.75	99.1	70.4
Corn DDGS, 7.5% Oil	0.00	66.6	190.9
Calcium carbonate	5.45	6.1	7.01
Calcium phosphate (monocalcium)	2.94	1.3	0.35
Sodium chloride	3.28	3.3	3.32
L-Lys-HCl	1.82	2.2	2.59
DL-Met	0.18	0.1	0.0
L-Thr	0.44	0.2	0.12
L-Trp	0.05	0.1	0.10
Vitamin premix with phytase	0.76	0.8	0.77
Trace mineral premix	0.76	0.8	0.77

Note: SBM: Soybean meal; DDG: Distiller-dried grains with solubles.

2.2. System Boundary and Functional Unit

The system boundaries for the LCA model were cradle to farm-gate, and the functional unit was '1 lb diet' at the feed production stage. The environmental footprint calculation in this study was for grow-finish swine diets in the U.S.A. Figure 1 shows a simplified process flowchart of the ingredients with their system boundaries.

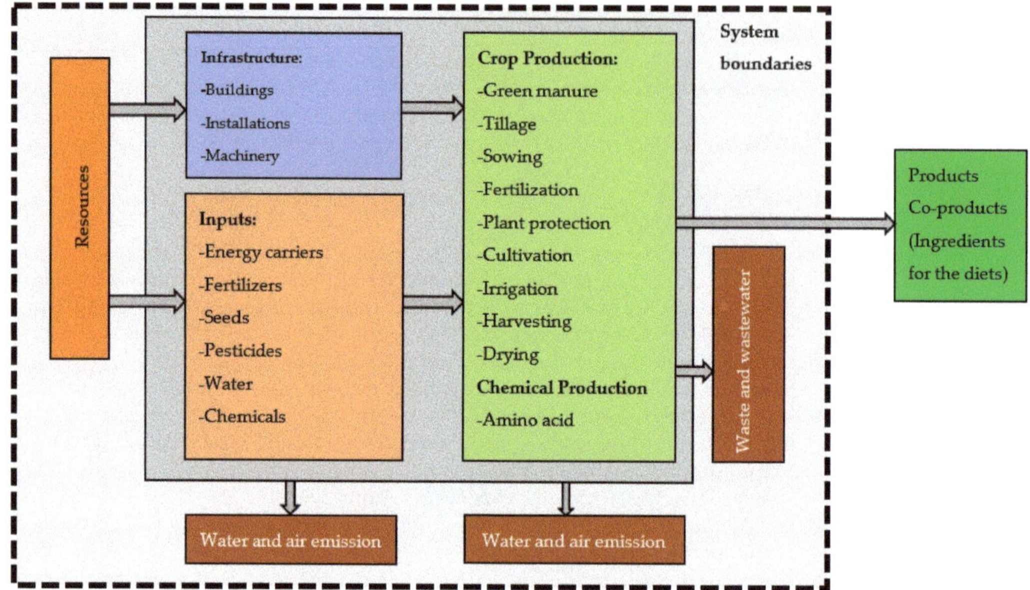

Figure 1. Process flowchart of the ingredients applied in the diets with their system boundary.

2.3. Life Cycle Inventory

Agricultural input data shown in Tables A2 and A3 in Appendix A for the grains (corn and soybean) production were from the USDA-NASS survey [17], Ecoinvent 3, and the Agri-footprint database (US EI 2.2) in SimaPro software version 8.5.2.0 unless otherwise stated [18]. It should be noted that necessary data for the feed ingredients' production correspond to the grains, and relevant product systems were within the United States unless otherwise stated. Both background and foreground processes included some available inventory data from the SimaPro Version 8.5.2.0 process library [18].

Foreground data including the yield of grains, fertilizer, and pesticide data were three-year-average data (2015, 2016, and 2017) in the United States. Synthetic fertilizers (N, P, K, and Sulphur) and pesticide processes were at the regional storehouse in the U.S.A., and the US-EI U database was followed from SimaPro version 8.5.2.0. The wastewater treatment process was selected from ELCD database 3 following Agri-footprint mass allocation.

The production cycle for corn and soybean was assumed to be one year, and the frequencies of fertilizer application were in accordance with the cultivation process throughout the year. All energy-consuming process data were from the USLCI database (DATASMART-2017, SimaPro 8.5.2.0). For LCA of amino acids, raw materials and inputs were collected from Marinussen and Kool [19], see Table A4 in Appendix A. The processing of raw materials and chemicals used for production, the transporting of materials to manufacturing plants, the emissions into the air and water from production, and the estimation of the energy demand and infrastructure of the plant (approximation) all followed the acrylic acid production model at the plant in the U.S.A. Methionine as an amino acid source for lysine production via the biosynthetic process was considered for Lysine production. For Threonine, Lysine was applied in the biosynthetic process.

2.4. Life Cycle Impact Assessment Method (LCIA)

An attributional LCA based on economic allocation was applied for all the ingredients and diets in this study (unless otherwise stated). The hierarchies perspective method of Recipe midpoints 2016 v1.06 was applied based on scientific consensus with regard to a 100-year period and the plausibility of the impact mechanism. Five impact categories, including

GWP, LU, WC, TE, and FR, were considered for the LCA studies. GWP was calculated as the CO_2 equivalent for a 100-year time scale; FR was from the higher heating value of fossil resources and expressed as kg oil-eq.; WC was expressed as m^3 water consumed; and LU was denoted as the relative species loss caused by a specified land-use type and expressed in m^2 annual crop eq.

2.5. Environmental Footprints by Mass and Economic Allocation

All the input data for feed ingredients (see Tables A2–A4 in Appendix A) and the processes associated with emissions were used for the LCA by SimaPro. The environmental footprint calculation by mass allocation was used according to their mass fractions, while for economic allocation, the economic fraction was derived using the following modified equation from Hossain et al. [20] (see Equation (1)).

$$B_i = P \times I \tag{1}$$

where B_i is the environmental impacts of the by-products i, P is the percent allocation, which refers to the fraction derived from the ratio of the main product and by-products according to their mass or economic value, and I is the total environmental impact of the final process products and co-products. When using economic allocation, the percent allocation for the products and co/by-products was estimated (Table 2) prior to the footprint calculation by SimaPro software using the following equation (see Equation (2)):

$$P = \frac{\text{Unit price of the product} \times \text{mass fraction of the product or byproducts} \times 100}{(\text{Unit price of product} \times \text{mass fraaction of the product}) + (\text{Unit price of the byproducts} \times \text{mass fraction of the byproducts})} \tag{2}$$

Table 2. The economic allocation of DDGS and SBM.

	Items	Unit Price ($/lb) [21–24]	Mass Allocation	Economic Allocation
DDGS	Ethanol	0.211	0.490	0.832
	DDGS	0.041	0.510	0.167
	Crude soy oil	0.271	0.217	0.492
SBM	Soy hulls	0.065	0.074	0.012
	SBM	0.146	0.709	0.496

Note: SBM: Soybean meal; DDG: Distiller-dried grains with solubles.

For all the ingredients, environmental footprint results were presented at a 1% cut-off, which meant the environmental load or contribution of a process less than 1% was discarded in the results.

3. Results and Discussion

3.1. Environmental Footprint of Individual Feed Ingredients

The environmental footprints of the byproduct DDGS by economic allocation was due to the relatively low prices of these byproducts. DDGS had the lowest environmental footprints compared to SBM and corn in all four categories of GWP, LU, WC, and FR, while SBM had the highest environmental footprints. The GWP, LU, and WC of DDGS via economic allocation were 22.1%, 81.4%, and 72.5% lower than that of corn, respectively (Table 3).

Table 3. Environmental footprints of SBM, DDGS, corn, bakery meal, and amino acids.

Ingredients		GWP kg CO_2 eq.	LU m^2 yr. Crop eq.	WC m^3	FR kg Oil eq.
SBM	Crude oil	0.390	0.346	0.160	0.028
	Soy hulls	0.054	0.135	0.062	0.010
	SBM	0.516	1.280	0.593	0.104
DDGS	DDGS	0.242	0.187	0.108	0.066
	Ethanol	1.200	0.932	0.535	0.328
Bakery meal		0.380 *	-	-	-
Corn		0.311	1.010	0.393	0.054
Amino acids					
L-Lysine-HCl		4.060	3.340	1.490	0.757
Methionine		9.060	0.728	4.930	2.940
Threonine		8.140	5.070	2.900	2.000
Tryptophan		9.620 *	-	-	-

Note: SBM: Soybean meal; DDG: Distiller-dried grains with solubles; GWP: Global warming potential; LU: Land use; WC: Water consumption; FR: Fossil resources. * ref. [10] (economic allocation).

The major contributors to the environmental footprints for individual feed ingredients are presented in Table 4. Fertilizer was the main contributor to GWP for corn and SBM. It accounted for 20% and 12.8%, respectively. The energy requirements for N fertilizer production ranged between 29 and 67 MJ/kg N, including values for both the low heating value and high heating value [25]. This energy mostly emanates from non-renewable energy and contributed to the dominant share of GWP to the corn and wheat middling. Non-renewable natural gas was the principal contributor to GWP for DDGS. For DDGS processing, when the non-renewable natural gas input energy was replaced with the renewable nuclear source, the GWP of DDGS could be further reduced by 39%, from 0.242 kg CO_2 eq. to 0.148 kg CO_2 eq. For amino acids, the production of raw materials (such as glucose) was the major contributor.

Table 4. Major contributors of environmental footprints for individual feed ingredients.

Ingredients	Major Contributing Factors	Contribution (%)			
		GWP	LU	WC	FR
Corn	Nitrogen ecoprofile at regional storehouse	19.7			
	Corn agricultural production		88.3	45.9	
	Natural gas, unprocessed, at extraction				26.9
SBM	Application lime ecoprofile at field	12.8			
	Soybean agricultural production at farm		94.3		
	Electricity, hydropower, at run-of river power plant			52.6	
	Natural gas, unprocessed, at extraction				22.6
DDGS	Natural gas burned at industrial furnace	41.3			
	Corn agricultural production at farm		79.7		
	Electricity, hydropower, at run-of river power plant			39.3	
	Natural gas, unprocessed, at extraction				74.0
Amino acid					
Lysine	Sugar, from sugar cane, from sugar production at plant	64.9	96.3	40.9	70.4
	Ammonium bicarbonate, at plant	26.1			38.8
Methionine	Electricity, natural gas, at power plant		71.0		
	Ammonia liquid at regional storehouse			84.6	
Threonine	Glucose global market for glucose at point of substitution unit process	54.0	92.3		50.2
	Ammonia, liquid, at regional storehouse			25.1	

Note: GWP: global warming potential; LU: land use; WC: water consumption; FR: fossil resources.

Between two DDGS diets, the displacement of the major ingredient, corn, was the highest with a 28.8% DDGS diet while for SBM it was a 10.1% DDGS diet (Table 5). DDGS inclusion in the diet did not only displace the major ingredients corn, SBM, and essential amino acids methionine and threonine, but also supplied a small amount of calcium carbonate and lysine in the diet. In the 28.8% DDGS diet, a unit of DDGS displaced 0.69 units of corn, which is higher than the national average by Arora et al. and RFA in refs. [26,27]. The per unit displacement of SBM by DDGS in all diets was within the range (0.2–0.3) reported in the literature.

Table 5. Displacement ratio for different DDGS inclusion in growing-finishing swine diet.

Ingredients	10.1%	28.8%
Corn	0.6532	0.6938
Soybean meal	0.31	0.2583
Calcium carbonate	−0.01	−0.0082
Calcium phosphate (monocalcium)	0.0252	0.0136
Sodium chloride	−0.0001	−0.0002
L-Lys-HCl	−0.006	−0.004
DL-Met	0.0017	0.0009
L-Thr	0.0031	0.0017
L-Trp	−0.0004	−0.0003
Vitamin premix with phytase	0.00	−0.0001
Trace mineral premix	0.00	−0.0001

Environmental footprint results from Table 6 demonstrated that GWP, LU, and TE were reduced with the increase in DDGS inclusion in the diets, while FR showed a trend of decline as compared with the standard diet. The highest environmental footprint reduction was attained with a 28.8% DDGS diet, which was 9.69% GWP, 22.97% LU, 2.36% WC, 20.3% TE, and 1.74% FR lower than the standard diet. A high FR footprint was attributed to the high energy requirement for the drying process of DDGS.

Table 6. Environmental footprint of growing-finishing swine diets—control/standard and with different DDGS inclusion for per kg diet at feed production stage.

Diet	GWP (kg CO_2 eq.)	LU (m^2 Area Crop eq.)	WC (m^3)	TE (kg 1,4-DCB)	FR (kg Oil eq.)
Standard (0% DDGS)	0.390	0.975	0.394	0.544	0.063
10.1% DDGS	0.374	0.898	0.365	0.502	0.063
28.8% DDGS	0.352	0.751	0.385	0.434	0.065

Note: GWP: Global warming potential; LU: Land use; WC: Water consumption; TE: Terrestrial Ecotoxicity FR: Fossil resources.

3.2. Sensitivity of Environmental Footprint to DDGS and Ethanol Price

The price of DDGS and ethanol historically in the US market has fluctuated over the decades (Figures 2 and 3). The price variation of DDGS could have a significant effect both economically and environmentally. Thus, a sensitivity test was conducted with the average (1.642), minimum (0.908), and maximum (3.898) price ratio from the historic DDGS and ethanol price data (Figure 4). The price ratio was computed from the average, minimum, and maximum prices of DDGS and ethanol from the historic data.

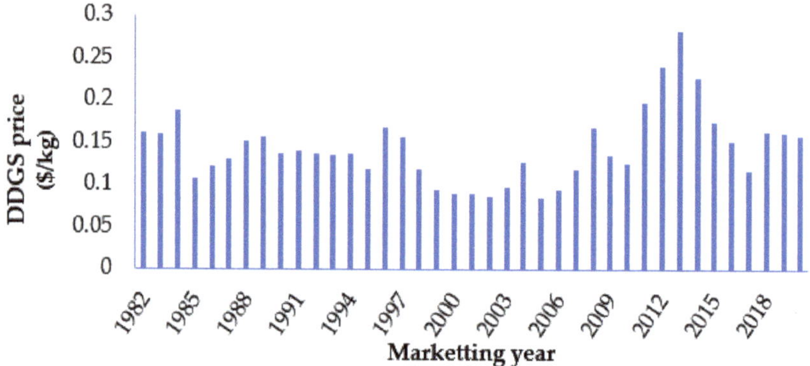

Figure 2. Historic price of DDGS ($/ton). Data from ref. [28].

Figure 3. Historic ethanol price in the US market. Data from ref. [29].

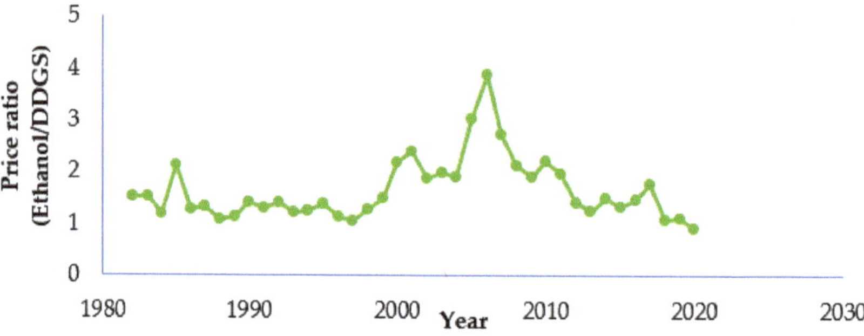

Figure 4. Historic price ratio of DDGS and ethanol in the U.S.A. market. Data from ref. [29].

Results from the DDGS and ethanol price sensitivity test demonstrated that with the historical average, minimum, and maximum price ratio between ethanol and DDGS, GWP and LU of the 28.8% DDGS diet is still below the standard diet. In contrast, the WC of the 28.8% DDGS diet is sensitive to the average, minimum, and maximum price ratios of ethanol and DDGS in comparison with the standard diet (Figure 5).

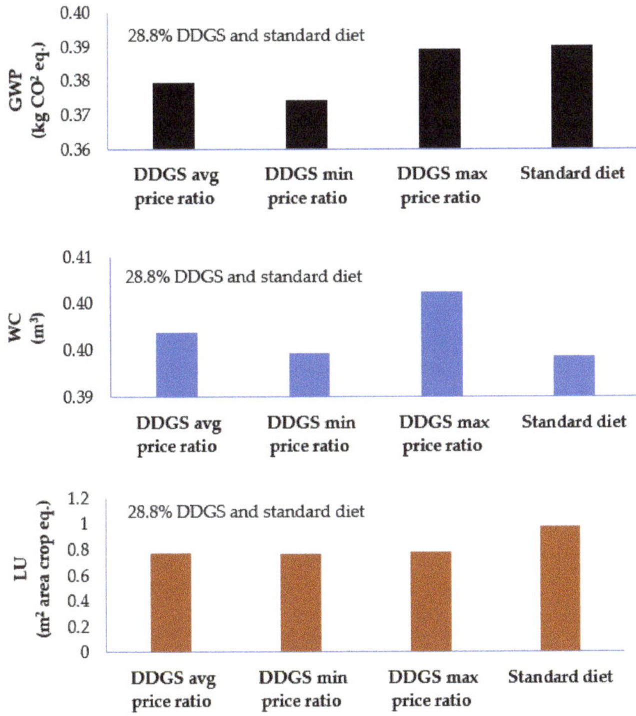

Figure 5. Sensitivity of environmental footprint in response to the price ratio of ethanol and DDGS.

3.3. Discussion

The displacement ratio of corn and SBM by DDGS reduced both the environmental footprint and the per dollar cost of the diet. The inclusion of DDGS compensated for a part of the amino acid requirement in the diet by providing an amino acid content three times greater than corn, thus playing a role in reducing the overall environmental footprint [4]. Besides amino acids, the addition of DDGS also supplies a portion of monocalcium phosphate in the diet. Another factor of the low environmental footprint of DDGS diets is that even with a high inclusion up to 28.8%, the low allocation to DDGS resulted in a low environmental footprint under economic allocation. Thus, from a check-and-balance observation, DDGS can reduce the overall environmental footprint in the diet as compared with the standard diet.

With an inclusion range of 10.1–28.8% of DDGS that corresponds to a corn displacement ratio range of 0.65–0.69 in the diet, we can save a GWP of 0.01–0.04 kg CO_2 eq. per lb feed at the feed production stage. In 2019, the total DDGS production in the US was 22.54 million metric tons, of which swine consumption was 3.6 million metric tons (16%) [27]. With the protein content of DDGS and corn of 28.15% and 8.24%, respectively, this amount of DDGS is equivalent to 12.33 million metric tons of protein equivalent of corn. Therefore, the current range of DDGS inclusion in the diet can save a GWP of up to 0.12–0.49 million metric tons of CO_2 eq. at the feed production stage on a national level. If the DDGS is not used for swine production, the crude protein and amino acid lysine content could be wasted. Based on the replacement ratio by Stein (2007), 28.8% DDGS inclusion can save 16.416% of corn and 12.24% of SBM [5]. This replacement can save the feed cost per pig by 28.65 $ and 3.88 $ for corn and SBM, respectively. Using such alternative diets, approximately 32.53 $ can be saved per pig, and this will reduce the overall swine production cost.

From the sensitivity results, with historical DDGS and ethanol price fluctuations from the base price, both 10.1% and 28.8% DDGS diets were demonstrated as environmentally benign diets compared to the standard diet. Although the 28.8% DDGS diet resulted in higher LU in comparison with the standard diet at the feed production stage, the discrepancy was, however, not too high at only 1.72%.

4. Conclusions and Limitations

An environmental footprint assessment was carried out on the DDGS diet as a substitute for standard corn–soybean for swine production in the U.S.A. Four impact categories, namely GWP, LU, WC, TE, and FR, are estimated for varying rates of DDGS inclusions (for instance, 10.1% and 28.8%) compared with the standard swine diet with 0% DDGS. Based on this assessment, the following conclusions were drawn:

- A DDGS displacement ratio of 0.65–0.69 can save a GWP of up to 0.12–0.49 million metric tons of CO_2 eq. at the feed production stage on a national level.
- Though the DDGS displacement ratio of 0.65–0.69 does not significantly impact WC and FR, it can save up to 26% LU and 8% TE.
- The historic price elasticity of DDGS and ethanol did not influence the diet's environmental footprint, indicating that the environmental footprint is not sensitive to the price of DDGS.
- With nutritional benefits and availability, DDGS remained one of the most important byproduct ingredients for the swine diet in the US.
- Although the amount of DDGS production is almost equal to the amount of ethanol in corn-based ethanol processing, with economic allocation, the environmental footprint of DDGS is lower than other ingredients in the diet.

The authors do acknowledge the current limitations of this study. For example, all the ingredients and diets were assumed to be produced in the geographic boundary of the U.S.A. The mass of the straw of the grain crops was not included in the system boundary. The industrial equipment's diesel and gasoline combustion data were attained from the USLCI database (SimaPro 8.5.2.0), and it may not be applicable to other parts of the world. We also acknowledge the uncertainties associated with this study due to the limited process data, such as methionine usage as an amino acid source for lysine fermentation. In contrast, for threonine, lysine is applied to the biosynthetic process. Lysine-producing microorganisms may not adapt to threonine (as amino acid source) in the medium for biosynthetic production.

Overall, from this study, we observed that DDGS inclusion in standard corn–soybean meal could potentially benefit the swine production sector and drive it towards being in line with sustainable standards. Furthermore, there has been a recent upward trend of DDGS use in the U.S.A. We believe it is high time to formulate policies that will accelerate the usage of DDGS at a massive level without compromising the growth performance of swine.

Author Contributions: Conceptualization, M.A.H. and Z.L.; methodology, M.A.H.; formal analysis, M.A.H.; investigation, Z.L.; resources, M.A.H. and Z.L.; data curation, M.A.H. and A.D.; writing—original draft preparation, M.A.H. and N.M.K.; writing—review and editing, M.A.H., Z.L. and N.M.K.; visualization, M.A.H. and N.M.K.; funding acquisition, Z.L. All authors have read and agreed to the published version of the manuscript.

Funding: This research was funded by the National Pork Board (NPB) Project 17-159.

Institutional Review Board Statement: Not applicable.

Informed Consent Statement: Not applicable.

Data Availability Statement: Not applicable.

Conflicts of Interest: The authors declare no conflict of interest.

Appendix A

Table A1. Nutrient budget of the diets at different phases of grow-finish swine in the U.S.A.

Nutrient Composition	Standard Corn-SBM Diet					10.1% DDGS Diet					28.8% DDGS Diet				
	Phase 1	Phase 2	Phase 3	Phase 4	Phase 5	Phase 1	Phase 2	Phase 3	Phase 4	Phase 5	Phase 1	Phase 2	Phase 3	Phase 4	Phase 5
	Weight Range (lb)					Weight Range (lb)					Weight Range (lb)				
	50–75	75–140	140–195	195–240	240–280	50–75	75–140	140–195	195–240	240–280	50–75	75–140	140–195	195–240	240–280
Gilt required SID Lys:NE Ratio	4.52	3.80	3.17	2.82	2.54	4.52	3.80	3.17	2.82	2.54	4.52	3.80	3.17	2.82	2.54
Calculated SID Lysine Required, %	1.10	0.94	0.80	0.72	0.65	1.10	0.94	0.80	0.71	0.65	1.09	0.93	0.79	0.70	0.65
PIC required SID Lys:NE Ratio	5.08	4.23	3.45	3.10	3.03	5.08	4.23	3.45	3.10	3.03	5.08	4.23	3.45	3.10	3.03
Calc. PIC SID Lysine Required, %	1.24	1.05	0.87	0.79	0.77	1.24	1.05	0.87	0.78	0.77	1.22	1.03	0.85	0.77	0.77
SID amino acids, %															
Lys	1.17	0.99	0.83	0.75	0.70	1.17	0.99	0.83	0.75	0.70	1.15	0.97	0.82	0.73	0.70
Ile:Lys	63	62	61	62	65	61	61	63	66	66	66	67	70	70	65
Leu:Lys	131	138	146	154	164	135	148	163	175	175	160	184	205	219	178
Met:Lys	32	31	30	28	30	31	29	29	31	32	29	33	36	38	32
Met & Cys:Lys	56	56	56	56	60	55	55	58	62	62	56	62	69	73	63
Thr:Lys	61	61	61	63	64	60	59	59	58	61	61	61	62	63	65
Trp:Lys	18.5	18.3	18.1	18.2	18.4	18.1	17.9	18.1	17.8	17.6	18.2	18.1	18.1	18.2	18.1
Val:Lys	69	69	70	72	76	68	70	74	78	78	76	80	86	89	78
His:Lys	42	42	43	44	47	41	42	44	47	47	45	47	50	51	47
Total Lys, %	1.31	1.11	0.94	0.85	0.80	1.32	1.13	0.96	0.87	0.81	1.34	1.15	0.99	0.89	0.81
ME, kcal/lb	1491	1497	1502	1506	1508	1490	1493	1500	1502	1506	1481	1483	1486	1488	1504
NE Noblet Grow/Finish, kcal/lb	1084	1107	1129	1138	1141	1007	987	1006	1011	1060	807	720	734	746	1032
NE Noblet Sow, kcal/lb	1125	1147	1167	1177	1179	1045	1022	1040	1045	1095	838	746	759	769	1066
DE NRC, kcal/lb	1554	1551	1549	1549	1550	1552	1548	1550	1549	1550	1551	1547	1544	1542	1548
NE NRC, kcal/lb	1103	1124	1144	1152	1155	1105	1123	1140	1146	1154	1092	1110	1122	1132	1152
SID Lys:NE, g/Mcal	4.81	3.99	3.29	2.95	2.75	4.80	4.00	3.30	2.97	2.75	4.78	3.97	3.32	2.92	2.76
CP, %	20.6	17.6	15.1	14.1	13.8	20.5	18.1	16.1	15.4	14.4	22.8	20.8	18.9	17.5	14.5
Ca, %	0.64	0.58	0.52	0.47	0.45	0.64	0.59	0.49	0.47	0.45	0.64	0.58	0.55	0.53	0.46
P, %	0.54	0.49	0.43	0.39	0.38	0.52	0.47	0.40	0.38	0.35	0.53	0.48	0.46	0.44	0.38
Available P w/o phytase, %	0.22	0.19	0.15	0.13	0.12	0.22	0.19	0.14	0.13	0.10	0.25	0.24	0.23	0.23	0.13
Available P, %	0.35	0.32	0.27	0.23	0.20	0.34	0.32	0.26	0.23	0.19	0.37	0.36	0.34	0.33	0.22
Avail P:calorie ratio g/mcal	1.05	0.96	0.81	0.70	0.62	1.05	0.96	0.78	0.71	0.56	1.14	1.10	1.05	1.00	0.67
Stand. Dig. P w/out phytase, %	0.30	0.26	0.21	0.19	0.18	0.28	0.24	0.19	0.18	0.15	0.29	0.26	0.24	0.23	0.18
Stand. Dig. P with phytase, %	0.40	0.37	0.31	0.28	0.26	0.39	0.35	0.29	0.27	0.23	0.40	0.36	0.34	0.33	0.26
STTD Ca, % without phytase	0.43	0.39	0.35	0.31	0.30	0.43	0.34	0.28	0.27	0.26	0.42	0.33	0.33	0.32	0.27
STTD Ca, % with phytase	0.47	0.43	0.38	0.34	0.33	0.47	0.43	0.35	0.34	0.32	0.46	0.41	0.39	0.37	0.33
Ca:P	1.17	1.18	1.22	1.19	1.18	1.23	1.26	1.22	1.22	1.26	1.20	1.20	1.20	1.20	1.21

Table A1. *Cont.*

Nutrient Composition	Standard Corn-SBM Diet					10.1% DDGS Diet					28.8% DDGS Diet				
	Phase 1	Phase 2	Phase 3	Phase 4	Phase 5	Phase 1	Phase 2	Phase 3	Phase 4	Phase 5	Phase 1	Phase 2	Phase 3	Phase 4	Phase 5
	Weight Range (lb)					Weight Range (lb)					Weight Range (lb)				
	50–75	75–140	140–195	195–240	240–280	50–75	75–140	140–195	195–240	240–280	50–75	75–140	140–195	195–240	240–280
STTD Ca:STTD P	1.17	1.16	1.22	1.22	1.28	1.21	1.22	1.21	1.25	1.38	1.17	1.13	1.13	1.15	1.29
Cost/ton	$194.04	$180.35	$167.03	$160.54	$156.90	$190.46	$175.40	$163.19	$156.48	$152.95	$183.73	$169.60	$159.41	$153.25	$154.60
Cost with processing	$206.04	$192.35	$179.03	$172.54	$168.90	$202.46	$187.40	$175.19	$168.48	$164.95	$195.73	$181.60	$171.41	$165.25	$166.60
Feed budget, lb/pig	50	155	158	148	145	50	155	158	149	145	51	157	161	150	146
Feed cost, $/pig	$5.17	$14.86	$14.13	$12.74	$12.27	$5.07	$14.49	$13.87	$12.52	$11.99	$4.96	$14.21	$13.79	$12.42	$12.13

SID = Standardized ileal digestible; STTD = Standardized total tract digestible; ME = Metabolizable energy; DE = Digestible energy; NE = Net energy.

Table A2. Inputs for agricultural production of corn grain in the U.S.A.

Inputs from Nature	
[1] Yield (lb/acre)	9699.2
*,[2] Water, unspecified natural origin, US (L)	77.5
*,[3] Occupation, annual crop (land-m^2a)	0.4047
Inputs from Technosphere: Materials/Fuels	
*,[3] Corn seed IP, at regional storehouse/US U (lb)	0.104020385
*,[4] Nitrogen ecoprofile, as N, at regional storehouse/US U (lb)	0.007423293
*,[4] Phosphate ecoprofile, as P, at regional storehouse/US U (lb)	0.005464368
*,[3] Manure, fertilizer, as applied N, at field/US U (lb)	0.001545702
*,[4] Potash ecoprofile, at regional storehouse/US U (lb)	0.007320191
*,[3] Lime ecoprofile, at factory/US U (lb)	0.000820022
Boron, at factory/US U (lb)	0
*,[4] Sulfur, at regional storehouse/US U(lb)	0.001340317
*,[5] Corn herbicides, at regional storehouse/US U (lb)	0.000409002
*,[5] Corn insecticides, at regional storehouse/US U (lb)	0.000119708
*,[6] Diesel produced and combusted, at industrial boiler/US U (gal)	0.00005480480
*,[6] Gasoline produced and combusted, at equipment/US U (gal)	0.000006094
*,[7] Fungicides, at regional storehouse/US- US-EI U (lb)	0.000047322
*,[5] Corn pesticides from NASS (emissions only)/US U (m^2)	0.4047
* Corn air, soil and water emissions ($PO_4 + NO_3$)/US U (m^2)	0
* Transport, lorry 16–32t, EURO3/US- US-EI U (kgkm)	45
Inputs from technosphere: electricity/heat	
*,[6] Natural gas produced and combusted, at industrial furnace/US U (cuft)	0.000243589
*,[6] Electricity, at grid, Western US NREL/US U (kwh)	0.00222624
*,[6] LPG production and combustion, at industrial boiler/US U_NPB_Wheat middling (lb)	0.0024239

[1] Average yield of 2015, 2016, and 2017 USDA-NASS survey. [2] Ecoinvent V 2.2, SimaPro 8.5.2.0. [3] Corn seed rate, manure and lime fertilizer, and occupation land data are taken from the US-EI U, SimaPro 8.5.2.0. [4] Average N, P, K, and S fertilizer data from USDA-NASS survey (2017, 2016, and 2015,). N, P, K, and S Ecoprofile at regional storehouse in the USA US-EI 2.2 (SimaPro 8.5.2.0). [5] Corn herbicides and insecticides data are collected from Camagro, 2013. Corn herbicides at regional storehouse in the USA US-EI 2.2 (SimaPro 8.5.2.0). [6] Diesel, natural gas, electricity, and LPG data are taken from (SimaPro 8.5.2.0). [7] Corn fungicides data collected from USDA-NASS survey, 2016. Corn fungicides at regional storehouse in the USA US-EI 2.2 (SimaPro 8.5.2.0). * Refers to the processes and associated data from (SimaPro 8.5.2.0).

Table A3. Inputs for agricultural production of soybean in the U.S.A.

Inputs from Nature	
Yield (lb/acre)	29,582
*,[1] Water, unspecified natural origin, US (L)	79.5
*,[2] Occupation, annual crop (m^2a)	0.76056338
Inputs from Technosphere: Materials/Fuels	
*,[1] Soybean seed IP, at regional storehouse/US U (lb)	0.03
*,[3] Nitrogen ecoprofile, as N, at regional storehouse/US U (lb)	0.006085193
*,[3] Phosphate ecoprofile, as P, at regional storehouse/US U (lb)	0.017579446
*,[3] Potash ecoprofile, at regional storehouse/US U (lb)	0.03076403
*,[1] Lime ecoprofile, at factory/US U (lb)	0.202713707
Boron, at factory/US U (lb)	0
*,[3] Sulfur, at regional storehouse/US U (lb)	0.005070994
*,[4] Soybean herbicides, at regional storehouse/US U (lb)	0.005551048
*,[4] Soybean insecticides, at regional storehouse/US U (lb)	0.00053854
*,[5] Diesel produced and combusted, at industrial boiler/US U (gal)	0.001680335
*,[5] Gasoline produced and combusted, at equipment/US U (gal)	0.000418155
*,[4] Soybean fungicides, at regional storehouse/US- US-EI U (gal)	0.000328938
*,[6] Soybeans pesticides from NASS (emissions only)/US U (m^2)	0.76056338
Soybean air, soil and water emissions ($PO_4 + NO_3$)/US U (m^2)	0

of the Bioeconomy Strategy in 2018 [6], has set the strategic approach to deployment of a sustainable European bioeconomy, to maximise its contribution towards the 2030 Agenda and its Sustainable Development Goals (SDGs) as well as the Paris Agreement. The Bioeconomy Strategy update proposes an action plan, which, inter alia, prioritises strengthening and scaling up the bio-based sectors, and unlocking investments and markets. Furthermore, other European policy priorities, in particular the Vision of the European industry in 2030 [10], the Circular Economy Action Plan [11], and the Communication on Accelerating Clean Energy Innovation [12], emphasise the importance of sustainably sourced and advanced renewable biomass materials to achieve their objectives. The EU bioeconomy already constitutes an important part of the European Union economy [13]. With European Green Deal priorities on a transition to net-zero greenhouse gas emissions by 2050, and the rapidly changing EU economy, the future of biomass use across the EU economy "looks to differ profoundly from what was imagined even three or four years ago", as stated in the latest study on the EU biomass use in a net-zero economy [14] (p. 3). The EU climate scenarios foresee a 70–80% increase in biomass use (ibid).

The transition of progressive European and world countries from fossil-based to bio-based economy fosters increasing demand for biomass within both domestic and global markets. According to OECD [15] projections, global demand for biomass in the baseline projection scenario will increase by 72% by 2060. Meanwhile, fossil fuel consumption will increase by 47% during the same period. This means that some of the fossil fuels will be replaced by biomass. Growth in biomass consumption is expected to be slower in OECD countries; it is expected to increase by 48% over the same period. The demand for biomass in emerging and developing countries will increase the most, by 136%. The biggest share of the extracted biomass globally is used as food for humans and feed for livestock, followed as raw materials for bio-based industries and bioenergy [16]. According to Carus and Dammer [17], feed predominates with a share of 60, food with 32, and energy use with 4% of the total biomass produced globally. Biomass for energy use, according to OECD [15], will increase by 1.5 times in OECD and by 2.4 times in the rest of the world by 2060. Globally, the use of biomass for energy will increase by 68%, whilst the use of fossil fuels will rise by 60%. Other uses of biomass are also becoming increasingly popular, though in many cases the growth of output is rather slow due to complex, inefficient, and costly manufacturing processes and decreasing economic viability of the products [18–20]. Ramos et al. [18] claim that there is a slight bias towards the development of bioproducts, such as bioplastics, a range of acids, surfactant resins, and biochemicals in Europe, whilst in North America, there is a clear tendency to produce biofuels. Moorkens et al. [21] observe that biopharmaceutical medicines represent a growing share of the global pharmaceutical market. Dos Santos et al. [22] find that biopharmaceuticals represent one-quarter of all pharmaceutical sales and provide suitable and efficient medical care for many previously untreatable diseases.

The aim of this study is to assess biomass self-sufficiency across the European Union member states. To achieve this, the self-sufficiency and import dependence ratios were applied using material flow data on the overall biomass, and separate types of biomass from economy-wide material flow accounts (EW-MFA). The analysis of spatial-temporal data was used to identify the long-lasting trend in terms of biomass self-sufficiency and import dependence across the target countries. We analyse previous trends of biomass self-sufficiency in the European Union, based on the domestic extraction-domestic consumption balance, both for the EU as a whole and for the individual member states (by the composition of the EU until 31 January 2020). Both spatial and temporal variability in the self-sufficiency is used to characterise the stability of capacity to satisfy our own requirements for biomass materials in the EU as a whole, and at the individual member state level. Moreover, a panel data analysis using GMM-SYS was performed to find the factors that have a statistically significant effect on the ratio of biomass self-sufficiency. The term biomass is used in correspondence to classification of biomass as a renewable material in the EW-MFA. In this regard, biomass is understood as the primary biological material derived from the natural environment and used in the economy.

2. Methodology

2.1. The Measure of Self-Sufficiency in Biomass

National self-sufficiency, according to Smith [23], is only a relative term, as in essence it refers to the balance between varieties of the constantly changing demand and adequacy of supply, which can never be in perfect equilibrium. A change in the actual equilibrium may lead to imports of a product, which has previously been exported. For a long time, self-sufficiency studies have been directed mainly at final products [24] and most research and discussion world-wide focused on food self-sufficiency [25].

However, the concept of self-sufficiency increasingly focuses on the issue of input self-sufficiency. For instance, Spero [26] examined the self-sufficiency in energy supply in the context of American national security policy as early as 1973. Recently, the concept of energy self-sufficiency is increasingly explored in scientific literature (e.g., country-level studies are performed by Welfle et al. [27], Saghir et al. [28], and Benti et al. [29]).

The paradigm of water self-sufficiency is increasingly being investigated by scientists; for example, Fragkou et al. [30] analysed the water self-sufficiency potential at the city-level in Mediterranean region, and Sarabi and Rahnama [31] performed a city level study of the potential for energy and self-sufficient water provision in Iran. Additionally, the paradigm of input self-sufficiency becomes increasingly used by researchers for conceptualising and analysing sustainable farming systems. Quite a few studies focus on relationships between input self-sufficiency (such as bioenergy, fodder, nutrients, and seed) and the sustainability or resilience of farms (e.g., Østergård, Markussen [32], Martin, Magne [33], Lebacq et al. [34], Soteriades et al. [35], Gaudino et al. [36], Jouan et al. [37], Masi et al. [38], and Kimming et al. [39]).

The analysis of a biomass-based, energy self-sufficient system for organic farms performed by Kimming et al. [39] focuses on energy balance, resource use, and greenhouse gas emissions. Vijay et al. [40] conducted a regional level study of biomass availability and the potential of energy self-sufficiency in rural areas. A similar study was conducted by Algieri et al. [41] in the southernmost region of the Italian peninsula in the Calabria region. Terrapon-Pfaff [42] assessed renewable energy self-sufficiency in the agricultural production and processing sector, using crop residues and wastes from processing. Harchaoui et al. [43] provided a framework for assessing net energy balances between food surplus, agricultural residues, and energy requirements to determine the potential for energy self-sufficiency in the agriculture sector.

Self-sufficiency can be defined in many ways [24,25]. In the present study, for a more pragmatic interpretation of national self-sufficiency in biomass materials, we used a similar understanding of self-sufficiency in food as suggested by the FAO [44] (p. 19): "the concept of food self-sufficiency is generally taken to mean the extent to which a country can satisfy its food needs from its own domestic production". In other words, self-sufficiency means the domestic food production is equal to or exceeds 100% of a country's food consumption [25]. As Clapp [25] stressed, trade is not ruled out within this definition of national self-sufficiency, as food self-sufficiency is defined by the ratio of food produced to food consumed at the domestic level, while both own-produced and imported products are used for domestic consumption. Food self-sufficiency usually indicates the extent to which a country relies on its own production resources. The higher the degree of self-sufficiency, the greater the ability of a country to satisfy domestic demand for food. Conversely, a lower self-sufficiency ratio indicates a higher dependence on food resources from outside the environment (Wang, 2009, cited in Luan et al. [45] (p. 395)).

Similarly, in this study, biomass self-sufficiency means the extent to which a country can satisfy its biomass material needs from its own domestic production. Biomass self-sufficiency is in this way understood as a country's capacity to satisfy domestic needs for biomass materials, using biological resources originating from the domestic natural environment. According to a broader understanding, biomass self-sufficiency means the extent of biomass material available to support economic activity in a spatial economic system (e.g., national, regional, local, etc.) based on domestic sustainable biomass potential.

Self-sufficiency can be measured in many ways; however, a ratio of production and consumption of individual products are the most common way of measuring self-sufficiency [24]. A key indicator for self-sufficiency measuring is the self-sufficiency ratio (SSR), which is defined "as the share of domestic production in total domestic use, excluding stock changes" [44] (p. 19). Our calculation of the biomass self-sufficiency ratio follows the statistical data on biomass domestically extracted from the natural environment and domestically used in the economy, developed by economy-wide material flow accounts (EW-MFA) [46]. The most basic nomenclature of material flow at different stages of the natural resource cycle used in EW-MFA will be applied. The biomass self-sufficiency ratio (SSR) was estimated based on the following algorithm:

$$SSR = \frac{DE \times 100}{DC} = \frac{DE \times 100}{(DE + IMP - EXP)},$$ (1)

where, DE denotes the domestic extraction that refers to the flows of biomass material extracted or harvested from the domestic natural environment, which physically enter the economic system for further processing or direct consumption. DC is the biomass domestic consumption that indicates the total amount of biomass consumed domestically in production and consumption activities. The biomass domestic consumption in Equation (1) is calculated as biomass domestic extraction plus biomass imports (IMP) minus biomass exports (EXP), i.e., $DC = DE + IMP - EXP$.

The SSR can be calculated for an individual biomass material (e.g., wheat, vegetables, fibres, straw, etc.), groups of biomass materials of similar origin (e.g., cereals, crop residues, wood, etc.), and for the aggregation of all biomass materials (i.e., biomass). The SSR indicates the extent to which a country relies on its own extraction of biomass, originating from the domestic natural environment, to meet domestic demand for biomass material, (i.e., the higher the ratio, the greater the national economy's self-sufficiency for biomass). An SSR over 100% indicates a national biomass material extraction surplus in relation to its domestic demand and therefore net exports. On the contrary, the lower the SSR (where the SSR is < 100%), the more the national economy's dependence upon biomass imports. An SSR < 100% indicates that domestic extraction of biomass from its own natural environment is less than demand for biomass quantity in a domestic market, and there is a demand for biomass imports to satisfy domestic needs. To analyse the time-space variability of biomass self-sufficiency across the European Union countries, the coefficient of variation (CV) of SSR was calculated.

National economy's dependence upon biomass imports could by mathematically expressed as follows:

$$IDR = \frac{IMP \times 100}{DI} = \frac{IMP \times 100}{(DE + IMP)},$$ (2)

where, IDR denotes the import dependence ratio expressed as a percentage, and DI denotes the direct input of biomass into the national economy. DI includes all biomass materials that are of economic value, and which are available for use in production and consumption activities. DI in Equation (2) is calculated as the sum of the domestic extraction of biomass plus its physical imports $(DI = DE + IMP)$. The IDR depicts the extent to which an economy relies upon imports to meet its biomass needs (i.e., the higher the ratio the greater the dependence on the import). IDR cannot be negative or higher than 100%. $IDR = 100\%$ indicates that there are no domestic extractions of biomass during the reference year. Based on the FAO [44] interpretation, the IDR measures the share of biomass imports in the domestic biomass consumption (both extracted locally and imported). Countries satisfying their domestic needs for biomass predominantly with domestic extractions will have $IDRs$ lower than 50%, while the countries relying more on imports than on domestic extraction will have $IDRs$ higher than 50%. However, it should be kept in mind that these ratios hold only if imports are mainly used for domestic consumption and are not re-exported. In the present study, the strength of correlation between SSR and *IDS* was analysed based on a whole sample of the EU countries.

2.2. Research Scope and Data Sources

The empirical analysis of biomass self-sufficiency is based on economy-wide material flow accounts (EW-MFA) data on the physical flows of biomass as a natural material at various stages of the flow chain, specifically domestic extraction, domestic material consumption, and imports. The data come from the Eurostat's "Material Flow and Productivity" database.

In EW-MFA, the indicator of domestic extraction of biomass is its flows from the domestic natural environment to the economy. Biomass extraction is defined as its amount in physical weight derived from the natural environment for use in the economy [46]. It is equivalent to the concept of used primary biomass harvest (such as primary crops, used crop residues, biomass harvested from grassland and grazed biomass, and wood harvest (wood removals)), as it is used in the Human Appropriation of Net Primary Production (HANPP) framework. "Biomass, the sum of recent, non-fossil organic material of biological origin, is one of the fundamental resources of any socioeconomic system" [47] (p. 471).

According to EW-MFA conventions [16], domestic extraction covers biomass that acquires the status of a product and is used as a natural materials input in further economic processes (or socioeconomic processes [47]). Meanwhile, biomass harvested from the natural environment without the intention of using it for economic needs is not included in flow of domestic extraction (e.g., felling losses in forests or crop residues remaining on field). In the ecosystem services literature, biomass remaining in natural or cultivated ecosystems after harvest denoted as "back-flows to nature" [48]. However, a systematic review of the literature on ecosystem services demonstrated the multiple environmental benefits of unused biomass remaining in ecosystems after extraction or harvest, for example: the benefits of returning crop residues to soils for climate regulation due to the formation of soil organic carbon and improving the relation between organic carbon and nitrogen in soil, cover crops and crop residues which are left on fields is central to biodiversity conservation in the agrarian landscape, and both cover crops and crop residues also contribute to water retention and the slow passage of water into deeper soil layers [48].

The analysis of national self-sufficiency and import dependency is targeted at two levels of detail of biomass materials: first, whole biomass corresponds to a 1-digit level of materials category in the EW-MFA classification (codes FM1), and second, two biomass groups according to OECD [49] classification, such as food materials (i.e., food crops, fodder crops and used crop residues, wild animals, essentially marine catches), small amounts of non-edible biomass (e.g., fibres, rubber), and related products including livestock, and woody materials (i.e., harvested wood and traded products essentially made of wood).

Both biomass groups according to OECD classification correspond to a 2-digit level of material classes in the EW-MFA classification (i.e., food materials by such codes: MF11, crops (excluding fodder crops), MF12, used crop residues, fodder crops, and grazed biomass, MF14, wild fish catch, aquatic plants and animals, hunting and gathering, and woody materials (code MF13)). Two additional biomass classes for livestock and livestock products (correspond, respectively, to codes MF15 and MF16) are not accounted as domestic extraction of biomass originating from the domestic natural environment but are considered as flows within the economic system. According to the EW-MFA principles [46], cultivated livestock (e.g., cows, pigs, sheep, etc.), and livestock products (e.g., milk, meat, eggs, animal leather, etc.) as well as livestock waste are not natural inputs and hence excluded from domestic extraction of biomass. Cultivated animals convert primary plant biomass into edible biomass for human consumption [48].

The data are from the Eurostat's "Material Flow and Productivity" database. The empirical study covers the period from 2000 to 2018 which is subdivided into two subperiods: 2000–2009 and 2010–2018, respectively. The spatial unit of analysis consists of the country. The study covers all 28 member states of the European Union (by the composition of the EU until 31 January 2020).

2.3. Econometric Model of Determinants of Biomass Self-Sufficiency

To explore the effect of some determinants on a degree of self-sufficiency in biomass, the two-step system generalised method of moments (GMM-SYS) was used in a dynamic panel model. This method helps to solve the problems of endogeneity, heteroscedasticity, and serial correlation [50]. The GMM-SYS estimator combines the regression in differences with the regression in levels [51]. This estimator is consistent if two tests are successful: (1) a Sargan test over-identification, which proves that instruments are valid (p-value > 0.05), and (2) an AR(2) test about no second-order autocorrelation (p-value > 0.05). A general form of a dynamic panel data model is as follows:

$$y_{it} = \gamma y_{i,t-1} + C + \beta' x_{it} + \mu_t + \alpha_i^* + \varepsilon_{it}, \tag{3}$$

where, $i = 1, 2, \ldots, N$ represents cross-sectional unit; $t = 1, 2, \ldots, T$ represents the time period; α^* reflects unobserved cross-sectional heterogeneity; μ refers to time-specific effect; ε means idiosyncratic error; y is the explained variable; $y_{(t-1)}$ is the lagged explained variable; C is constant; x stands for explanatory variables; γ and β are parameters that reflect the impact of right-hand side variables on regression [52,53].

The analysis of world region patterns [47] revealed that the level of biomass use, measured by its consumption per capita, is determined by patterns of land use evolved historically as well as by population density, rather than by affluence or economic development status. The regional peculiarities of the land use system can lead to significant differences in the quantitative and qualitative structure of biomass harvest, as noted in the UNEP document [16]. In light of the above information, the core explanatory variable of this research is agricultural and forest land share (%). To make the results of the econometric model more convincing, other control variables are included in the model. The variables used in this paper and the expected correlation with the explained variable are presented in Table 1. The data used for calculations have been collected from the Eurostat's and FAOSTAT's databases.

Table 1. Research variables.

	Variables	Expected Correlation [a]	Variable Description
Explained variable	Biomass self-sufficiency ratio (*B_SSR*)	-	%Ratio between domestic extraction and domestic consumption.
Core explanatory variable	Agricultural and forest land share, % (*land*)	+ive	%Ratio between the sum of both agricultural and forest land areas and the total land area (excluding area under inland waters and coastal waters).
Control variables	Biomass domestic extraction per ha (*biomas_extr*)	−ive/+ive	Ratio between biomass domestic extraction and the sum of both agricultural and forest land areas.
	Share of bioenergy in renewable energy, % (*bioen_renw*)	+ive	%Ratio between bioenergy and primary production of total renewables.
	Share of bioenergy in total primary energy production, % (*bioen_prim*)	+ive	%Ratio between bioenergy and total primary production.
	Energy imports dependency, % (*en_imp*)	−ive	%Ratio between net import and gross available energy.
	Biomass materials intensity, kg per GDP (PPS [b]) (*biomas_int*)	+ive	Ratio between biomass direct inputs and GDP.

Table 1. *Cont.*

Variables	Expected Correlation [a]	Variable Description
Resource productivity, GDP PPS per tonne (*res_prod*)	−ive/+ive	Ratio between GDP and domestic material consumption.
Population density, persons per km² (*pop_dens*)	−ive	Ratio between the number of population and the land area.
Employment in total knowledge-intensive activities, % (*empl_kia*)	−ive	%Ratio between employment in total knowledge-intensive activities and total employment.

Notes: [a] Expected correlation with the explained variable; +ive stands for "positive"; −ive stands for "negative"; [b] PPS stands for purchasing power standard.

Based on Equation (3), our econometric model can be specified as follows:

$$\begin{aligned} B\_SSR_{it} = {} & \gamma B\_SSR_{i,t-1} + C + \beta_1 land_{it} + \beta_2 biomas\_extr_{it} + \\ & \beta_3 bioen\_renw_{it} + \beta_4 bioen\_prim_{it} + \beta_5 en\_imp_{it} + \beta_6 biomas\_int_{it} + \\ & \beta_7 res\_prod_{it} + \beta_8 pop\_dens_{it} + \beta_9 empl\_kia_{it} + \mu_t + \alpha_i^* + \varepsilon_{it}, \end{aligned} \tag{4}$$

where, *B_SSR* refers to biomass self-sufficiency ratio; *B_SSR*(-1) means lagged biomass self-sufficiency ratio variable; *land* is agricultural and forest land share, %; *biomas_extr* is biomass domestic extraction per ha; %; *bioen_renw* refers to share of bioenergy in renewable energy; *bioen_prim* means share of bioenergy in total primary energy production; *en_imp* denotes energy imports dependency; *biomas_int* is biomass materials intensity, kg per GDP (PPS); *res_prod* is resource productivity, GDP PPS per tonne; *pop_dens* means population density, persons per km²; *empl_kia* refers to employment in total knowledge-intensive activities, %.

The descriptive statistics of the variables are presented in Table 2.

Table 2. Descriptive statistics.

Variable	Observations	Mean	St. Dev.	Min	Max
B_SSR	532	103.03	41.69	6.86	417.19
land	532	76.82	11.17	29.22	94.14
biomas_extr	532	5.80	4.19	1.43	21.00
bioen_renw	530	67.14	21.84	0.00	99.92
bioen_prim	532	22.85	20.59	0.00	92.27
en_imp	532	55.55	27.83	−50.60	104.14
biomas_int	532	0.26	0.14	0.06	0.97
res_prod	532	1.53	0.75	0.43	4.18
pop_dens	532	171.28	245.88	17.00	1548.30
empl_kia	308	35.36	6.62	19.20	60.40

3. Results

3.1. The Profile of the EU's Biomass Extraction–Consumption Balance

As an aggregated biomass, the European Union is rather self-sufficient in biomass originating from the domestic natural environment and has the potential to meet all domestic needs for biomass, with sufficient domestic extraction, as illustrated by the SSR curve in Figure 1. The Figure also displays a slightly higher amount of biomass domestic consumption compared to its domestic extraction throughout the entire study period; domestic consumption increased by 7.7% (i.e., from 1718.4 million tonnes in 2000 to 1850.6 million tonnes in 2018), while domestic extraction grew by 8.2% (i.e., from 1659.3 to

1796.1 million tonnes). The obtained results indicate that the EU's biomass self-sufficiency has been higher in the present decade than in the previous decade but has slightly decreased in recent years (i.e., from almost 98.9% in 2016 (when it was at its highest degree for the past nineteen years) to 97.1% by 2018).

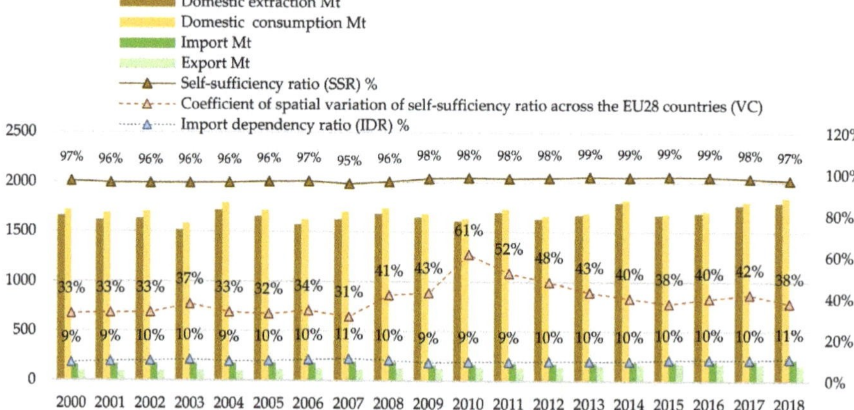

Figure 1. Trends of the biomass extraction, consumption, and self-sufficiency in the EU-28. Source: Own composition based on the material flow accounts data from the Eurostat database.

Biomass self-sufficiency varied spatially across the EU countries, as illustrated by the spatial variation coefficient of self-sufficiency ratio (VC) in Table 3. In this regard, a previous study [54] has revealed that nearly two thirds of the EU member states specialised in primary production (in other words: biomass as renewable resources production) and can meet all domestic requirements for biomass material and have the potential to export biomass to the EU internal or world markets. By contrast, in the remaining third of the EU member states (Luxembourg, United Kingdom, Belgium, Ireland, Denmark, Malta, Austria, Sweden, and Germany) the primary production sector is under-represented and biomass materials production is insufficient to meet domestic demand, requiring the import of biomass. However, the degree of self-sufficiency in biomass varies significantly across the EU member states, as the coefficient of spatial variation (VC) in Figure 1 (see also Table 3) indicates. The trend revealed by Figure 1 is a steadily decreasing difference in biomass self-sufficiency between the EU countries in the years following the global economic crisis.

Despite being almost 100% biomass self-sufficient, the EU economy also depends on imports of biomass materials. The EU is both importing and exporting significant amounts of biomass material. Biomass accounts for more than one-tenth of total material import and a quarter of total material export of the EU in the years following the global economic crisis. The EU is a net importer of biomass materials, just as are most of the member states. In 2018, imports exceeded exports by 33% (Figure 1). During the 2000–2018 period, biomass import increased by 38% (i.e., from 158.9 to 217.3 million tonnes) and export increased by 63% (i.e., from 99.8 to 162.9 million tonnes).

Thus, the import dependence of the EU on biomass currently is above 11%, and it shifted to a higher level throughout the study period. The highest and steadily increasing dependence on imports since 2009 was observed in overall woody materials (13% in 2018). The import dependence on the total food material category is 6%, and it increased after the economic crisis, possibly due to slower growth in extraction or harvesting compared to total consumption growth (on average 1.5% and 1.6% per year in 2010–2018, respectively). The import dependence on the general nutrient category is 6% and it increased to a higher level after the economic crisis, possibly due to slower extraction or yield growth compared to total consumption growth (on average 1.5% and 1.6% respectively, 2010–2018).

Table 3. Temporal variation of the biomass self-sufficiency ratios in the EU countries.

| | Average 2016–2018 (SSR) | 2000–2009 | | | 2010–2018 | | | ±%p SSR$_{2SP}$ Less SSR$_{1SP}$ | ±%p CTV$_{2SP}$ Less CTV$_{1SP}$ |
		Average (SSR$_{1SP}$)	Standard Deviation	Coefficient of Temporal Variation (CTV$_{1SP}$)	Average (SSR$_{2SP}$)	Standard Deviation	Coefficient of Temporal Variation (CTV$_{2SP}$)		
EU(28)	97.9	96.1	0.77	0.8	98.2	0.56	0.6	2.1	−0.2
LV Latvia	225.2	214.0	25.67	12.0	279.7	70.74	25.3	65.8	13.3
EE Estonia	183.9	150.8	20.31	13.5	173.7	9.81	5.7	22.9	−7.8
CZ Czechia	157.0	131.1	13.63	10.4	159.1	7.79	4.9	28.0	−5.5
SI Slovenia	147.5	106.0	20.56	19.4	142.9	7.05	4.9	36.8	−14.5
BG Bulgaria	135.9	109.7	6.43	5.9	135.6	6.72	5.0	25.8	−0.9
FI Finland	126.2	114.9	6.40	5.6	124.0	3.27	2.6	9.0	−2.9
HU Hungary	123.7	115.0	6.94	6.0	125.8	6.19	4.9	10.8	−1.1
HR Croatia	123.5	109.9	5.18	4.7	122.3	6.08	5.0	12.5	0.3
LT Lithuania	121.2	109.5	2.37	2.2	120.0	4.63	3.9	10.5	1.7
SK Slovakia	117.1	109.3	3.81	3.5	114.3	6.34	5.5	5.0	2.1
RO Romania	111.5	100.1	2.32	2.3	112.2	3.40	3.0	12.1	0.7
FR France	110.9	109.5	3.05	2.8	111.4	1.16	1.0	1.9	−1.8
SE Sweden	110.7	112.5	4.11	3.7	110.5	3.19	2.9	−2.0	−0.8
PL Poland	99.9	97.8	1.23	1.3	99.2	2.30	2.3	1.4	1.1
ES Spain	96.6	88.1	2.41	2.7	96.2	1.28	1.3	8.1	−1.4
DE Germany	94.2	101.1	1.87	1.9	95.3	1.16	1.2	−5.8	−0.6
AT Austria	91.8	98.3	2.54	2.6	93.1	1.90	2.0	−5.2	−0.5
DK Denmark	88.9	89.5	2.92	3.3	91.7	3.08	3.4	2.3	0.1
IE Ireland	88.2	93.7	0.96	1.0	90.8	2.71	3.0	−2.9	2.0
EL Greece	86.8	86.8	3.66	4.2	88.7	2.12	2.4	2.0	−1.8
NL Netherlands	84.6	83.1	3.82	4.6	81.8	2.91	3.6	−1.4	−1.0
UK United Kingdom	79.1	80.9	1.46	1.8	79.9	1.60	2.0	−0.9	0.2
PT Portugal	78.5	83.8	3.54	4.2	81.6	2.98	3.6	−2.2	−0.6
IT Italy	77.9	82.8	2.04	2.5	80.5	3.27	4.1	−2.3	1.6
BE Belgium	70.0	71.9	3.13	4.3	69.9	2.46	3.5	−2.0	−0.8
LU Luxembourg	69.4	62.9	5.62	8.9	70.6	7.19	10.2	7.7	1.2
CY Cyprus	32.8	40.4	16.15	40.0	37.7	5.43	14.4	−2.6	−25.6
MT Malta	15.0	21.1	3.51	16.7	19.0	3.52	18.5	−2.0	1.8
Stand. deviation	41.9	34.0	-	-	47.0	-	-	-	-
Coefficient of spatial variation (CSV)	39.8	34.3	-	-	43.7	-	-	-	9.4

Source: Own calculation based on the material flow accounts data from the Eurostat database.

As for the total food materials group, including small amounts of non-edible biomass, the degree of the EU's self-sufficiency was 1–2 percentage points lower than 100% over the nineteen study years, except for 3 percentage points in 2007 (Figure 2). A similar result was observed for SSR in materials of crops origin (see the curve of crop materials sub-group in Figure 2). This means that the EU harvests more biomass for food (including feed, fibre, biofuels, and biogas as well) than it consumes, thus avoiding a supply side problem in recent decades [55]. In contrast, the EU's self-sufficiency in wild fish and aquatic materials (including small amounts of hunting and gathering materials) was noticeably lower, owing to its limited natural resources. Indeed, the SSR of this biomass sub-group decreased from 75% to 61% in the past two decades. Accordingly, the EU's import dependency on this biomass sub-group shifted from 34% to almost 67% in the last nineteen years.

As for overall woody materials group, the EU's self-sufficiency increased from 92.9% to 96.9% in 2000–2009 but fell to 95.2% in 2010–2018 (Figure 2). The EU is "approximately self-sufficient" ("approximately food self-sufficient" is proposed by O'Hagan (1975, cited in Clapp [25], p. 4), meaning the self-sufficiency ratio interval between 95 and 105 percent) in woody fuel and domestic needs, mainly relying on its own production. The import dependency ratio of the latter material category was low and stable (i.e., about 5%) during the two decades in the study period, thus the RSS in wood fuel is as high as 95% over the last two decades. The lowest degree of self-sufficiency is illustrated by industrial roundwood; however, its SSR increased from below 92% in 2000 to above 95% in 2018.

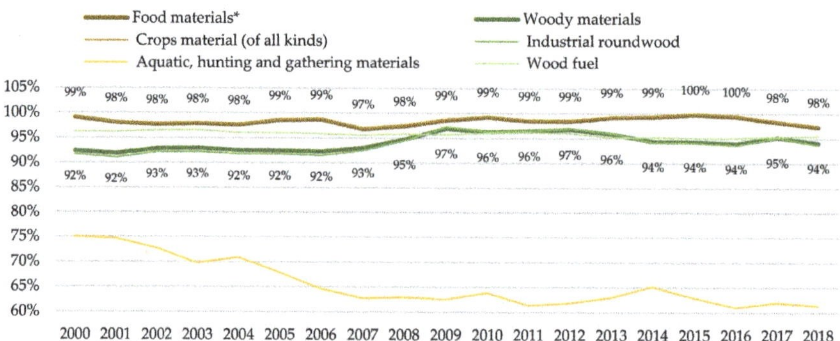

Figure 2. Trends of the self-sufficiency in biomass groups and sub-groups in the EU-28. Notes: * including small amounts of non-edible biomass; SSR value presented by food and woody materials groups. Source: Own composition based on the material flow accounts data from the Eurostat database.

3.2. The Profile of Biomass Self-Sufficiency across the EU Countries

Based on average SSR value and its variation coefficient, Table 3 presents the temporal and spatial distribution characteristics of the biomass self-sufficiency ratio for both the EU as a whole and at the level of member states for the two sub-periods from 2000 to 2009 and 2010 to 2018. The data compiled in the table below reveals that both spatial and temporal variability of SSR is very uneven in the EU. In the EU as a whole, very small coefficients of temporal variation over both sub-periods (approx. 0.8% and 0.6%, respectively CTV_{1SP} and CTV_{2SP}) display stability in all regions for the last nineteen years considered. However, the biomass self-sufficiency varies greatly across members states. The high coefficient of spatial variation (CSV) during both sub-periods illustrates the significant differences in the degree of biomass self-sufficiency among countries. Additionally, the difference among the EU countries increased remarkably over the last decade, as evidenced in the change of SSR variation, from 34.3% in the first sub-period to 43.7% in the second sub-period.

The data in Table 3 illustrates the asymmetry between the degree of biomass self-sufficiency and its temporal variation at the country level. The highest coefficient of the SSR variation was found in countries that display the highest or lowest SSR. Moderate or higher variations of SSR (>10%) during the first sub-period 2000–2009 were found in Latvia, Estonia, and the Czech Republic, which demonstrated the highest SSR (214%, 150%, and 131% on average in 2016–2018, respectively), and in Malta and Cyprus, which displayed the lowest SSR (21%, 150%, and 131% on average, respectively), meaning that the nature of the biomass domestic extraction-consumption system in these countries was the most unstable in the whole EU region. The decreased SSR variation in the second half (2010–2018) displays more stability in Estonia, the Czech Republic, and Cyprus, whereas the increased variation in Latvia and Malta demonstrates more instability. The decreased variation coefficient in the second half indicates more stability of biomass domestic extraction-consumption systems in the other thirteen countries as well.

Nearly half of the EU member countries (i.e., the first thirteen listed in Table 3) were largely self-sufficient in biomass materials originating from the domestic natural environment. They had a ratio of self-sufficiency of over 109% since 2000. In addition, five countries (Latvia, Estonia, the Czech Republic, Slovenia, and Bulgaria) had a very high degree of biomass self-sufficiency, with biomass yields exceeding domestic consumption by more than 1.35 to 2.35 times at the end of the study period. Additionally, the SSR of all these countries increased in the last decade compared to the previous decade, as indicated by a positive difference between the SSR values in both sub-periods (Table 3). Poland and Spain were "approximately biomass self-sufficient", meaning that they had a ratio of self-sufficiency between 95 and 105% during the last decade. Germany was "approximately biomass self-sufficient" until 2016 as well; however, its self-sufficiency in biomass materials declined to almost 94% by 2018. This implies that all the sixteen countries can meet all domestic requirements for biomass materials, and have potential (except Poland, Spain, and Germany) to export biomass to satisfy the growing demand for biomass, not only on the EU internal market, but also in global markets.

In contrast, an SSR below 95% was found in twelve of the EU member states (these include Austria and the rest of the countries listed in Table 3, above) during both considered decades. This indicates that the degree of biomass self-sufficiency is low, and these countries were less than self-sufficient. Moreover, in most countries (i.e., nine out of twelve, except for Denmark, Greece, and Luxembourg), biomass self-sufficiency declined over the last decade compared to the previous one, indicating the deterioration of their capacity to meet their own domestic demand for biomass materials. As a result, all these countries faced a significant increase in their dependence on biomass imports from the rest of the EU countries or other countries (see Figure 3).

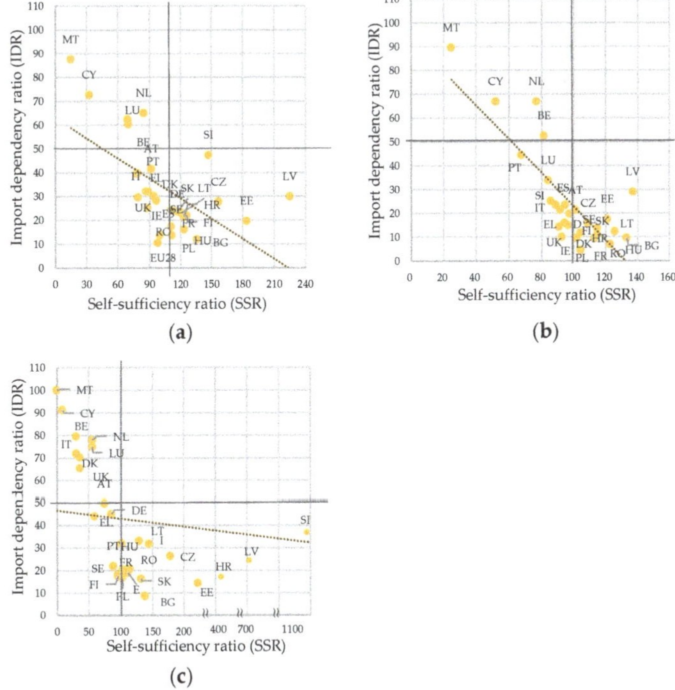

Figure 3. Self-sufficiency and import dependency relationship graph for biomass materials across the EU member states. Notes: In the whole sample of the EU−28 countries, a moderate negative correlation was found between the self-sufficiency and import dependency ratio (r = −0.62; correlation is significant at the 0.01 level (2-tailed)), a strong negative correlation was found between the self-sufficiency and import dependency ratio for food materials (r = −0.81; correlation is significant at the 0.01 level (2-tailed)), and a moderate negative correlation between the self-sufficiency and import dependency ratio for woody materials (r = −0.31; correlation is significant at the 0.01 level (2-tailed)). Source: Own composition based on the material flow accounts data from the Eurostat database. (a) Biomass materials (total); (b) food materials; (c) woody materials.

The exceptionally low biomass self-sufficiency ratio for the entire considered period was observed in the two island nations: Cyprus and Malta. This means that both countries satisfied their domestic needs for biomass materials by relying more on imports than on local production. As illustrated in Figure 3, in both countries, the import dependency ratio was higher than 50% (i.e., nearly 90% and 67% on average, in 2016–2018). Indeed, the degree of biomass self-sufficiency decreased in both countries (i.e., from 50% to below 33% in Cyprus and from above 25% to 15% in Malta, in the past half-century). This means that dependence on biomass imports has increased in both countries.

Some countries lack the necessary natural resource base or face ecological asset scarcity to be self-sufficient [56]. Such countries (e.g., Cyprus, Malta, Netherlands, Belgium, and Luxembourg; see Figure 2) tend to rely on imports to meet their needs and may become import dependent. At the same time, however, biomass import dependency is not simply a result of natural resource constraints; this may be due to several other factors, including drops in domestic production, changing production and demographic shifts, subsidies, and export earnings (interpretation extended from food imports to overall biomass imports, based on Luan et al. [57]).

Figure 3a indicates the relative position of the EU countries in terms of their ability to satisfy domestic needs for biomass materials, based on the domestic extraction and import. Both the self-sufficiency and import dependency ratios are expressed through average value in 2016–2018. A moderate negative correlation between SSR and IDR was found for the whole biomass materials group ($r = -0.62$, $p < 0.01$), which means a comparatively higher dependence on imports at a lower degree of self-sufficiency. Figure 3 illustrates that only in five EU countries is a larger portion of biomass domestic demand satisfied through imports (i.e., Malta (88%), Cyprus (73%), Netherlands (65%) Luxembourg (88%), and Belgium (60%)). In the remaining EU countries, the greater share of domestic needs for biomass was satisfied through domestic production imports, ranging from 12% for Bulgaria up to 47% for Slovenia.

A strong negative correlation ($r = -0.81$, $p < 0.01$) between SSR and IDR was found for the food materials group, including small amounts of non-edible biomass. An export share larger than 50% appears for Malta (90%), Cyprus and the Netherlands, (67%), and Belgium (52%), revealing their dependence on food imports (Figure 3b). As for the whole woody materials group, the correlation analysis indicates a moderately negative relation between the self-sufficiency and import dependency variables (Figure 3c). More than half of domestically consumed woody biomass was imported in seven EU countries, such as Luxembourg (75%), Netherlands (78%), United Kingdom (65%), Denmark (70%), Italy (72%), Belgium (79%), and Cyprus (91%). The imported woody materials amounted to 50% of Austria's domestic biomass needs, while Malta used only imported wood materials for domestic consumption.

3.3. The Effect of Determinants on Biomass Self-Sufficiency

To obtain valid results of statistical analysis, a log (logarithm) transformation of variables is performed. Also, outliers are removed from the dataset as they can affect the results of the model. The verification of the dependency between explained (y) and each of the explanatory variables (x) was checked, and a linear relation between each pair of variables was revealed. For this reason, a linear form of each variable is included in the model. A bivariate correlation between each pair of different explanatory variables was examined and the results revealed that there is a strong correlation (higher than |0,7|; based on Ratner [57]), between two pairs of variables: l ("l" denotes the log transformation of a variable) pop_dens & l_biomas_extr (0,8) and l_biomas_int & l_res_prod (−0,8). For this reason, the variables that are highly correlated are included in the model separately, and four cases of the model are performed. The results of the estimations are presented in Table 4. The program used to run the model was GRETL.

All estimations in Table 4 passed AR(2) and Sargan over-identification tests and thus the results are not affected by the second-order correlation, and the instruments are valid. The results reveal that the impact of the indicator of major interest (i.e., the agricultural and forest land share variable) on the biomass self-sufficiency ratio is positive and statistically significant in all cases of the model. A positive effect indicates that an increase in agricultural and forest land increases the biomass self-sufficiency ratio,; this confirms the expected effect of the core explanatory variable on the explained variable.

Table 4. Estimation results of the model.

	I	II	III	IV
l_B_SSR(-1)	0.5724 *** (0.0000)	0.5239 *** (0.0000)	0.5011 *** (0.0000)	0.5110 *** (0.0000)
const	1.9929 ** (0.0113)	1.8644 *** (0.0004)	2.5105 *** (0.0000)	1.7343 *** (0.0027)
l_land	0.2360 *** (0.0000)	0.2999 *** (0.0000)	0.2453 *** (0.0000)	0.3083 *** (0.0000)
l_biomas_extr	−0.0523 *** (0.0050)	−0.0416 *** (0.0000)		
l_bioen_renw	0.1365 *** (0.0009)	0.0577 ** (0.0180)	0.1954 *** (0.0000)	0.0826 *** (0.0002)
l_bioen_prim	0.0357 *** (0.0080)	0.0186 (0.2123)	−0.0127 (0.3090)	0.0050 (0.7510)
l_en_imp	−0.2717 *** (0.0009)	−0.2033 *** (0.0022)	−0.1329 ** (0.0353)	−0.1408 ** (0.0000)
l_biomas_int		0.1166 *** (0.0000)		0.0720 *** (0.0012)
l_res_prod	0.0169 (0.5860)		0.1186 *** (0.0000)	
l_pop_dens			−0.1288 *** (0.0000)	−0.0595 *** (0.0000)
l_empl_kia	−0.0800 (0.4050)	−0.0016 (0.9815)	−0.2380 *** (0.0003)	−0.0249 (0.7221)
AR(2) test	0.0233 (0.9814)	−0.0643 (0.9488)	0.3468 (0.7287)	0.2999 (0.7643)
Sargan test	20.5052 (1.0000)	19.1705 (1.0000)	21.0238 (1.0000)	19.9745 (1.0000)
Number of countries	27 [a]	27 [a]	27 [a]	27 [a]
Number of observations	270	269	273	272

Notes: [a] Malta is not included because of a lack of some data. *** and ** indicate statistical significance at the 1%, 5%, and 10% significance level, respectively. z-values are presented in parentheses of estimates. *p*-values of AR(2) and Sargan tests are provided in parentheses. All estimations are two-step GMM-SYS.

The expected effect of biomass domestic extraction on the biomass self-sufficiency ratio was positive or negative. In both cases of the model, when this variable was included, the negative direction of exposure was obtained, and the results were statistically significant. The coefficient of this variable indicates that a one per cent increase in biomass domestic extraction decreases the biomass self-sufficiency ratio by about 0.05 per cent. The estimation results also indicate that an increase in the share of bioenergy in renewable energy contributes positively to the biomass self-sufficiency ratio, in all four constructed cases of the model; this is in line with previous expectations. Energy import dependency proved to be statistically significant. The magnitude and direction of this variable indicate that an increase in energy import dependency by 1 per cent reduces the biomass self-sufficiency ratio by about 0.13–0.27 per cent; the effect of this variable is as expected.

The estimates suggest that the biomass materials intensity variable has a positive and statistically significant effect on the explained variable, indicating that an increase in biomass material intensity affects an increase in the biomass self-sufficiency ratio.; the results are in line with a priori expectations. This research has demonstrated that population density has a statistically significant effect on the biomass self-sufficiency ratio in all cases of the model. This effect is negative (i.e., when population density increases, biomass self-sufficiency decreases). Further results of the model suggest that three variables are statistically significant only in one case of the model. These variables are the share of bioenergy in total primary energy production, resource productivity, and employment in total knowledge-intensive activities.

4. Discussion

In this section, the study results on biomass self-sufficiency and import dependence, as well as its determinants in the EU countries and the United Kingdom, are discussed in the context of ecological constraints of biomass extraction and harvesting, the EU Green Deal, and biomass supply-demand trends at a global scale.

The obtained results suggest that the European Union is rather self-sufficient in terms of biomass and has the potential to meet all domestic requirements for biomass, with sufficient domestic natural resources. The EU's biomass self-sufficiency has been higher in the present decade than in the previous decade but has recently slightly diminished. It is emphasised that, although the domestic demand and domestic supply of biomass is currently balanced in the European Union, increasing the supply of biomass that originates from its own sources, to match the large increase in demand, would be very difficult [14]. A range of ecological boundaries place limitations. Biomass is finite in nature and functionally; time for regrowing or recovering is needed. Biomass harvest depends on ecosystem health, regeneration rate, and the availability of land, soil fertility, and productivity, etc. Exceeding ecological boundaries can lead to ecosystem degradation [58]. Due to ecological boundaries, the EU's bioeconomy strategy requires that biomass be used only within safe ecological limits, to strengthen the resilience of both terrestrial and aquatic ecosystems and ensure their contribution to climate mitigation and sustainability of their biodiversity [6]. Moreover, biomass harvesting or extraction has multifaceted negative implications for natural ecosystems, such as declines in biodiversity, the reduction in natural carbon stocks and sinks, nitrate pollution, and GHG emissions from biomass production. Consequently, biomass continues to play a role in the EU's low-carbon transition; its use will need to be limited and targeted towards the most added value to climate and environmental objectives [58].

Despite a high degree of self-sufficiency, the EU economy also depends on imports of biomass materials. The EU is a net importer of biomass materials, as are most of the member states; the EU's imports exceeded exports by 33% (2018). Based on the EU-28 states' sample, a moderate negative correlation between self-sufficiency and import dependency ratios was found for the whole biomass material group and for the wood materials sub-group, and a strong negative correlation was found for the food material sub-group, including small amounts of non-edible biomass. As several studies have found, on average globally, the largest share of biomass extracted from cultivated and non-cultivated (wild) natural biological resources is used domestically [16,47,59]. Currently, only one tenth of the globally extracted biomass from the natural environment is traded internationally; however, in some countries the share of biomass international trade may be large [16]. Using the simulated interregional flows of goods within the world economy, based on the global multi-regional input-output table [59], it was found that most of the local biomass extraction is consumed locally. On the contrary, only a small part of biomass use, embodied in final consumption, originates from domestic sources. The reasons substantiate the need for biomass self-sufficiency studies to uncover the extent to which a country can satisfy its biomass needs from its own domestic production, based on its natural resource base.

Additionally, the latest research [5] demonstrates that the extraction of raw materials (including biomass) in Europe is under threat due to various reasons, such as the lack of knowledge on raw materials, insufficient awareness of the true ecological boundaries of the planet, competing land use and urbanisation, erosion and degradation of soils, as well as the reluctant and at times hypocritical attitude towards the sustainability of raw materials used in imported products. The "spatial disconnect" between biomass production and consumption has been recognised as pivotal, regarding global environmental degradation [60].

Several studies ([61–63]) demonstrate that the availability of biomass as a renewable resource arising from living ecosystems is unevenly distributed, in regards to its demand and availability. Some of the regions with the greatest demand have a relatively low availability of local biomass resources. Our findings, based on the spatio-temporal analysis of biomass self-sufficiency, illustrate that the ability of EU countries to meet their domestic biomass demand, at the expense of local resources, varies greatly. The degree of biomass self-sufficiency varies significantly across the EU member states, although the differences increased remarkably over the last decade. The asymmetry between the degree of biomass self-sufficiency and its temporal variation at the country level is higher in the countries that display the highest or lowest self-sufficiency ratio. Nearly half of the EU member countries (i.e., Latvia, Estonia, Czechia, Slovenia, Bulgaria, Finland, Hungary, Croatia, Lithuania, Slovakia, Romania, France, and Sweden) were largely self-sufficient in biomass materials (SSR above 105%). Poland and Spain were "approximately biomass self-sufficient". The rest of the EU member countries were under self-sufficiency (SSR below 95%). Exceptionally, the lowest biomass self-sufficiency ratio for the entire considered period was observed in the two island nations: Cyprus and Malta (SSR below 33% and 15%, respectively, at present). Also, four EU member countries (Portugal, Italy, Belgium, and Luxembourg) and the United Kingdom are rather poorly self-sufficient in biomass (from 70% to 80%), and in Austria, Denmark, Ireland, Greece, and the Netherlands, the degree of biomass self-sufficiency is quite low (80–90%). The import dependency on of all these countries is much higher than the EU average, ranging from 30% in Greece and Denmark to 86% in Malta, on average over the last five years. Therefore, it can be expected that a considerable number of challenges for biomass's supply and demand can be expected, regarding the ambitious targets of the Green Deal and increasing competition in the biomass export markets.

From the point of view of the European Green Deal's ambitious target to reduce net greenhouse gas emissions to zero by 2050, the EU climate scenarios foresee a 70–80% increase in biomass use demand, while its supply will lag behind, a 40–100% gap relative to the large increases in demand [14]. In addition, the launched strategies of Farm to Fork and Biodiversity, which are a part of the Green Deal, are likely to influence the reduction of agricultural production (production of biomass of agricultural origin in the context of this study) in the EU. According to EW-MFA statistics, biomass of agricultural origins account for 82% of total biomass extraction in the EU-27 on average, over the period 2016–2019. A few studies were performed to analyse the possible impact of the targets of these strategies on the EU's agricultural production. The quantitative results of the impact studies are different. Henning et al. [64] identify a decline in agricultural production in the EU from 2.6% to 45%. Bremmer et al. [65] conclude a decrease in agricultural production, up to 30%. Beckman et al. [66] find a decrease in agricultural production from 7% to 12%, and Barreiro-Hurle et al. [67] identify up to a 15% reduction in the supply of agricultural production. In any case, all these impact studies acknowledge that the supply changes of various agricultural products can dramatically decline in 2030 in the EU-27. In addition, on a global level, the consumption of primary biomass is projected to almost double over the coming four decades (rising from 22,5 Gt in 2020 to 41 Gt in 2060), if the new policies to improve resource use efficiency and stimulate the transition to a circular economy are not developed [15].

The main purpose of the econometric analysis in this study is to test the effect of the share of agricultural and forest land in total land area on biomass self-sufficiency. A panel data analysis is proposed in this study. The two-step GMM method is used for econometric analysis, controlling for potential factors, including bioenergy development, and social and economic characteristics. The estimation results indicate that the core explanatory variable (i.e., agricultural and forest land share) has a positive effect on the biomass self-sufficiency ratio; this means that an increase in the share of agriculture and forestry in the total land area increases the biomass self-sufficiency ratio. This finding is in line with the findings of Ladanai and Vinterbäck [68], Field et al. [69], and Benti et al. [29], who stated that land availability is crucial for the future since energy from biomass depends on it. Biomass is defined as, potentially, a main source of energy. As UNEP [16] mentions, regional distinctiveness of the land use system can affect the framework of biomass, both in terms of quantity and quality. Vera et al. [70] note, "the biomass potentials and environmental impacts strongly depend on location of specific biophysical conditions, land use/cover prior to conversion, and feedstock type" (p. 5). Thus, land use is an important factor for biomass accessibility.

In addition, a negative impact of population density on biomass self-sufficiency is present in all cases of the model. This indicator, as a land use characteristic as well, is widely discussed in the scientific literature. A growing population (directly related to population density) increases the fear that, within a few decades, agricultural production will have to increase, but a large portion of the land is not useful, as it is degraded; "there is a deepening awareness of the long-term consequences of the loss of biodiversity, with the prospect of climate change" [71] (p. 560). As UNEP [16] asserts, a high population density is usually related to a high dependence on biomass imports at a country level, while countries with a low population density are usually net exporters of biomass. Also, population density is associated with the domestic extraction of biomass. As Krausmann et al. [47] stress, the highest domestic extraction is in countries where population density is low.

As expected, the econometric model displays a negative effect of energy import dependency on the biomass self-sufficiency ratio. In this respect, reducing the energy import dependency requires increasing biomass production. As Field et al. [69] state, there are many opportunities for greater energy independence. The potential of bioenergy depends on the protection of forests and agricultural land against such processes as degradation, desertification, etc. [72]. On the contrary, a positive effect of the share of bioenergy in renewable energy on the biomass self-sufficiency ratio is found by the model regression, as was expected while formulating the model. Bioenergy is viewed as the most important option of renewable energy in the future as well as at present. As stated before, closely related to the production of bioenergy is land, which is a crucial element for the bioenergy industry.

5. Conclusions

This study serves as an overview of biomass self-sufficiency in the European Union as a whole and in individual member state levels. It provides an essential framework for this topic and a detailed analysis of the effect of some determinants on the biomass self-sufficiency ratio at the EU level. The analysis covers the 2000–2018 period and includes 28 EU countries (the composition of the EU until 31 January 2020). A pragmatic interpretation of the national self-sufficiency in biomass materials was used in the present study. Biomass self-sufficiency is calculated as a percentage ratio of biomass domestic extraction to its domestic consumption. Thus, biomass self-sufficiency indicates the extent to which a country can meet its needs for biomass materials, using resources coming from domestic extraction or harvest.

Our study has some limitations, mainly related to the narrowed definition of biomass as the primary biological material domestically extracted from the natural environment and domestically used in the national economy, as developed by the economy-wide material flow accounts (EW-MFA). Secondly, data on the biomass domestic extraction and consumption from the EW-MFA database was used in the analysis, keeping in mind that the NEW-MFA accounts for the physical flows of primary biomass from the natural environment to the economy. Due to this limitation, the biological waste generated by primary production and subsequent economic processes (manufacturing, trade, and final consumption), and returned to the production and consumption processes, were not included in the analysis. Despite the limitations, this study can be used by governments, policy makers, bioeconomists, and even macroeconomists, since the concept of self-sufficiency features is prominently used in many regulatory guidelines, policies, and recommendations on food security, anti-poverty, energy security, renewable energy, circularity of economy, sustainable development, transition to a low carbon economy, etc. Further research is needed to explore the contribution of biomass from waste streams into biomass self-sufficiency and to assess the country's ability to meet domestic demand for biomass. There is a need to explore ecological, economic, and social constraints to biomass use and harvesting from natural and cultivated ecosystems.

Author Contributions: Conceptualization, V.V.; methodology, V.V. and A.A.; software, A.A.; validation, V.V. and A.A.; formal analysis, V.V. and A.A.; investigation, V.V. and A.A.; resources, V.V., A.A. and N.R.; data curation, V.V. and A.A.; writing—original draft preparation, V.V., A.A. and N.R.; writing—review and editing, V.V.; visualization, V.V. and A.A.; supervision, V.V.; project administration, V.V.; funding acquisition, V.V., A.A. and N.R. All authors have read and agreed to the published version of the manuscript.

Funding: This research was supported by the European Union's Horizon 2020 research and innovation programme, under grant agreement No 862699.

Institutional Review Board Statement: Not applicable.

Informed Consent Statement: Not applicable.

Data Availability Statement: Data are available in the Eurostat's and FAOSTAT's databases.

Conflicts of Interest: The authors declare no conflict of interest.

References

1. United Nations. *World Population Prospects 2019: Highlights*; Department of Economic and Social Affairs, Population Division (ST/ESA/SER.A/423): New York, NY, USA, 2019; p. 143.
2. Saunders, J.T.; Adenauer, M.; Brooks, J. Analysis of long-term challenges for agricultural markets. *OECD Food Agric. Fish. Pap.* **2019**, *131*, 42. [CrossRef]
3. Van Meijl, H.; Havlik, P.; Lotze-Campen, H.; Stehfest, E.; Witzke, H.; Perez Dominguez, I.; Bodirsky, B.; van Dijk, M.; Doelman, J.; Fellmann, T.; et al. *Challenges of Global Agriculture in a Climate Change Context by 2050 (AgCLIM50)*; Publications Office of the European Union: Luxembourg, 2017; p. 70. [CrossRef]
4. McCormick, K.; Kautto, N. The bioeconomy in Europe: An overview. *Sustainability* **2013**, *5*, 2589–2608. [CrossRef]
5. VERAM. Vision for Raw Materials in Europe and for Europe Part II. Report on Raw Material Research and Innovation Vision for 2050 European Union's, Project Funded from the European Union's Horizon 2020 Research and Innovation Programme Under Grant Agreement No 690388 H2020, Ref. Ares(2018)4131269-06/08/2018.2018. 2018. Available online: http://veram2050.eu/wp-content/uploads/2018/10/D4.2-Report-on-raw-material-research-and-innovation-vision-for-2050.pdf (accessed on 15 October 2021).
6. European Commission. *A Sustainable Bioeconomy for Europe: Strengthening the Connection Between Economy, Society and the Environment (COM(2018) 673)*; Publications Office of the European Union: Brussels, Belgium, 2018; p. 14.
7. Cristóbal, J.; Matos, C.T.; Aurambout, J.P.; Manfredi, S.; Kavalov, B. Environmental sustainability assessment of bioeconomy value chains. *Biomass Bioenergy* **2016**, *89*, 159–171. [CrossRef]
8. Scarlat, N.; Dallemand, J.F.; Monforti-Ferrario, F.; Nita, V. The role of biomass and bioenergy in a future bioeconomy: Policies and facts. *Environ. Dev.* **2015**, *15*, 3–34. [CrossRef]
9. European Commission. *Innovating for Sustainable Growth: A Bioeconomy for Europe. Communication from the Commission to the European Parliament, the Council, the European Economic and Social Committee and the Committee of the Regions*; European Commission: Brussels, Belgium, 2012; p. 9.
10. European Commision. *A Vision for the European Industry Until 2030: Final Report of the Industry 2030 High Level Industrial Roundtable*; Publications Office of the European Union: Luxembourg, 2019; p. 48.
11. European Commission. *Closing the Loop—An EU Action Plan for the Circular Economy*; European Commission: Brussels, Belgium, 2015; p. 21.
12. European Commission. *Accelerating Clean Energy Innovation*; European Commission: Brussels, Belgium, 2016.
13. Joint Research Centre. *Bioeconomy Report 2016*; JRC Scientific and Policy Report; Joint Research Centre: Brussels, Belgium, 2017; p. 124.
14. Material Economics. *EU Biomass Use in a Net-Zero Economy—A Course Correction for EU Biomass*; Material Economics Sverige AB: Stockholm, Sweeden, 2021.
15. OECD. *Global Material Resources Outlook to 2060: Economic Drivers and Environmental Consequences*; OECD Publishing: Paris, France, 2018; p. 214.

16. UNEP. *The Use of Natural Resources in the Economy: A Global Manual on Economy Wide Material Flow Accounting*; UNEP: Nairobi, Kenya, 2021; p. 143.
17. Carus, M.; Dammer, L. Food or non-food: Which agricultural feedstocks are best for industrial uses? *Ind. Biotechnol.* **2013**, *9*, 171–176. [CrossRef]
18. Ramos, J.L.; García-Lorente, F.; Valdivia, M.; Duque, E. Green biofuels and bioproducts: Bases for sustainability analysis. *Microb. Biotechnol.* **2017**, *10*, 1111–1113. [CrossRef]
19. Budzianowski, W.M. High-value low-volume bioproducts coupled to bioenergies with potential to enhance business development of sustainable biorefineries. *Renew. Sustain. Energy Rev.* **2017**, *70*, 793–804. [CrossRef]
20. Schoenung, S.; Efroymson, R.A. *Algae Production from Wastewater Resources: An Engineering and Cost Analysis*; Oak Ridge National Lab (ORNL): Oak Ridge, TN, USA, 2018; p. 11.
21. Moorkens, E.; Meuwissen, N.; Huys, I.; Declerck, P.; Vulto, A.G.; Simoens, S. The market of biopharmaceutical medicines: A snapshot of a diverse industrial landscape. *Front. Pharmacol.* **2017**, *8*, 12. [CrossRef] [PubMed]
22. Santos, N.V.D.; de Carvalho Santos-Ebinuma, V.; Pessoa, A., Jr.; Pereira, J.F.B. Liquid–liquid extraction of biopharmaceuticals from fermented broth: Trends and future prospects. *J. Chem. Technol. Biotechnol.* **2017**, *93*, 1845–1863. [CrossRef]
23. Smith, G.O. Theory and Practice of National Self-Sufficiency in Raw Materials. *Proc. Acad. Political Sci. City N. Y.* **1926**, *12*, 116–122. [CrossRef]
24. Kettunen, L. Self-sufficiency of agriculture in Finland in 1970–1983. *Agric. Food Sci.* **1986**, *58*, 143–150. [CrossRef]
25. Clapp, J. Food self-sufficiency: Making sense of it, and when it makes sense. *Food Policy* **2017**, *66*, 88–96. [CrossRef]
26. Spero, J.E. Energy self-sufficiency and national security. *Proc. Acad. Political Sci.* **1973**, *31*, 123–136. [CrossRef]
27. Welfle, A.; Gilbert, P.; Thornley, P. Increasing biomass resource availability through supply chain analysis. *Biomass Bioenergy* **2014**, *70*, 249–266. [CrossRef]
28. Saghir, M.; Zafar, S.; Tahir, A.; Quadi, M.; Siddique, B.; Homung, A. Unlocking the Potential of Biomass Energy in Pakistan. *Front. Energy Res.* **2019**, *7*, 1–18. [CrossRef]
29. Benti, N.E.; Gurmesa, G.S.; Argaw, T.; Aneseyee, A.B.; Gunta, S.; Kassahun, G.B.; Aga, G.S.; Asfaw, A.A. The current status, challenges and prospects of using biomass energy in Ethiopia. *Biotechnol. Biofuels* **2021**, *14*, 1–24. [CrossRef] [PubMed]
30. Fragkou, M.C.; Vincent, T.; Gabarelli, X. An ecosystemic approach for assessing the urban water self-sufficiency potential: Lessons from the Mediterranean. *Urban Water J.* **2015**, *13*, 663–675. [CrossRef]
31. Sarabi, S.G.; Rahnama, M.R. From self-sufficient provision of water and energy to regenerative urban development and sustainability: Exploring the potentials in Mashhad City, Iran. *J. Environ. Plan. Manag.* **2021**, *64*, 2459–2480. [CrossRef]
32. Østergård, H.; Markussen, M.V. Energy Self-sufficiency from an Emergy Perspective Exemplified by a Model System of a Danish Farm Cooperative. In Proceedings of the 6th Biennial Emergy Research Conference, The Center for Environmental Policy, Gainesville, FL, USA, 14–16 January 2010; Brown, M.T., Sweeney, S., Eds.; University of Florida: Gainesville, FL, USA, 2010; pp. 311–322.
33. Martin, G.; Magne, M.A. Agricultural diversity to increase adaptive capacity and reduce vulnerability of livestock systems against weather variability—A farm-scale simulation study. *Agric. Ecosyst. Environ.* **2015**, *199*, 301–311. [CrossRef]
34. Lebacq, T.; Baret, P.V.; Stilmant, D. Role of input self-sufficiency in the economic and environmental sustainability of specialised dairy farms. *Animal* **2015**, *9*, 544–552. [CrossRef] [PubMed]
35. Soteriades, A.D.; Stott, A.W.; Moreau, S.; Charroin, T.; Blanchard, M.; Liu, J.; Faverdin, P. The relationship of dairy farm eco-efficiency with intensification and self-sufficiency. Evidence from the French dairy sector using life cycle analysis, data envelopment analysis and partial least squares structural equation modelling. *PLoS ONE* **2016**, *11*, 21. [CrossRef] [PubMed]
36. Gaudino, S.; Reidsma, P.; Kanellopoulos, A.; Sacco, D.; van Ittersum, M.K. Integrated assessment of the EU's Greening reform and feed self-sufficiency scenarios on dairy farms in Piemonte, Italy. *Agriculture* **2018**, *8*, 137. [CrossRef]
37. Jouan, J.; Ridier, A.; Carof, M. Legume production and use in feed: Analysis of levers to improve protein self-sufficiency from foresight scenarios. *J. Clean. Prod.* **2020**, *274*, 123085. [CrossRef]
38. Masi, M.; Vecchio, Y.; Pauselli, G.; di Pasquale, J.; Adinolfi, F. A typological classification for assessing farm sustainability in the Italian bovine dairy sector. *Sustainability* **2021**, *13*, 7097. [CrossRef]
39. Kimming, M.; Sundberg, C.; Nordberg, Å.; Baky, A.; Bernesson, S.; Norén, O.; Hansson, P.A. Life cycle assessment of energy self-sufficiency systems based on agricultural residues for organic arable farms. *Bioresour. Technol.* **2011**, *102*, 1425–1432. [CrossRef]
40. Vijay, V.; Subbarao, P.M.; Chandra, R. An evaluation on energy self–sufficiency model of a rural cluster through utilization of biomass residue resources: A case study in India. *Energy Clim. Change* **2021**, *2*, 100036. [CrossRef]
41. Algieri, A.; Andiloro, S.; Tamburino, V.; Zema, D.A. The potential of agricultural residues for energy production in Calabria (Southern Italy). *Renew. Sustain. Energy Rev.* **2019**, *104*, 1–14. [CrossRef]
42. Terrapon-Pfaff, J.C. Linking energy-and land-use systems: Energy potentials and environmental risks of using agricultural residues in Tanzania. *Sustainability* **2012**, *4*, 278–293. [CrossRef]
43. Harchaoui, S.; Chatzimpiros, P. Can agriculture balance its energy consumption and continue to produce food? A framework for assessing energy neutrality applied to French agriculture. *Sustainability* **2018**, *10*, 4624. [CrossRef]
44. FAO. FAO Statistical Pocketbook 2012—World Food and Agriculture. Available online: http://www.fao.org/docrep/015/i2490e/i2490e00.htm (accessed on 5 September 2019).
45. Luan, Y.; Cui, X.; Ferrat, M. Historical trends of food self-sufficiency in Africa. *Food Secur.* **2013**, *5*, 393–405. [CrossRef]

46. EUROSTAT. *Economy-Wide Material Flow Accounts, Handbook 2018 Edition*; Publications Office of the European Union: Luxembourg, 2018; p. 142.
47. Krausmann, F.; Erb, K.H.; Gingrich, S.; Lauk, C.; Haberl, H. Global Patterns of Socioeconomic Biomass Flows in the Year 2000: A Comprehensive Assessment of Supply, Consumption and Constraints. *Ecol. Econ.* **2008**, *65*, 471–487. [CrossRef]
48. Mayer, A.; Kaufmann, L.; Kalt, G.; Matej, S.; Theurl, M.C.; Morais, T.G.; Leip, A.; Erb, K.H. Applying the Human Appropriation of Net Primary Production framework to map provisioning ecosystem services and their relation to ecosystem functioning across the European Union. *Ecosyst. Serv.* **2021**, *51*, 101344. [CrossRef]
49. OECD. Environment Database—Material Resources: Concepts and Classifications. Available online: https://stats.oecd.org/# (accessed on 16 August 2019).
50. Leitão, N. GMM Estimator: An Application to Intraindustry Trade. *J. Appl. Math.* **2012**, *2012*, 1–12. [CrossRef]
51. Das, D.K. Determinants of current account imbalance in the global economy: A dynamic panel analysis. *J. Econ. Struct.* **2016**, *5*, 2–24. [CrossRef]
52. Matuzeviciute, K.; Butkus, M.; Karaliute, A. Do Technological Innovations Affect Unemployment? Some Empirical Evidence from European Countries. *Economies* **2017**, *48*, 48. [CrossRef]
53. Brañas-Garza, P.; Bucheli, M.; García-Muñoz, T. Dynamic Panel Data: A Useful Technique in Experiments. 2011. Available online: http://www.ugr.es/~{}teoriahe/RePEc/gra/wpaper/thepapers10_22.pdf (accessed on 19 November 2021).
54. Vitunskiene, V.; Ramanauske, N. Spatial Concentration of Biomass Production Sector in the European Union. Current Analysis on Economics & Finance. 2019. Available online: https://mesford.ca/wp-content/uploads/2019/04/Spatial-Concentration-of-Biomass-Production-Sector-in-the-European-Union.pdf (accessed on 16 August 2019).
55. Candel, J.J.L.; Breeman, G.E.; Stiller, S.J.; Termeer, C.J.A.M. Disentangling the consensus frame of food security: The case of the EU Common Agricultural Policy reform debate. *Food Policy* **2013**, *44*, 47–58. [CrossRef]
56. Galli, A.; Halle, M.; Grunewald, N. Physical limits to resource access and utilisation and their economic implications in Mediterranean economies. *Environ. Sci. Policy* **2015**, *51*, 125–136. [CrossRef]
57. Ratner, B. The Correlation Coefficient: Its Values Range Between +1/−1, Or Do They? *J. Target. Meas. Anal. Mark.* **2009**, *18*, 139–142. [CrossRef]
58. Andersen, S.P.; Allen, B.; Domingo, G.C. *Biomass in the EU Green Deal: Towards Consensus on the Use of Biomass for EU Bioenergy, (Policy Report)*; Institute for European Environmental Policy (IEEP): Brussels, Belgium, 2021; p. 69.
59. Ji, X.; Liu, Y.; Meng, J.; Wu, X. Global supply chain of biomass use and the shift of environmental welfare from primary exploiters to final consumers. *Appl. Energy* **2020**, *276*, 115484. [CrossRef]
60. Kalt, G.; Kaufmann, L.; Kastner, T.; Krausmann, F. Tracing Austria's biomass consumption to source countries: A product-level comparison between bioenergy, food and material. *Ecol. Econ.* **2021**, *188*, 107129. [CrossRef]
61. Welfle, A. Balancing growing global bioenergy resource demands-Brazil's biomass potential and the availability of resource for trade. *Biomass Bioenergy* **2017**, *105*, 83–95. [CrossRef]
62. Gough, C.; Garcia-Freites, S.; Jones, C.; Mander, S.; Moore, B.; Pereira, C.; Röder, M.; Vaughan, N.; Welfle, A. Challenges to the use of BECCS as a keystone technology in pursuit of 1.5 °C. *Glob. Sustain.* **2018**, *1*, 1–9. [CrossRef]
63. Ying, H.P.; Phun Chien, C.B.; Yee Van, F. Operational management implemented in biofuel upstream supply chain and downstream international trading: Current issues in southeast Asia. *Energies* **2020**, *13*, 1799. [CrossRef]
64. Henning, C.; Witzke, P. Economic and Environmental Impacts of the Green Deal on the Agricultural Economy: A Simulation Study of the Impact of the F2F-Strategy on Production, Trade, Welfare and the Environment Based on the CAPRI-Model. 2021. Available online: https://grain-club.de/fileadmin/user_upload/Dokumente/Farm_to_fork_Studie_Executive_Summary_EN.pdf (accessed on 15 December 2021).
65. Bremmer, J.; Gonzalez-Martinez, A.; Jongeneel, R.; Huiting, H.; Stokkers, R. Impact Assessment Study on EC 2030 Green Deal Targets for Sustainable Food Production. 2021. Available online: https://edepot.wur.nl/555349 (accessed on 15 December 2021).
66. Beckman, J.; Ivanic, M.; Jelliffe, J.L.; Baquedano, F.G.; Scott, S.G. *Economic and Food Security Impacts of Agricultural Input Reduction Underthe European Union Green Deal's Farm to Fork and Biodiversity Strategies*; Agricultural Economic Reports; United States Department of Agriculture (USDA), Economic Research Service: Washington, DC, USA, 2020; p. 59. [CrossRef]
67. Barreiro-Hurle, J.; Bogonos, M.; Himics, M.; Hristov, J.; Pérez-Domínguez, I.; Sahoo, A.; Salputra, G.; Weiss, F.; Baldoni, E.; Elleby, C. *Modelling Environmental and Climate Ambition in the Agricultural Sector with the CAPRI Model*; JRC121368; Publications Office of the European Union: Luxembourg, 2021. [CrossRef]
68. Ladanai, S.; Vinterbäck, J. *Global Potential of Sustainable Biomass for Energy*; SLU Report 013; Department of Energy and Technology: Stockholm, Sweden, 2009; ISSN 1654-9406.
69. Field, C.B.; Campbell, H.E.; Lobell, D.B. Biomass energy: The scale of the potential resource. *Trends Ecol. Evol.* **2008**, *23*, 65–72. [CrossRef] [PubMed]
70. Vera, I.; Hilst, F.V.D.; Hoefnagels, R. *Regional Specific Impacts of Biomass Feed-Stock Sustainability: D4.3 Report on Biomass Potentials and LUC—Related Environmental Impact*; Utrecht University: Utrecht, The Netherlands, 2020; p. 70.
71. Popp, J.; Lakner, Z.; Harangi-Rákos, M.; Fári, M. The effect of bioenergy expansion: Food, energy and environment. *Renew. Sutainable Energy Rev.* **2014**, *32*, 559–578. [CrossRef]
72. WBA. Global Biomass Potential Towards 2035. 2016. Available online: http://www.worldbioenergy.org/uploads/Factsheet_Biomass%20potential.pdf (accessed on 15 December 2021).

 *sustainability*

Article

Web-Based Decision Support System for Managing the Food–Water–Soil–Ecosystem Nexus in the Kolleru Freshwater Lake of Andhra Pradesh in South India

Meena Kumari Kolli [1,2,]*, Christian Opp [3], Daniel Karthe [4,5,6] and Nallapaneni Manoj Kumar [7,]*

1. National Academy of Agricultural Sciences, NASC, Pusa Campus, New Delhi 110012, India
2. Division of Agricultural Physics, Indian Agricultural Research Institute, Pusa Campus, New Delhi 110012, India
3. Faculty of Geography, Philipps-Universität Marburg, Deutschhausstraße 10, 35037 Marburg, Hesse, Germany; opp@mailer.uni-marburg.de
4. Institute for Integrated Management of Material Fluxes and of Resources, United Nations University, Ammonstr. 74, 01067 Dresden, Saxony, Germany; Karthe@unu.edu
5. Faculty of Environmental Sciences, Technische Universität Dresden, Helmholtzstr. 10, 01069 Dresden, Saxony, Germany
6. Environmental Engineering Section, German-Mongolian Institute for Resources and Technology, Nalaikh District, Ulaanbaatar 12790, Mongolia
7. School of Energy and Environment, City University of Hong Kong, Kowloon, Hong Kong
* Correspondence: meenu.rgukt@gmail.com (M.K.K.); mnallapan2@cityu.edu.hk (N.M.K.)

Citation: Kolli, M.K.; Opp, C.; Karthe, D.; Kumar, N.M. Web-Based Decision Support System for Managing the Food–Water–Soil–Ecosystem Nexus in the Kolleru Freshwater Lake of Andhra Pradesh in South India. *Sustainability* **2022**, *14*, 2044. https://doi.org/10.3390/su14042044

Academic Editor: Idiano D'Adamo

Received: 29 December 2021
Accepted: 7 February 2022
Published: 11 February 2022

Publisher's Note: MDPI stays neutral with regard to jurisdictional claims in published maps and institutional affiliations.

Abstract: Most of the world's freshwater lake ecosystems are endangered due to intensive land use conditions. They are subjected to anthropogenic stress and severely degraded because of large-scale aquafarming, agricultural expansion, urbanization, and industrialization. In the case of India's largest freshwater lake, the Kolleru freshwater ecosystem, environmental resources such as water and soil have been adversely impacted by an increase in food production, particularly through aquaculture. There are numerous instances where aqua farmers have indulged in constructing illegal fishponds. This process of aquafarming through illegal fishponds has continued even after significant restoration efforts, which started in 2006. This underlines the necessity of continuous monitoring of the state of the lake ecosystem in order to survey the effectiveness of restoration and protection measures. Hence, to better understand the processes of ecosystem degradation and derive recommendations for future management, we developed a web mapping application (WMA). The WMA aims to provide fishpond data from the current monitoring program, allowing users to access the fishpond data location across the lake region, demanding lake digitization and analysis. We used a machine learning algorithm for training the composite series of Landsat images obtained from Google Earth Engine to digitize the lake ecosystem and further analyze current and past land use classes. An open-source geographic information system (GIS) software and JavaScript library plugins including a PostGIS database, GeoServer, and Leaflet library were used for WMA. To enable the interactive features, such as editing or updating the latest construction of fishponds into the database, a client–server architecture interface was provided, finally resulting in the web-based model application for the Kolleru Lake aquaculture system. Overall, we believe that providing expanded access to the fishpond data using such tools will help government organizations, resource managers, stakeholders, and decision makers better understand the lake ecosystem dynamics and plan any upcoming restoration measures.

Keywords: Kolleru Lake; land use; aquafarming; fishponds; illegal fishponds; food–water–soil–ecosystem nexus; Google Earth Engine; freshwater ecosystem web model; India's largest freshwater lake

1. Introduction

Lakes are of considerable value to humankind: they provide drinking water and form the basis for commercial fishery and agriculture, they are linked to energy production and constitute important transportation pathways, and they often have cultural and recreational significance. Moreover, lakes and wetlands play a significant role in regional biodiversity and are invaluable along the migratory routes of birds [1,2]. The specific characteristics of lakes vary significantly according to their origin and geographic location. Despite these differences, many of the world's lakes are under acute threat. According to Mammides [3], one-third of global lakes are subject to such considerable human pressure that they are existentially threatened. As the human population increases exponentially, many of the world's lakes are affected by land reclamation for agricultural expansion, settlements, and industry [4–6]. Economic benefits degrade most of the world's lakes by exploiting their resources, productivity, and identity [7–9].

Meanwhile, single direct and multiple diffuse sources significantly cause pollution and introduce many impairments, which leads to water quality deterioration by eutrophication and algae [10,11]. Whenever lakes are exposed to multiple adverse impacts, lake ecosystems may become more sensitive and vulnerable to changes in climate and hydrology, water quality, or land use. Changes in the lakes themselves can have significant effects on the regional climate and riparian ecosystems [12–15]. However, prioritizing food security as a political goal may adversely affect other environmental resources, including the hydrosphere, the pedosphere, and the biosphere. Therefore, the resource nexus concept aims at integrated approaches that consider food security as a development goal in the contexts of water and soil security and the preservation of viable ecosystems [16,17].

Recent advancements in digital platforms, remote sensing, and GIS (Geographic Information System) technologies have increased and widened their potential for environmental applications [18–23], such as monitoring and modelling environmental resources, such as water and soils, and the biosphere's states, processes, and fluxes [18–22]. However, web modeling services on hydrological catchment applications are a relatively new research area, and to date, the uses of location-based service (LBS) systems are limited. LBS delivers real-time data and information services where the content is illustrated to the user's current or projected location and context [24]. It will be more efficient to model with the combination of field and remote sensing data methods. Furthermore, it is useful for the determination of any ongoing changes with real-time datasets.

In this paper, we present a case study on the recent degradation of Lake Kolleru, India, focusing on the food–water–soil–ecosystem nexus and the integration of ground-based and remote sensing data for monitoring water and soil fluxes as well as the general ecological state of the lake. The concept of the water–food–soil nexus was first popularized by the 2011 Bonn Nexus conference. It has since developed into one of the most widely applied approaches considering the interrelations between different environmental compartments and processes which are exposed to multiple human impacts [25]. From a management perspective, the nexus does not only look at synergies between different objectives, but also at potential trade-offs [26]. Trade-offs may become particularly problematic when single resources or development goals are prioritized by decision makers. An example of this is food security, which is a basic prerequisite for human health and socio-economic development, and therefore defined as the second Sustainable Development Goal (SDG).

In the present study, we report on the first application of a web-based decision support system for monitoring protection and restoration efforts in the Lake Kolleru Basin. Over the past four decades, this lake has suffered significantly under the illegal construction of fishponds, leading to significant nutrient pollution and sedimentation problems [27,28]. Particular focus is directed at land use changes before and after the "Operation Kolleru" restoration program [29]. Our previous studies have shown a coherent picture of the massive land use changes in the Kolleru Lake ecosystem [27]. Clearwater areas in the lake have completely vanished through human interference by constructing fishponds [30]. It can be argued that one of the reasons why restoration and protection efforts were only partly successful in the past is that traditional monitoring methods such as field surveys were a very laborious way of identifying expansions of aquaculture. This study describes a web mapping system using open-source software for the location of the lake region's fishponds, based on the data extracted from machine learning algorithms. The application helps the readers carry out their own assessment of any new illegal construction of fishponds across the lake. It allows the user to update onsite data to a web model. This helps the stakeholders and state government authorities in their decision-making processes for the future development of lake management because they become able to identify new illegal fishponds and resolve arising conflicts.

2. Study Area

Kolleru is the largest freshwater lake in India, located in Andhra Pradesh (Figure 1). Geographically, it is situated between $16°33'10''$ and $16°47'44''$ northern latitude, and $80°4'5.5''$ and $81°24'27.5''$ eastern longitude. It has a distinctive ecosystem that supports biodiversity, and it is rich in flora and fauna. It was recognized as a wetland of international importance by the Ramsar Convention act in November 2002 [31]. It is located between the delta regions of southern India's largest perennial rivers, the Krishna and the Godavari, and serves as a natural flood-balancing reservoir between these two river basins. The lake is fed by seasonal rivers such as Budameru and Tammileru, and additionally, 68 minor irrigation canals flow into the lake. The lake's average water spread area is 902 km^2, falling below the 3.05 m contour level during the southwest monsoon period. The minimum and maximum water depths are 1 and 3 m, and the average annual precipitation is 1094 mm [27]. Agriculture and aquaculture are the major economic activities in this wetland region, where approximately 14,000 families live. As they illegally encroach lake areas for aquaculture expansion, Lake Kolleru's open water area has shrunk, lake water quality deteriorated, and its ecosystem has come under threat.

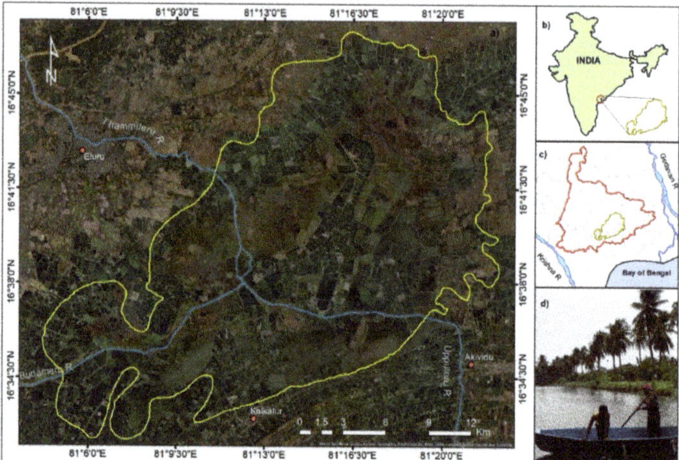

Figure 1. (**a**) Kolleru Lake aquaculture; (**b**) location of the study area in India map; (**c**) Kolleru Lake catchment; (**d**) practicing of aquaculture (photo: Monika Mandal).

The increasing encroachment of Kolleru Lake has led to increasing disputes between environmental authorities and the public. The illegal expansion of aquaculture degraded the lake to an extent where no trace of clear water could be recorded over the past three decades [27]. Despite efforts to restore the lake, irregular lake monitoring activities effectively permitted aquaculture to grow even after restoration measures were implemented. In 2018, fishponds occupied 136 km^2 of the lake area, and weed infestations covered about 152 km^2, together spanning a total of 58.6% of the lake's sanctuary, and the rest of the area was occupied by marshy lands, paddy fields, and built-up areas [27].

Andhra Pradesh, particularly the massive distribution of inland aquaculture formed around the Kolleru Lake freshwater ecosystem, developed into India's most important region of inland fishery [32]. A once-significant lake area was thereby transformed into fishponds, gradually replacing other landcover classes in the wetland ecosystem. To protect the lake from illegal construction, the state government's initial efforts were made to restore the lake area in 2006 through the "Operation Kolleru" program [29]. However, a mixture of high population density and the absence of other employment options in the region induced villagers in the Lake Kolleru Basin to aggressively encroach the lake area for aquaculture farming. A single restoration program was not sufficient to effectively stop this process and protect the lake area from illegal fishponds. Therefore, continuous monitoring of fishpond dynamics, particularly the creation of new ponds, is an essential component of future lake restoration measures.

Apart from the aquaculture threats, Lake Kolleru is subjected to multiple external pressure sources from non-point source pollution, particularly agricultural runoff, soil erosion, and sedimentation. In the catchment region, the massive application of chemical fertilizers, including various nitrogen (N) and phosphorus (P) compounds, and their mobilization by agricultural runoff, cause severe water quality problems [28]. The accumulation of nitrate–nitrogen (NO_3_N) deposited near the lake downstream has led to eutrophication and proliferating weeds. Therefore, pollution abatement measures focusing on nutrient loading are necessary for lake water protection.

3. Materials and Methods

3.1. Input Data

In this study, we used the Landsat-8 satellite series composition with 30 m spatial resolution to prepare a land use classification of the Kolleru Lake in 2018. The Landsat data for 2018, comprising 43 images, were aggregated into a single image by applying the median function for the Random Forest (RF) classification model in Google Earth Engine (GEE). To achieve accurate results, the observed fishpond data were extracted by applying the Normalized Difference Water Index (NDWI), which was calculated based on the spectral indices of Landsat-8. Nearly 70 training samples were used to distinguish between six different land use classes: weed infestations, paddy fields, marshy land, the open lake area, built-up land, and fishponds. The polygon featured training samples collected from high-resolution Google Earth images. For each category, approximately more than 10 samples were collected. In 2018, the fishponds occupied a 136 km^2 (27.9%) area in the Kolleru wildlife sanctuary. The overall accuracy and Kappa coefficients were 88% and 0.84, respectively. After the "Operation Kolleru" restoration program, a fast-growing distinctive land use class was recognized, which turned into a biodiversity threat to Kolleru Lake's natural fauna and flora. Kolli et al. (2020), Pattanaik et al. (2010), and Barman (2004), in their previous studies, showed a clear picture of biodiversity loss with extensive land cover changes for economic profits by constructing fishponds [27,33,34]. This study determined the fishpond data, facilitated them to users, and upgraded the latest identification of fishponds across the lake. The fishpond data were extracted from the 2018 land use image in ArcGIS software, the model's primary data input. The 2018 land use image is mainly used to separate fishponds from other land use classes. Our main objective is to create a web-based application for a better understanding of lake management problems and solutions for any case of secondary restoration measures.

The methodological workflow comprised four stages: problem definition, land use classification and fishpond data extraction, database on a WebGIS, and client–server architecture interface. Figure 2 is an exemplary block diagram representing an inflow of data to the client-side or server-side web server. The first stage includes identifying critical lake factors by communicating with government authorities, stakeholders, research communities, and Kolleru Lake Development Committees (KLDCs). Additional information was obtained from reports in newspapers, magazines, articles, and local news channels. The second stage includes the preparation of a land use map for the year 2018, based on a machine learning algorithm in GEE, and fishpond data extraction to prepare the primary data input for modeling. The third stage is devoted to working with data storage in a database. It also involves data files published on a web server. The final stage implements the web model for protecting the lake ecosystem against the illegal construction of fishponds. Furthermore, the model will be discussed with the researchers, stakeholders, and state government authorities.

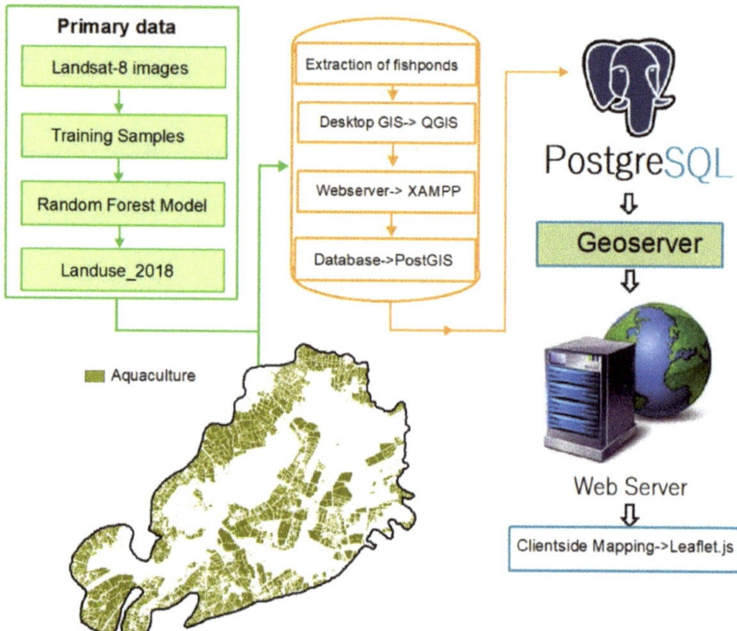

Figure 2. Methodology flowchart.

The model development was divided into two parts: database preparation and web mapping application. The database consists of fishpond data containing each fishpond's spatial information (point location) with a column for the X and Y coordinates. The fishpond polygon layer was converted into the geometric location (i.e., medium of each fishpond boundary) to a point shapefile for better mapping. Furthermore, the fishpond data were converted into the GeoJSON (Geographic JavaScript Object Notation) file format. GeoJSON is a required spatial data format for the map library to display the web server's spatial data [35]. It is an open standard format designed to represent simple geographical features and their non-spatial attributes, based on JavaScript Object Notation [36]. GeoJSON is supported by numerous mapping and GIS software packages, including OpenLayers, Leaflet, and MapServer [37].

3.2. WebGIS Database

The GIS users require map data maintained by other sources. Therefore, data sharing and updating are crucial. Current advanced technologies such as WebGIS can address GIS data issues, including sharing, processing, manipulation, visualization, and updating in the web server domain to widen the adoption to a larger number of potential users [38]. GeoServer is an open-source Java-based web mapping service that enables sharing geospatial data and publishes them on the network [39]. It supports a wide variety of spatial data extensions that handle various datasets, as well. One of the greatest advantages of GeoServer is that it complies with OGC (Open Geospatial Consortium) standards that established a series of data exchange protocols such as Web Map Service (WMS), Web Feature Service (WFS), and Web Coverage Service (WCS) [40–42]. There are certain prerequisites to use a GeoServer such as Java, XAMPP, Apache, and Tomcat. This study used the GeoServer database to publish the Kolleru Lake aquaculture data on a web map service to facilitate the public source fishpond data. PostgreSQL extended with PostGIS is an open-source geospatial database with an object-relational database model installed to store all kinds of primary datasets [43]. In order to maintain data management and data consistency, PostGIS and PostgreSQL, which store both spatial data and attribute data in one database, were used [44]. The fishpond data were uploaded into the PostgreSQL database through the PostGIS server. Figure 2 shows the PostgreSQL database, which can be used to design the fishpond modeling using the web server architecture system. The fishpond data contain both the spatial data and attribute data information. In the first phase of development, the configuration of fishpond data stored in the PostGIS database was published in the GeoServer. For that, a new workspace named "Kolleru_fishponds" was created, and then a new store comprising all kinds of geospatial data was added. The PostGIS database was facilitated in a new store that previously loaded the fishpond shapefile into the PostgreSQL server. Finally, then the fishpond data could be published on the web GeoServer. The fishponds' information can be monitored, visualized, or edited by any user from the web server.

3.3. Client–Server Architecture

The web model application has a three-layer architecture. PostgreSQL was extended with PostGIS, used as a backend to store the fishpond data. The GeoServer was used to create layer services and to allow the publication of the PostGIS data in a web server, while the Leaflet library was used to create the Graphical User Interface (GUI). The Leaflet, created by Vladimir Agafonkin, is the leading open-source JavaScript library for web-based interactive maps, and it is updated continuously [45]. It has been well documented and supported for different applications with large amounts of plugins. For this study, the Leaflet library for a working environment for programming was used. We used it for a web browser user interface to develop a map request entry webpage. For example, if a user has entered the construction of a new illegal fishpond location, the webpage receives a request from the user by selecting a "request map" button. After that, new fishpond data are typically generated in the remote server from the user's device to synchronize the new data into an existing database.

The client–server module interface affords the individual users to configure and manipulate the fishpond mapping data remotely. The data layer provides access to the database through web services. The web service is a gateway between the data layers to allow the client application and server application to access the database. Furthermore, the published fishpond data layer in GeoServer is accessed through the leaflet. The overall client–server architecture of a generated web mapping application is shown in Figure 3.

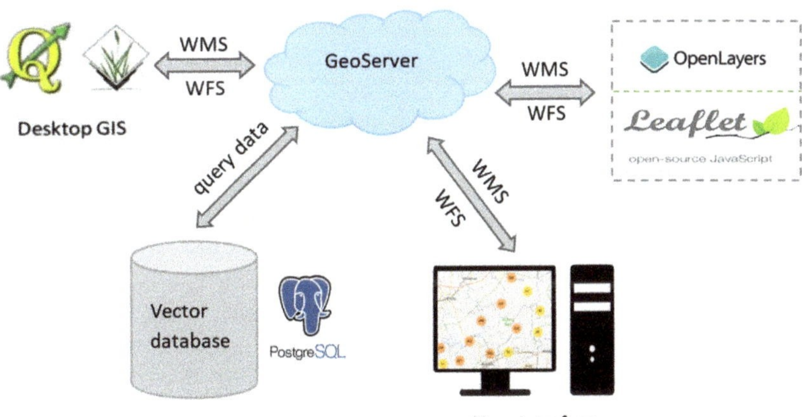

Figure 3. The client–server architecture of a generated web mapping application.

4. Results

The web-based GIS interface between the server and the client-side module can design both static and dynamic datasets. Thus, the web-based module was developed and integrated with the PostGIS database to store the input data and to model the fishponds. They are required to map ongoing changes and control structures for development activities, designed with XAMPP, machine learning datasets, PostgreSQL database, and GeoServer. At the same time, client-side mapping is facilitated with the Leaflet Java plugin. This module allows users to define the fishpond's location, which computes input parameter values for the web system. Then, it generates the URL (Uniform Resource Locator) to transfer the input data parameters to the PostGIS database service. It computes all possible dimensions and provides enough storage space for the output data interface. The links to source codes are given in Appendix A.

4.1. Displaying of Fishpond Data on a Webpage

In Figure 4, the displaying of fishpond data on a webpage is shown. Here, the fishpond data stored in a PostGIS database were added to the Leaflet guide through the WMS layer published on a WebGIS platform. The fishpond data are permanently stored in a PostgreSQL database. Figure 4a depicts the fishponds' locations on a web page shown as a marker cluster layer in the Kolleru Lake ecosystem. A total of 2770 fishponds were identified in the 2018 classified image that was overlayed on a Google Earth image. Since thousands of fishponds were dug into mere fish drains, we could show the data as a cluster marker layer to better visualize the map in close proximity with other marker icons. However, in the case of the maximum zoom in the clustered fishpond area, each fishpond can be separated with a unique marker icon and represented as the area's center, as shown in Figure 4b. The user can interact with the data of each fishpond location on a web interface. In addition, a layer control panel was provided that allows users to switch between the different base layers for a larger and better visualization of the study area.

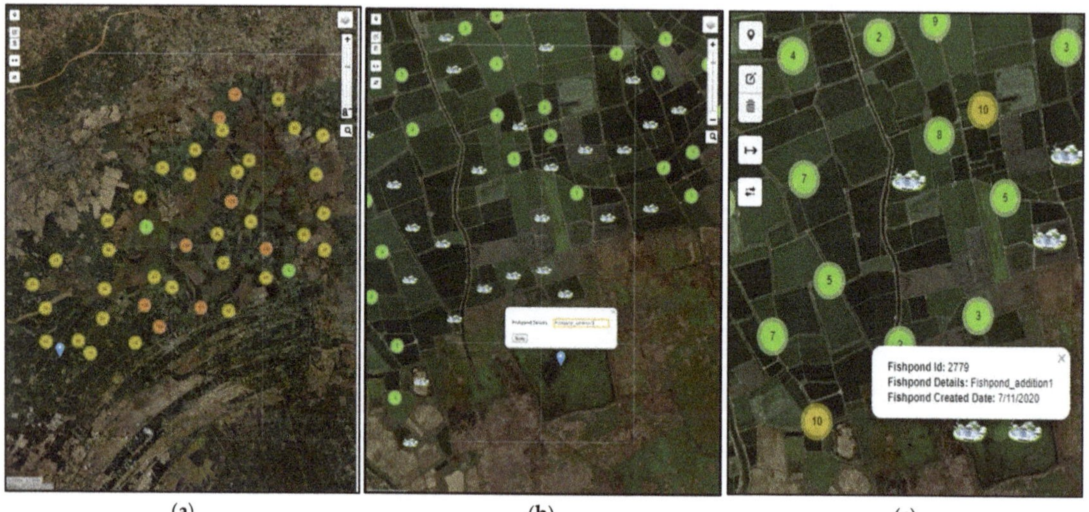

(a) (b) (c)

Figure 4. (**a**). Fishpond data as shown in the form of a marker cluster layer on a web application in the Kolleru Lake ecosystem. (**b**) User generates the URL to add the new fishpond location. (**c**) Result of attribute query based on a user required location.

4.2. Web-Based Server and a Client-Side Module

Figure 4b depicts an exemplary map displayed on a webpage on a web browser user interface. The user sends a request of a fishpond's location from a client-side computing device to a web server (https://webgis.in/fishponds/index.html, accessed on 7 November 2020). The user's desired location then shows as a marker with created details "name: Fishpond_addition1". After entering the specific location of a fishpond to be mapped or added to the database, the user then requests a map by selecting the "Save" button. Before submitting a request to a web server for storing in a database, the user should verify with ground truth information about whether fishponds exist or not. This information is generated on a remote server on the user's computing device, transmitted to the web server, and eventually displayed on a webpage.

Figure 4c illustrates the user's desired location of a fishpond saved in a database identified on a webpage with a marker icon (i.e., fishpond ID, latitude and longitude, and date). The newly entered fishpond displayed on a web page shows a new ID, name, and created date. This helps to assess uncertainties related to older and potentially out-of-date information.

Figure 5 shows that new fishpond credentials with the name "Fishponds_addition1" were saved in a PostGIS database and reflected on a map with current time and detail. This application allows any new fishpond entry to be displayed on a webpage and registered in the existing database.

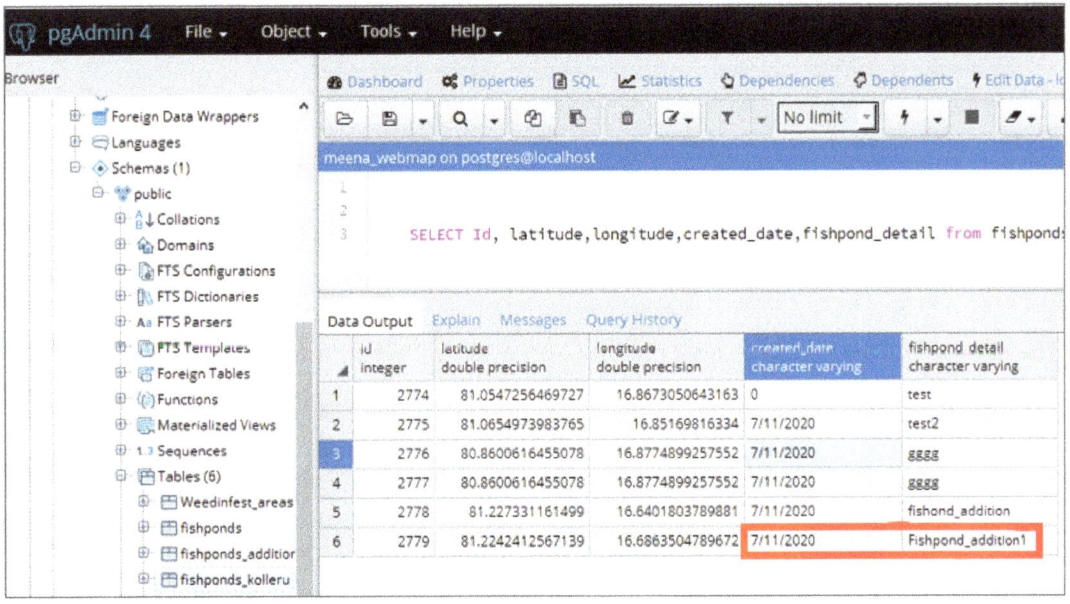

Figure 5. Storing of new fishpond data in the PostGIS database.

5. Discussion

Figure 6 depicts the displacement of the fishponds' occupation area before and after the Kolleru wildlife sanctuary restoration measures. The fishpond data for 1999 are derived from the Kolli et al. (2020) land use classification map for further analysis [27].

Figure 6a shows that fishponds occupied 29.7% of the overall lake area. This was the highest dominant land use class. The majority of other land use classes of 1999 were paddy fields, marshy lands, and weed-infested areas that had entirely disturbed the lake ecosystem. Floods were aggravated within the fishponds due to the construction of high-rise embankments that polluted the surrounding lake areas [30]. Restoration processes were initiated to dismantle the fishponds in 2006 through the volunteer "Operation Kolleru" program [29]. "Operation Kolleru" brought the solution to stop the illegal expansion of fishponds [33]. The success of this operation was only temporary, as evidenced by the significant development of fishponds across the lake area observed in the 2018 land use image.

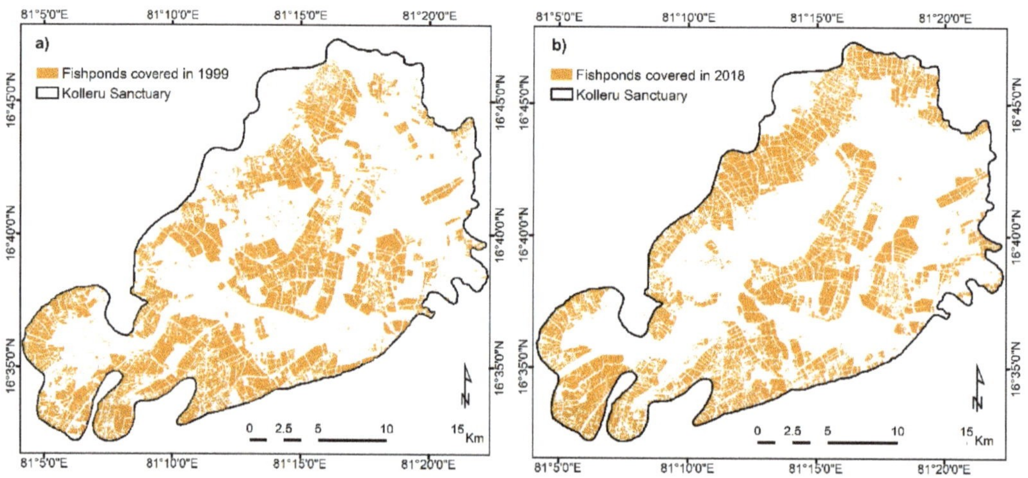

Figure 6. Fishpond-occupied area in (**a**) 1999; (**b**) 2018.

Figure 6b illustrates that the 2018 distribution of fishponds in the Kolleru Lake ecosystem resembles—and in some parts, exceeds—the fishpond regions in the 1999 image. The fishpond-occupied area was 27.7%, about 2% less than before the lake restoration processes took place. This indicates that the lake remained stable for a specific period, and the rate of encroachment was relatively faster after restoration. However, during the "Operation Kolleru" program, the affected fishpond areas were turned into marshy lands, accounting for about 59.8% immediately after the restoration program [27]. Therefore, people targeted these marshy areas for further expansion of fishponds. Additionally, we compared monthly satellite image features and identified that February was well suited for monitoring the fishponds' development. Since the Kolleru Lake ecosystem marshy areas dug into fishponds, especially in February, the lake is dried and easy for digging. Therefore, February is a suitable time to monitor the illegal construction of fishponds. Our model will be useful for both monitoring and decision-making solutions for stakeholders.

Figure 7 illustrates the loss and gain of fishponds between governmental and human activities across the Kolleru Lake ecosystem. The lake endured intensive stress due to frequent land use changes before and after restoration practices. At the time of restoration in 2006, a significant number of fishponds were destroyed.

Figure 7a depicts the fishpond loss area after the restoration program, which is about 5.17% of the lake area. Most destroyed fishponds were concentrated in the middle lake area, especially within the 3 ft contour level. Thus, the local people used the 5 ft contour area for aquaculture farming and were not interested in returning the land to the government. Furthermore, they encroached the 3 ft lake area, completely degrading the lake ecosystem. The state government of Andhra Pradesh intended to protect the lake area up to the 3 ft contour (Figure 7a), which resulted in the "Operation Kolleru" program.

According to Figure 7b, the fishponds' growth showed in the direction of the 5 ft contour area (i.e., around the 3 ft contour level) of the 2018 image. The fishponds gained after the restoration program account for 4.82% of the lake area. We compared the loss and gain of fishponds between 1999 and 2018, or before and after the restoration program, for a better interpretation analysis. Due to the "Operation Kolleru" program, the lake's 3 ft area was not encroached by the local people for aquaculture practice (Figure 7b). However, it was mostly covered with weeds, marshy areas, and less area with paddy fields. Better lake measurement practices are essential for lake protection.

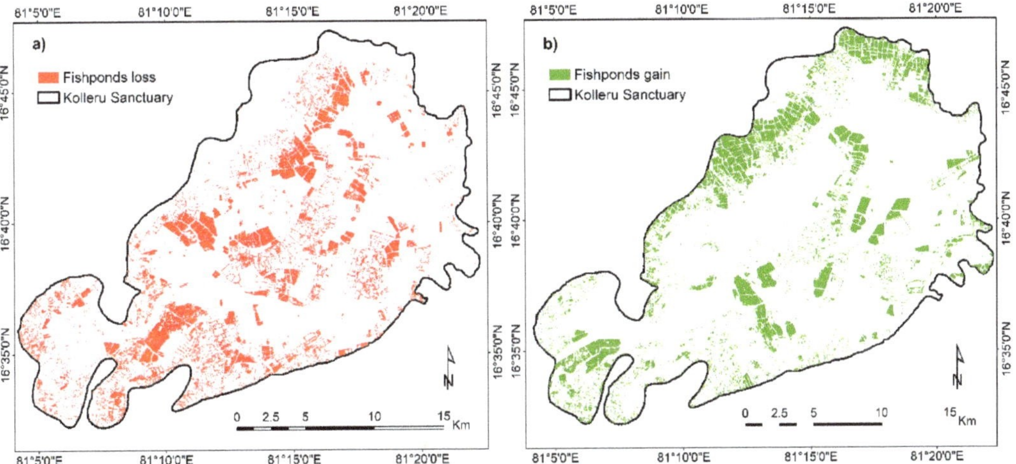

Figure 7. (**a**) Fishpond loss after 1999; (**b**) fishponds gained in 2018.

Human economic growth activities degrade the lakes, and Kolleru Lake is the best example of India's largest freshwater aquaculture expansion. The state government has formulated committees for Kolleru development activities. However, lake encroachment by illegal activities has dominated the lake ecosystem.

6. Conclusions, Limitations, and Future Work

Here, we described the development of an interactive web mapping application for monitoring aquaculture dynamics in Kolleru Lake. In addition to the values of stakeholders and decision makers, this methodological tool plays a crucial role in surveying land use dynamics in the lake basin. Recently, the lake has experienced land degradation due to the expansion of fishponds across the lake region. According to the results and analysis in this study, environmental managers and authorities can strongly benefit from a tool that combines a remote sensing approach with crowd-sourced field mapping.

A machine learning-based algorithm was used to prepare the land use categorized map in Google Earth Engine. In the context of the present work, the prototype version of the web-based services system constituted an exemplary but pragmatic approach for monitoring lake degradation by aquaculture. This was achieved through a GIS interface to the web-based system to model the fishponds. This model is fully automated through the SQL programming, the PostGIS database, the server, and the client-side web interface. The model helps the government, based on the lake's present land use conditions, where the user can update the illegal construction of fishpond location details to the web server. Thus, decision makers can employ this easy-to-use system for identifying the most affected areas by aquaculture growth and establish better lake management activities. This application demonstrates how the web–client interaction can be easily used for minimizing the expansion of illegal fishponds in the Kolleru Lake ecosystem.

It is important to acknowledge the current limitations of this study. For instance, the fishpond data have not been cross-checked with the ground truth data but only validated with high-resolution Google Earth images. Therefore, further investigations should include detailed ground-based data collections. A second potential issue is data redundancy, as any registration of new fishpond data by a user results in the generation of new ID and latitude/longitude coordinates in the database. As individual ponds may be counted several times, the number of fishponds in the Kolleru Lake wetland may be overestimated. In such instances, the use of automated and manual consistency checks and an advancement of the variables considered for fishpond registration must be considered in the future. Finally, in this study, we developed a web-based decision support system using open-source technologies. This web-based approach obviously requires an internet connection, which is not always available to stakeholders living in rural areas, e.g., due to their socio-economic status or limited telecommunication networks. Therefore, the integration of offline mobile applications can be considered in the future.

Author Contributions: Conceptualization, D.K. and C.O.; methodology, M.K.K.; software, M.K.K.; validation, M.K.K. and N.M.K.; formal analysis, C.O.; investigation, M.K.K. and N.M.K.; resources, D.K.; data curation, M.K.K.; writing—original draft preparation, M.K.K. and N.M.K.; writing—review and editing, D.K., C.O. and N.M.K.; visualization, M.K.K.; supervision, C.O. and D.K.; project administration, M.K.K. and N.M.K.; funding acquisition, N.M.K. All authors have read and agreed to the published version of the manuscript.

Funding: This research received no external funding.

Institutional Review Board Statement: Not applicable.

Informed Consent Statement: Not applicable.

Data Availability Statement: The data used in this study can be made available upon request to the corresponding author.

Acknowledgments: The authors would like to thank Rahul Gupta for assisting in developing the model and Monika Mandal for providing field photos and valuable information.

Conflicts of Interest: The authors declare no conflict of interest.

Appendix A

Source code for the developed web-based model for the sustainable management of aquaculture in the Kolleru Lake is available at: https://github.com/aneemkolli/Kolleru_fishponds.git. The weblink for the Kolleru Lake fishponds is available at https://webgis.in/fishponds/index.html.

References

1. Fluet-Chouinard, E.; Messager, M.; Lehner, B.; Finlayson, C. Freshwater Lakes and Reservoirs. In *The Wetland Book. Dordrecht*; Finlayson, C., Milton, G., Prentice, R., Davidson, N., Eds.; Springer: Berlin/Heidelberg, Germany, 2018.
2. Sterner, R.W.; Keeler, B.; Polasky, S.; Poudel, R.; Rhude, K.; Rogers, M. Ecosystem services of Earth's largest freshwater lakes. *Ecosyst. Serv.* **2020**, *41*, 101046. [CrossRef]
3. Mammides, C. A global assessment of the human pressure on the world's lakes. *Glob. Environ. Chang.* **2020**, *63*, 102084. [CrossRef]
4. Ding, L.; Chen, K.-L.; Cheng, S.-G.; Wang, X. Water ecological carrying capacity of urban lakes in the context of rapid urbanization: A case study of East Lake in Wuhan. *Phys. Chem. Earth Parts A/B/C* **2015**, *89-90*, 104–113. [CrossRef]
5. Wang, Y.; Peng, B.; Xion, J.; Zhang, H. Study on the spatial pattern and influencing factors of population urbanization of Dongting Lake area. *Grograph. Res.* **2013**, *32*, 1912–1922.
6. Zeballos-Velarde, C.; Yamaguchi, K. Impacts of land reclamation on the landscape of Lake Biwa, Japan. *Procedia-Soc. Behav. Sci.* **2011**, *19*, 683–692. [CrossRef]
7. Lienhoop, N.; Messner, F. The Economic Value of Allocating Water to Post-Mining Lakes in East Germany. *Water Resour. Manag.* **2009**, *23*, 965–980. [CrossRef]
8. Jayanthi, M.; Rekha, P.N.; Kavitha, N.; Ravichandran, P. Assessment of impact of aquaculture on Kolleru Lake (India) using remote sensing and Geographical Information System. *Aquac. Res.* **2006**, *37*, 1617–1626. [CrossRef]
9. Hickley, P.; Muchiri, M.; Boar, R.; Britton, R.; Adams, C.; Gichuru, N.; Harper, D. Habitat degradation and subsequent fishery collapse in Lakes Naivasha and Baringo, Kenya. *Ecohydrol. Hydrobiol.* **2004**, *4*, 503–517.
10. Schindler, D.W.; Carpenter, S.; Chapra, S.C.; Hecky, R.E.; Orihel, D. Reducing Phosphorus to Curb Lake Eutrophication is a Success. *Environ. Sci. Technol.* **2016**, *50*, 8923–8929. [CrossRef]
11. Ekholm, P.; Mitikka, S. Agricultural Lakes in Finland: Current Water Quality and Trends. *Environ. Monit. Assess.* **2006**, *116*, 111–135. [CrossRef]

12. Cooper, S.D.; Lake, P.S.; Sabater, S.; Melack, J.M.; Sabo, J.L. The effects of land use changes on streams and rivers in mediterranean climates. *Hydrobiologia* **2013**, *719*, 383–425. [CrossRef]
13. Mao, D.; Cherkauer, K.A. Impacts of land-use change on hydrologic responses in the Great Lakes region. *J. Hydrol.* **2009**, *374*, 71–82. [CrossRef]
14. Pham, S.V.; Leavitt, P.R.; McGowan, S.; Peres-Neto, P. Spatial variability of climate and land-use effects on lakes of the northern Great Plains. *Limnol. Oceanogr.* **2008**, *53*, 728–742. [CrossRef]
15. Hecky, R.E.; Bootsma, H.A.; Kingdon, M.L. Impact of Land Use on Sediment and Nutrient Yields to Lake Malawi/Nyasa (Africa). *J. Great Lakes Res.* **2003**, *29*, 139–158. [CrossRef]
16. Bleischwitz, R.; Spataru, C.; Vandeveer, S.D.; Obersteiner, M.; Van Der Voet, E.; Johnson, C.; Andrews-Speed, P.; Boersma, T.; Hoff, H.; Van Vuuren, D.P. Resource nexus perspectives towards the United Nations Sustainable Development Goals. *Nat. Sustain.* **2018**, *1*, 737–743. [CrossRef]
17. Lal, R. The Nexus Approach to Managing Water, Soil, and Waste under Changing Climate and Growing Demands on Natural Resources. In *Governing the Nexus: Water, Soil, and Waste Resources Considering Global Change*; Kurian, M., Ardakanian, R., Eds.; Springer: Berlin/Heidelberg, Germany, 2015; pp. 39–60.
18. Jain, S.K.; Lohani, A.K.; Singh, R.D.; Chaudhary, A.; Thakural, L.N. Glacial lakes and glacial lake outburst flood in a Himalayan basin using remote sensing and GIS. *Nat. Hazards* **2012**, *62*, 887–899. [CrossRef]
19. Mergili, M.; Schneider, J.F. Regional-scale analysis of lake outburst hazards in the southwestern Pamir, Tajikistan, based on remote sensing and GIS. *Nat. Hazards Earth Syst. Sci.* **2011**, *11*, 1447–1462. [CrossRef]
20. Leblanc, M.; Favreau, G.; Tweed, S.; Leduc, C.; Razack, M.; Mofor, L. Remote sensing for groundwater modelling in large semiarid areas: Lake Chad Basin, Africa. *Appl. Hydrogeol.* **2006**, *15*, 97–100. [CrossRef]
21. Ye, Q.; Zhu, L.; Zheng, H.; Naruse, R.; Zhang, X.; Kang, S. Glacier and lake variations in the Yamzhog Yumco basin, southern Tibetan Plateau, from 1980 to 2000 using remote-sensing and GIS technologies. *J. Glaciol.* **2007**, *53*, 673–676. [CrossRef]
22. Zainab, N.; Tariq, A.; Siddiqui, S. Development of Web-Based GIS Alert System for Informing Environmental Risk of Dengue Infections in Major Cities of Pakistan. *Geosfera Indones.* **2021**, *6*, 77–95. [CrossRef]
23. D'Amico, G.; Szopik-Depczyńska, K.; Beltramo, R.; D'Adamo, I.; Ioppolo, G. Smart and Sustainable Bioeconomy Platform: A New Approach towards Sustainability. *Sustainability* **2022**, *14*, 466. [CrossRef]
24. Li, C.; Longley, P. A Test Environment for Location-Based Services Applications. *Trans. GIS* **2006**, *10*, 43–61. [CrossRef]
25. Constanza, R.; Kubiszewski, I. A Nexus Approach to Urban and Regional Planning Using the Four-Capital Frame-work of Ecological Economics. In *Environmental Resource Management and the Nexus Approach*; Hettiarachchi, H., Ardakanian, R., Eds.; Springer: Cham, Switzerland, 2016; pp. 79–111.
26. Kurian, M.; Ardakanian, R. The Nexus Approach to Governance of Environmental Resources Considering Global Change. In *Governing the Nexus: Water, Soil and Waste Resources Considering Global Change*; Kurian, M., Ardakanian, R., Eds.; Springer: Dordrecht, The Netherlands, 2014; pp. 3–13. [CrossRef]
27. Kolli, M.K.; Opp, C.; Karthe, D.; Groll, M. Mapping of Major Land-Use Changes in the Kolleru Lake Freshwater Ecosystem by Using Landsat Satellite Images in Google Earth Engine. *Water* **2020**, *12*, 2493. [CrossRef]
28. Kolli, M.K.; Opp, C.; Groll, M. Identification of Critical Diffuse Pollution Sources in an Ungauged Catchment by Using the Swat Model- A Case Study of Kolleru Lake, East Coast of India. *Asian J. Geogr. Res.* **2020**, *3*, 53–68. [CrossRef]
29. Azeez, P.; Kumar, A.; Choudhury, B.; Sastry, V.; Upadhyay, S.; Reddy, K.; Rao, K. *Report on the Proposal for Downsizing the Kolleru Wildlife Sanctuary (+5 to +3 Feet Contour)*; The Ministry of Environment and Forests Government of India: New Delhi, India, 2011.
30. Rao, K.; Krishna, G.; Malini, B. Kolleru lake is vanishing—A revelation through digital image processing of IRS-1D LISS III sensor data. *Curr. Sci.* **2004**, *86*, 1312–1316.
31. Harikrishna, K. Land Use/Land Cover patterns in and around Kolleru Lake, Andhra Pradesh, India Using Remote Sensing and GIS Techniques. *Int. J. Remote Sens. Geosci.* **2013**, *2*, 2319–3484.
32. Belton, B.; Padiyar, A.; Ravibabu, G.; Rao, K.G. Boom and bust in Andhra Pradesh: Development and transformation in India's domestic aquaculture value chain. *Aquaculture* **2017**, *470*, 196–206. [CrossRef]
33. Pattanaik, C.; Prasad, S.; Nagabhatla, N.; Sellamuthu, S. A case study of Kolleru Wetland (Ramsar site), India using remote sensing and GIS. *IUP J. Earth Sci.* **2010**, *4*, 70–77.
34. Barman, R.P. The fishes of the Kolleru Lake, Andhra Pradesh, India, with comments. *Rec. Zool. Sur. India* **2004**, *103*, 83–89.
35. Butler, H.; Daly, M.; Doyle, A.; Gillies, S.; Hagen, S.; Schaub, T. *The Geojson Format*; Technical Report; Internet Engineering Task Force (IETF): Fremon, CA, USA, 2016.
36. Piedrafita, R.; Béjar, R.; Blasco, R.; Marco, A.; Zarazaga-Soria, F.J. The digital 'connected' earth: Open technology for providing location-based services on degraded communication environments. *Int. J. Digit. Earth* **2017**, *11*, 761–782. [CrossRef]
37. Horbiński, T.; Lorek, D. The use of Leaflet and GeoJSON files for creating the interactive web map of the preindustrial state of the natural environment. *J. Spat. Sci.* **2020**, 1–17. [CrossRef]
38. Masetti, G.; Kelley, J.G.W.; Johnson, P.; Beaudoin, J. A Ray-Tracing Uncertainty Estimation Tool for Ocean Mapping. *IEEE Access* **2017**, *6*, 2136–2144. [CrossRef]
39. Dai, Y.; Duan, Z.; Ai, D. Construction and Application of Field Investigation Support Platform for Land Spatial Planning Based on GeoServer. *J. Phys. Conf. Ser.* **2020**, *1621*. [CrossRef]

40. Gao, S.; Mioc, D.; Anton, F.; Yi, X.; Coleman, D.J. Online GIS services for mapping and sharing disease information. *Int. J. Heal. Geogr.* **2008**, *7*, 8. [CrossRef] [PubMed]
41. Yu, X.W.; Liu, H.Y.; Yang, Y.C.; Zhang, X.; Li, Y.W. GeoServer Based Forestry Spatial Data Sharing and Integration. *Appl. Mech. Mater.* **2013**, *295–298*, 2394–2398. [CrossRef]
42. Boulos, M.N.K.; Honda, K. Web GIS in practice IV: Publishing your health maps and connecting to remote WMS sources using the Open Source UMN MapServer and DM Solutions MapLab. *Int. J. Heal. Geogr.* **2006**, *5*, 6. [CrossRef] [PubMed]
43. Bordogna, G.; Kliment, T.; Frigerio, L.; Brivio, P.A.; Crema, A.; Stroppiana, D.; Boschetti, M.; Sterlacchini, S. A Spatial Data Infrastructure Integrating Multisource Heterogeneous Geospatial Data and Time Series: A Study Case in Agriculture. *ISPRS Int. J. Geo-Inf.* **2016**, *5*, 73. [CrossRef]
44. Zhang, L.; Yi, J. Management methods of spatial data based on PostGIS. In Proceedings of the 2010 Second Pacific-Asia Conference on Circuits, Communications and System, Beijing, China, 1–2 August 2010; Volume 1, pp. 410–413.
45. Peterson, G. GIS cartography. In *A Guide to Effective Map Design*, 2nd ed.; CRC Press: Boca Raton, FL, USA, 2014.

 *sustainability*

Article

Potential of NTFP Based Bioeconomy in Livelihood Security and Income Inequality Mitigation in Kashmir Himalayas

Ishtiyak Ahmad Peerzada [1,2,*], Mohammad A. Islam [1], James Chamberlain [3], Shalini Dhyani [2,4], Mohan Reddy [5] and Somidh Saha [6,7,*]

[1] Faculty of Forestry, Sher-e-Kashmir University of Agricultural Sciences and Technology of Kashmir, Srinagar 191201, India; ajaztata@gmail.com
[2] Commission on Ecosystem Management (CEM), International Union for the Conservation of Nature (IUCN), 1196 Gland, Switzerland
[3] Forest Service, Southern Research Station (USDA), Blacksburg, VA 24060, USA; james.l.chamberlain@usda.gov
[4] Critical Zone Research Group, CSIR National Environmental Engineering Research Institute (NEERI), Nagpur 440020, India; shalini3006@gmail.com
[5] Nurture Agtech Private Limited, Bellandur, Bangaluru 560103, India; nimmalamohan@gmail.com
[6] Research Group Sylvanus, Institute for Technology Assessment and Systems Analysis (ITAS), Karlsruhe Institute of Technology (KIT), 76133 Karlsruhe, Germany
[7] Institute of Forest Sciences, Albert Ludwig University of Freiburg, 79085 Freiburg, Germany
* Correspondence: drishtiyak@skuastkashmir.ac.in (I.A.P.); somidh.saha@kit.edu (S.S.)

Citation: Peerzada, I.A.; Islam, M.A.; Chamberlain, J.; Dhyani, S.; Reddy, M.; Saha, S. Potential of NTFP Based Bioeconomy in Livelihood Security and Income Inequality Mitigation in Kashmir Himalayas. *Sustainability* **2022**, *14*, 2281. https://doi.org/10.3390/su14042281

Academic Editors: Nallapaneni Manoj Kumar, Md Ariful Haque and Sarif Patwary

Received: 31 December 2021
Accepted: 14 February 2022
Published: 17 February 2022

Publisher's Note: MDPI stays neutral with regard to jurisdictional claims in published maps and institutional affiliations.

Abstract: The contribution of non timber forest products (NTFPs) has been acknowledged globally for their role in conservation, income generation, livelihood improvement and rural development. The potential of a NTFP-based bioeconomy has given a new dimension to the forest sector, and NTFPs are now considered favourably by the resource rich developing economies. The actual contribution of NTFPs has never been adequately estimated due to lack of sufficient baseline information on extraction, consumption patterns and traded quantities in Kashmir, India. Complicated management frameworks and fragmented value chains have eclipsed their diverse social life cycle in Kashmir. Therefore the present study investigates the bioeconomic transformation, livelihood contribution, income inequality mitigation and determinant socioeconomic factors of NTFP extraction in the Kashmir Himalayas. A multistage random sampling technique was employed to collect data through participatory household-based surveys from different villages. Data were collected through structured in-depth interviews, non-participant observation and focussed group discussions. Descriptive and analytical statistics were used for data analysis. The Lorenz curve and Gini index were used to evaluate the influence of household NTFP incomes on income inequality mitigation, and econometric models were developed to identify key factors that influence the level of household income from NTFPs to determine their potential for supporting livelihood security and bioeconomy in the region.

Keywords: NTFP; NWFP; Kashmir Himalayas; bioeconomy; livelihood; income inequality; Lorenz curve; Gini coefficient

1. Introduction

The term non-timber forest products (NTFPs) and allied terms such as "minor", "secondary" and non-wood forest products have emerged as umbrella expressions for a range of plant and animal resources other than timber (or wood, in the case of non-wood forest products) derived from forests or forest species. DeBeer and McDermot in 1989 [1] defined NTFPs as all the biological materials, other than timber, extracted from forests for human use. This definition excludes minerals and includes fuelwood, bamboo and animal products. By contrast, the FAO in 1999 defined non-wood forest products (NWFPs) as "goods of biological origin other than wood that are derived from forests as well as other wooded land that also includes trees outside forests" [2].The FAO's 2015 Forest Resource

Assessment [3] suggests that NWFPs are "Goods derived from forests that are tangible and physical objects of biological origin other than wood", in order to increase consistency in country reportings. In India, researchers have defined the products obtained from plants of forest origin, as well as insects, animals, animal parts and items of mineral origin except timber, as minor forest products (MFPs) or NTFPs or NWFPs [4]. The Indian constitution, the central policy "National Forest Policy 1988" and "The Scheduled Tribes and Other Traditional Forest Dwellers (Recognition of Forest Rights) Act, 2006" use the term MFPs. However, variations of this term, such as secondary or NTFPs, are now more frequently used by international and national organisations, governments, foresters and academics, depending on the requirements as well as objectives. Hence, for the objective and scope of this study, we have used the term non-timber forest products (NTFPs).

NTFPs have played an essential role in sustaining the livelihoods, income generation, food and nutritional security, fuelwood, fodder and traditional medicine as subsistence support to the rural communities since time immemorial [5–7]. Around 1.6 billion people throughout the world are reported to consume and trade NTFPs [8]. About 80% people across developing countries are reported to use plants for nutritional security [9] and traditional medicine [10]. More than two billion people use biomass-based fuels, mostly fuelwood for cooking and heating purposes [11]. NTFPs comprise a significant component of food security in the developing countries, and communities consume them more for subsistence than trade, as NTFPs are considered a relevant safety net and economic buffer [12] to support them during agricultural shortfalls or lean periods [13].

Research on the role of NTFPs in income generation and rural livelihoods has greatly increased in the last few decades, and has been reported to contribute to 20–60% of the rural household income [9,14] of forest fringe communities globally. In India, the annual contribution of NTFPs to income corresponds to US$2.7 billion, supporting more than 55% of the total employment in the forest sector. One third of India's rural population is reported to derive substantial household incomes from NTFPs [15]. More than 60% population of the Jammu and Kashmir (J&K) UT of India harvests NTFPs for food, nutrition, medicine, income and employment generation purposes [16,17].

The renewed interest in the management of NTFPs has improved extensively due to their significant contribution to addressing income inequality mitigation and supporting sustainable development. Conservative estimates indicate that a large portion of total forest products' value comes from NTFPs, although the magnitude of this value may vary from site to site. In diversified value chains, NTFPs provide medicine, aroma, spices, flavours, phytonutrients and nutritional variety in contemporary diets.

Despite maximum local consumption, trade and poor representation of details in national and international statistics [10,18–20], NTFPs are being increasingly recognized for their significant roles in supporting local and state economies. Comprehensive investigation into the dynamics of NTFPs in rural livelihoods, as well as trends of production, collection, consumption, trade and sustainability, are essential. In order to enhance the local subsistence as well as to support regional bioeconomy with NTFPs, it is important to examine the site-specific potential of different NTFPs. This means that site-specific assessments regarding the consumption patterns and potential of NTFPs with a greater focus on forest-based livelihoods are crucial and relevant. Moreover, the collection, consumption and trade of freely available forest-based NTFPs are influenced by the context and site-specific household characteristics. Among household characteristics, gender and age are more important than numbers. Therefore, this study was conceived to understand the synergy of local community households' dependency on these resources, as well as to understand the links with adaptation to various stresses and the potential contribution to bioeconomy.

1.1. NTFPs in Income Inequality Mitigation

Poverty has been described as an evident deprivation in well-being or living below a defined threshold of income [21]. However, forest fringe communities living in poverty are particularly vulnerable to adverse events beyond their control. They are often treated badly by the state, system and society, and are excluded from voice as well as power [21,22]. Poverty is a complex material scarcity, lack of access to basic needs (education, health, nutrition and food security), absence of political autonomy, lack of freedom of choice and effect of social inequality, among others. In addition, the occurrence of poverty and the intensity and extent of inequality—i.e., the distribution of income between the poor and rich—also helps in differentiation [22].

The livelihoods of forest-dependent communities are intrinsically delicate and exposed to an array of jolts and seasonal instability; hence, rural households maintain diversified livelihood approaches, such as the harvesting and trade of NTFPs, for both subsistence and cash income. The significance of NTFPs to the livelihood of marginalized communities can help offset inequalities [23]. Forest fringe communities do not have sufficient productive lands or access to formal employment opportunities, which forces them to extract NTFPs for subsistence consumption and income generation. Incomes from NTFPs are most important to poor and less educated people compared to rich and educated people [23], having substantial use within the households. The contribution of NTFPs to rural households and local economies is ignored by poverty estimation surveys, due to insufficient information on their income-balancing impact of reducing inequalities among rural households. Hence, NTFPs are not adequately considered in the poverty reduction strategies of most developing and underdeveloped countries, as poverty analysis based on income or material use discounts the role of forests [9].

With the onset of gloomy economic circumstances due to increasing population, demand for food, water and healthcare requirements, resource exhaustion and climate-induced disasters, bioeconomy is expected to provide opportunities for environment-friendly raw material sourcing, mainly based on renewable and recycled resources. The forest vis-à-vis NTFP sector is an efficient renewable bioresource base for meeting the growing demands, if managed sustainably. The increasing consumer demand for diversified bioactive compounds extracted from various NTFPs for pharmaceutical, nutraceutical, cosmeceutical, food and beverage industries can transform local economies into a bioeconomy. In bioresource-rich countries, NTFPs are more relevant to forest fringe communities for income generation, livelihood improvement and modelling the rural development by promoting new bio-products through the NTFP-based bioeconomy. Despite the huge dependence of rural people on NTFPs in India, there is little emphasis in its national policies and research priorities on the forest bioeconomy. However, a bioeconomy created through biotechnological transformations of the energy and pharmaceutical sectors has been envisioned by the Ministry of Science and Technology of the Government of India [16].

1.2. Closing the Data Gaps: On NTFP-Based Livelihood for Bioeconomy

Feeding a growing world population sustainably is a key challenge for the 21st century and is well acknowledged and highlighted by the United Nations Sustainable Development Goals of 2015. Bioeconomy has gained urgency due to financial crisis, inflation and loss of livelihoods caused by pandemics, resource exhaustion and fossil-fuel-induced climate change. Bio-based products are innovative elements of a bioeconomy, and can materialize only if the flow of resources in the economy is well understood. Given the fact that most forest products are consumed by local households, and do not enter the formal markets, very little is known about the value of these products contributing to the bioeconomy. Only wood and its supply chain are accounted for and considered as essential pillars of a forest-based bioeconomy. Globally, NTFPs have played a vital role in ensuring human well-being, by efficiently supporting local livelihoods, businesses, culture and indigenous practices through the diversification of income from formal as well as informal forest sectors, such as trees outside forests. However, NTFP-based production systems, management as well as value chains fall within a very diverse set of socio-ecological and socio-economic complexities. This results in crucial challenges as well as opportunities that require attention, to explore and understand the importance of NTFPs to bioeconomies and human wellbeing, and to harness their potential—from the local level to supporting international sustainable development goals. Moreover, most of the NTFPs and wild edibles consumed and traded by local communities are never formally reported or accounted for. The actual contribution of forest products to livelihoods is difficult to understand without better data [24]. The potential role of NTFPs in income diversification and uplifting rural economies has been hindered by the lack of clear baseline data, analytical frameworks and inclusive value chains in India. Despite their significance to trade, NTFP markets are mostly informal and scattered, with no formal records maintained, leading to an inadequate information flow on the contribution of NTFP trade at the local as well as national level. Hence, NTFP are poorly acknowledged, despite their significant contribution to local income and livelihood generation [25]. Data on the valuation of NTFPs, as well as their entry in government records for production and exports, are also limited and inconsistent in India [26].

Proper NTFP management offers a sustainable basis for livelihoods once diversified and developed as tradable products. Various studies have proved that NTFP production has contributed to higher-than-average income compared to the national income [27] if managed sustainably. The development of NTFPs or NTFP-based value-added products could improve the living standards of several local communities if promoted and facilitated through transformations of local economies into bioeconomies. However, the literature on bioeconomy has focussed more on the technological aspects of developing new biobased products and the policy process that supports transition to bioeconomy [16,27]. There is insufficient information to demonstrate the impact of biobased products vis-à-vis NTFPs on local livelihood upliftment and economic benefits. Therefore, there is a significant data gap to investigate and understand the impact of NTFPs on local livelihoods and the potential impacts of local economies largely on bioeconomies.

The contemporary biobased industry offers tremendous opportunities for indigenous people and local communities by endorsing products that are consistent with, and acceptable to, traditional ways of life, values and cultures. This can help with the creation of sustainable and culturally meaningful employment in local communities. Moreover, indigenous people and local communities are familiar with, and skilled at, identifying, harvesting and using NTFPs; hence, they are well-suited to this type of work [28]. Once all of these varied forms of income generation are considered, the value of NTFPs could increase significantly [29]. Therefore, accurate baseline data on the potential value and contribution of NTFPs to local communities at the household level may significantly help with designing appropriate policy interventions. Collecting such information is also critical to understand the importance of NTFPs to protecting indigenous and traditional cultural values and practices.

1.3. Research Aim

The Kashmir part of the Indian Himalayan Region (IHR) is a unique mountain ecosystem which harbours rich floral, faunal and cultural diversity, in addition to the largest source of freshwater resources harnessed by both India and Pakistan. The forests of the region have diverse NTFPs, with a substantial contribution to its rural livelihood and local economy. Hence, the present study attempted to analyse the determinant factors that influence the extent of households' dependence on NTFPs for their livelihoods and income generation. We also attempted to categorise the potential of NTFPs to mitigate income inequality. The study aimed to address the following questions:

- What is the role of NTFPs in household subsistence and income generation in supporting a bioeconomy?
- How do different sources of income and socio-economic factors influence the contribution of NTFPs to the livelihood strategies of households?
- What is the potential impact of the local economy on the bioeconomy?

2. Materials and Methods

2.1. Study Site Description

This study was conducted in the Langate Forest Division in the frontier district of the Kupwara, Jammu and Kashmir (J&K) Union Territory (UT) of India (Figure 1). The Langate Forest Division is situated between the northern latitude at 34°13′ to 34°30′ and eastern longitude at 73°56′ to 74°26′, with an altitudinal range of 1590 to 4308 m asl (meters above sea level). However, the principal forest cover extends up to 3500 m asl only in the dominant eastern aspect. This area faces severe cold during winter and pleasant weather during summer months. The temperature ranges between −5 °C minimum in winter and up to 22 °C maximum during summers; while the mean annual rainfall is 1270 mm. This area has temperate, sub-alpine and alpine climatic conditions with rich biodiversity, including many rare species of NTFP, such as *Morchella esculenta, Aconitum heterophyllum, Saussurea lappa, Taxus wallichiana* and *Trillium govanianum*, among others. Out of the total geographical area (2744 km^2), district Kupwara has about 1534.52 km^2 (55.92%) of forest, of which 760.06 km^2 is very dense, 423.61 km^2 is moderately dense and 350.85 km^2 is open forest [30]. According to the 2011 census, district Kupwara has a population of around 875,564, of which 776,322 people live in rural areas and 99,242 people live in urban settings.

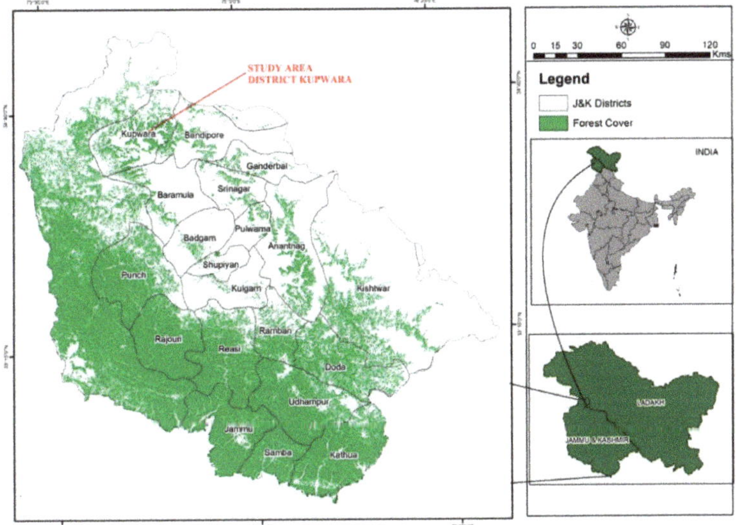

Figure 1. Location map of the study area. Raw image source: APA format, Landsat-8 (USGS Earth Explorer). Time period: January–February 2021 [31].

2.2. Methodology

The purpose of the household survey was to map the availability of NTFPs, livelihood dependency of people on NTFPs, economic valuation, current extraction, consumption patterns, prioritization and usage patterns.

2.2.1. Sampling Procedure

A multistage sampling technique [32] was used to select the ranges and villages from Langate Forest Division. In the first stage, out of the four forest ranges in the division, two prominent forest ranges (i.e., Mawar and Rajwar), having the maximum forest cover and forest fringe villages, were selected for this study after a proper reconnaissance survey. As the target groups selected for the household interviews were people living in closer proximity to the forests; therefore, at the second stage, ten sample villages, five villages from each range, were selected on the basis of livelihood (forest/agriculture), dependency on

NTFPs, density of forest, ratio of forest-dependent population and proximity of household to forests. At the third stage, a total of one hundred households extracting, consuming and selling NTFPs were selected randomly, which was 8% of the total households in each village (Table 1). Household heads or eldest members of the families were considered for focussed and in-depth interviews, as they are generally the main earners, decision makers and future planners of the households' transactions.

Table 1. Sampling framework and sample size distribution in the study area.

S. No.	Range	Village	Sampling Frame	Sample Size	Sampling Intensity
1.	Rajwar	Dardahaji	100	8	8%
2.	Rajwar	Satkhoji	389	31	8%
3.	Rajwar	Briniyal	100	8	8%
4.	Rajwar	Uthroosa	90	7	8%
5.	Rajwar	Shilthara	68	6	8%
6.	Mawar	Reshwari	60	5	8%
7.	Mawar	Puthwari	75	6	8%
8.	Mawar	Monabal	90	7	8%
9.	Mawar	Lahikot	126	10	8%
10.	Mawar	Bandi	146	12	8%
Total	2	10	1244	100	8%

2.2.2. Data Collection

The data were collected from the sample households through interviews using a structured interview schedule and focussed group discussions guided by a checklist of questions [33]. The questions asked through the interview schedule included socioeconomic characteristics of households; the collection, consumption and trade of NTFPs; the quantity of NTFPs marketed, various sources of household income; and the economic contribution of NTFPs. The socioeconomic variables of the households included age, education level, social membership, household size, household labour, farm size, livestock ownership, main occupation, wealth status, gross annual income, proximity to forest and forest visits. The variables were measured using a socioeconomic status scale [34] after modification. The focussed group discussions were held with 10–12 participants, including village elders with good knowledge of the identification and use of NTFPs. The observations extracted from the focussed group discussions were used to triangulate and validate the data collected through the household surveys, and also to interpret the results and draw inferences.

2.2.3. Data Analysis

Descriptive statistics including the percentage, average, standard deviation and range [35] were applied to summarize the socioeconomic characteristics; NTFP collection; and the consumption, trade, income generation and contribution of NTFPs to household incomes. The Lorenz curve [36] and Gini coefficient [37] were applied to evaluate the distribution of household NTFP incomes and their impact on income inequality mitigation [38]. The Lorenz curve was generated in MS Excel by drawing a line chart with cumulative share of population on the horizontal axis and cumulative share of income on the vertical axis. The Gini coefficient was calculated using the following formula:

$$G = A/A + B = 2A = 1 - 2B \tag{1}$$

where

G is Gini coefficient

A is an area between the line of perfect equality and the Lorenz curve

B is the area under the Lorenz curve

The data collected in terms of local units were converted into International System Units (ISU) and analysed using statistical analytical package SPSS Ver. 21.0. The results are displayed through various tables and graphs.

2.2.4. Analytical Framework

Multiple regression analysis [39] determined the socioeconomic variables that influenced the household NTFP incomes. It was hypothesized that household NTFP income is inextricably influenced by the socioeconomic characteristics of the household. Here, the household NTFP income was the regress and socioeconomic characteristics were the regressors. The b-values in the analysis were the impact multipliers, which explain the magnitude of the effect of the unit change on the quantity of a household NTFP income. The conceptual model based on the multivariate function is given below:

$$Y = a + b_1x_1 + b_2x_2 + \ldots + b_{10}x_{10} + e \qquad (2)$$

where Y is the household NTFP income (INR/year, also indicated in USD);

x_1–x_{10} are socioeconomic characteristics; a is the constant or intercept; b_1–b_{10} are regression coefficients and e is an error term.

3. Results

3.1. Diversity and Use Pattern of NTFPs

The use pattern of NTFPs was characterised by a range of factors, including the access to resources, diversity of species available in the nearby forest areas and availability of markets. However, the use pattern varied from area to area and even between households within a village or community. Therefore, the NTFPs extracted by local community households were classified into different categories based on the use pattern (Table 2) of each species (for example, medicine, fuelwood, fodder, vegetable, spice or wild fruit), and part harvested (leaves, fruits, fruiting body, roots or stem). The investigation further revealed that around 50 species of NTFPs distributed across 36 families (Figure 2) were used by the households of the study area for different purposes. Among these, 65% were herbs, 17% were trees, 12% were shrubs, 4% were fungi and 2% were climbers (Figure 3). Apparently, the tubers, roots, rhizomes (28%), and leaves (28%) were the highest-exploited parts, followed by fruits and seeds (8%), bark (8%) and the whole plants and branches (6% each); other parts, such as flowers, nuts, wickers and fruiting bodies, were the least-exploited parts (Figure 4). As the use categories of the species were concerned, about 23 species were used for medicines; 11 for vegetables; 5 for fuelwood; 4 species each for fodder, wild fruit and spice; and 3 species for wicker (Figure 5).

Table 2. Diversity and use pattern of NTFPs in the study area.

S.No.	Species/Habit	Family	Local Name	English/Common Name	Part Used	Uses
1.	*Abies pindrow* Royle (tree)	Pinaceae	Budul	Himalayan fir	Branch/bark	Fuel wood
2.	*Achillea millefolium* Linn(herb)	Compositae	Berguer	Yarrow	Leaf	Medicine
3.	*Aconitum heterophyllum* Wall (herb)	Ranunculaceae	Atis	Aconite	Tuber	Medicine
4.	*Acorus calamus* Linn (herb)	Araceae	Vai	Sweetflag	Rhizome	Medicine
5.	*Allium humile* Kunth (herb)	Amaryllidaceae	Jangli-piaz	Allium	Whole plant	Vegetable/spice
6.	*Angelica glauca* Edgew. (herb)	Apiaceae	Chohore	Angelica	Root	Medicine/spice
7.	*Arnebia benthamii* Wall ex G. Don (herb)	Boraginaceae	Kahzaban	Arnebia	inflorescence/root	Medicine
8.	*Artemisia absinthium* Linn (herb)	Asteraceae	Tethwan	Artimisia	Leaf, flower	Medicine

Table 2. *Cont.*

S.No.	Species/Habit	Family	Local Name	English/Common Name	Part Used	Uses
9.	*Atropa accuminata* Royle ex Lindl (herb)	Solanaceae	Jal-kafal	Atropa	Root, leaf	Medicine
10.	*Berberis lycium* Royle (shrub)	Berberidaceae	Kawdach	Berberis	Root/fruit	Medicine
11.	*Bergenia ciliata* (Haw.) Sternb (herb)	Saxifragaceae	Zakhmi-hayat	Berginia	Root/whole plant	Medicine
12.	*Betula utilis* D.Don (tree)	Betulaceae	Burza	Birch	Leaf, bark	Medicine
13.	*Bunium persicum* (Boiss). Fedts (herb)	Apiaceae	Kala zeera	Cumin	Seed	Spice
14.	*Capsella bursa-pastoris* (L.) Medic (herb)	Brassicaceae	Kralmond	Shepherds purse	Leaf	Vegetable
15.	*Castanea sativa* Mill (tree)	Fagaceae	Gour	Sweet chestnut	Nut	Wild fruits
16.	*Cedrus deodara* G.Don. (tree)	Pinaceae	Deodar	Himalayan cedar	Branch/bark	Fuel wood
17.	*Cichorium intybus* Linn. (herb)	Asteraceae	Kasini	Chicory	Whole plant	Vegetable
18.	*Corylus jacquemontii* Decne (shrub)	Betulaceae	Hazel nut	Indian tree hazel	Nut/leaf	Wild fruits
19.	*Dactylis glomerata* Linn. (herb)	Poaceae	Ghass	Orchard grass	Leaf	Fodder
20.	*Dioscorea deltoidea* Wall. ex Griseb (climber)	Dioscoreaceae	Krish	Dioscoria	Tuber	Medicine
21.	*Diplazium esculentum* (Retz.) Sw. (herb)	Athyriaceae	Kasrod/Dade	Vegetable fern	Leaf	Vegetable
22.	*Dipsacus inermis* Wall. (herb)	Caprifoliaceae	Wopalhakh	Himalayan teasel	Leaf	Vegetable
23.	*Fritillaria roylei* Hook. (herb)	Lilaceae	Sheedkhar	Himalayan fritillary	Bulb	Medicine
24.	*Helvella crispa* (Scop.) Fr. (fungi)	Helvellaceae	Shajkan	Common helvel	Fruiting body	Vegetable
25.	*Indigofera heterantha* Wall ex. Brandis (shrub)	Fabaceae	Krats	Himalayan indigo	Wicker/leaf	Wicker/kangri making
26.	*Inula racemosa* Hook. f. (herb)	Compositae	Poshkarmool	Inula	Root	Medicine
27.	*Juglans regia* Linn. (tree)	Juglandaceae	Doon	Walnut	Nut/branch/bark	Wild fruits
28.	*Jurinea dolomiaea* Boiss. (herb)	Asteraceae	Guggal	Jurinea	Root	Medicine
29.	*Mentha longifolia* Linn. (herb)	Lamiaceae	Pudina	Wild mint	Leaf	Spice/medicine
30.	*Morchella esculenta* (Linn.) Pers. (fungi)	Morchellaceae	Guchi	Wild Morel	Fruiting body	Vegetable
31.	*Origanum vulgare* Linn. (herb)	Lamiaceae	Wanbaber	Oregano	Leaf	Medicine
32.	*Parrotiopsis jacquemontiana* (Decne) Rehd (shrub)	Hamamelidaceae	Pohu	Parrotia	Wicker/leaf	Wicker for kangri making/fodder
33.	*Picrorhiza kurrooa* Royle ex Benth (herb)	Scrophulariaceae	Kutki	Picrorhiza	Rhizome	Medicine
34.	*Pinus wallichiana* A.B. Jacks (tree)	Pinaceae	Kail	Blue pine	Branch/bark	Fuel wood
35.	*Plantago lanceolata* Linn (herb)	Plantaginaceae	Gul	Plantago	Leaf	Vegetable

Table 2. *Cont.*

S.No.	Species/Habit	Family	Local Name	English/Common Name	Part Used	Uses
36.	*Poa pratensis* Linn (herb)	Poaceae	Ghass	Meadow grass	Leaf	Fodder
37.	*Podophyllum hexandrum* Royle (herb)	Podophyllaceae	Wanwangun	Podophyllum	Root/fruit	Medicine
38.	*Polygonatum verticillatum* Linn (herb)	Liliaceae	Salam-mishri	Polygonatum	Root	Medicine
39.	*Punica granatum* Linn (tree)	Lythraceae	Anar	Pomegranate	Fruit	Wild Fruit
40.	*Rosa webbiana* Wallich ex Royle (shrub)	Rosaceae	Jangli-gulab	Wild rose	Flower	Medicine
41.	*Rheum webbianum* Royle (herb)	Polygonaceae	Pambhaakh	Himalayan rhubarb	Leaf/root	Vegetable/Medicine
42.	*Rumex nepalensis* Spreng (herb)	Polygonaceae	Obej	Dock	Leaf	Vegetable
43.	*Salix alba* Linn (tree)	Salicaceae	Vir	Salix	Wicker/leaf	Wicker/Kangri making/Fodder/Fuelwood
44.	*Saussurea costus* C.B. Clarke (herb)	Asteraceae	Kuth	Costus	Root	Medicine
45.	*Taraxacum officinale* Weber (herb)	Compositae	Handh	Taraxacum	Whole plant	Vegetable
46.	*Thymus serpyllum* Linn (shrub)	Lamiaceae	Javend	Thyme	Leaf	Spice
47.	*Trillium govanianum* Wall. ex. D. Don (herb)	Melanthiaceae	Tripatri	Himalayan trillium	Root	Medicine
48.	*Valeriana jatamansi* Jones (herb)	Valerianaceae	Mushkbala	Valeriana	Root	Medicine
49.	*Viola odorata* Linn (herb)	Violaceae	Bunafsha	Viola	Flower	Medicine
50.	*Ziziphus jujube* (L.) Mill (tree)	Rhamnaceae	Breyi	Common jujube	Fruit	Medicine

3.2. Household Socioeconomic Variables

The results of this study revealed that the NTFP collectors were between the age group of 20 to 84 years, with mean age of 48.84 years. The middle-aged people were generally economically active, hard-working and the main earner group of the society. The mean score of the education level of the NTFP collectors was 1.55, which is equivalent to the primary school level. To understand the literacy rate in the area, six categories were defined, which ranged from illiterate, below primary, primary school, middle school, high school and graduation and above. The literacy levels in terms of formal education were observed to be quite low. The prevalence of low literacy among NTFP collectors was due to the remoteness of the area, lack of higher educational facilities, low socio-economic conditions and higher involvement of young people in livelihood earnings. The proportion of uneducated persons was found to be higher than that of other categories. The mean social membership of the NTFP collectors was only 1.33, which indicates that they had membership of at least one organization; the majority of the NTFP collectors had no social memberships. Low social participation shows a grousing magnitude of interest and willingness of the NTFP collectors towards membership in various formal and informal organisations. The mean value (1.68) of the household family size indicates that the NTFP collectors had a household composition or number of family members above five. Considering children as added assets, a need for family labour and a lack of knowledge of family planning are the key reasons for large families. The majority of the sampled households contributed as workers, with a mean of 3.13 workers per household.

This proved that a considerable number of workers in the surveyed households accounted for the large quantity of extraction, consumption and marketing of NTFPs. The larger section of NTFP collectors were marginal and small landholders, with a mean landholding size of only 1.68 ha. Almost all the sampled households possessed livestock. The mean value of the livestock unit (2.29) shows that they owned livestock ranging from 5 to 10 per household. The mean score of the main occupation was 3.27, indicating that agriculture was the prevalent main occupation among the sampled households. The main occupations to compensate the household income were both wage labour and non-farm labour (NTFP collection). Agriculture and allied activities, such as the cultivation of vegetables and fruits, constitute a considerable proportion of the livelihood portfolio of the households. The NTFP collectors were mostly poor, with a very low wealth status with a mean of only 1.40. This clearly indicates widespread poverty in the forest fringe villages around the study area. The gross annual income of the NTFP collectors ranged from INR 5400.00 (USD 71.93) to INR 309,523.00 (USD 4122.99) with a mean of INR 67,122.44 (USD 894.09). The majority of the NTFP collectors had a significantly low income status. The NTFP collectors lived very close to forests and had to walk around only 0.30 to 1.90 km. The sampled households would frequently visit forests, with a mean of 1.52 visits (Table 3). People living closer to the forest had a higher dependency on NTFPs to meet their daily livelihood needs, which implies frequent forest visits.

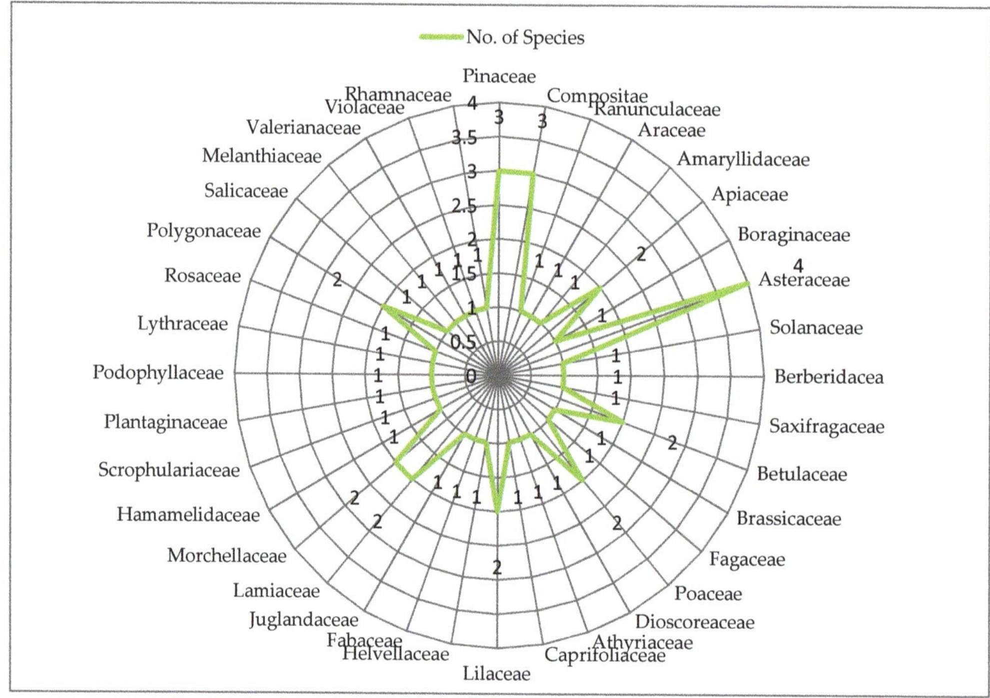

Figure 2. Proportion of NTFP species by family (N = 100).

3.3. Extraction and Consumption of NTFPs

The NTFP collectors extracted a total of 506.43 tons of fuel wood, 90.52 tons of fodder, 2.62 tons of wild fruits, 4.42 tons of wild vegetables, 1.99 tons of mushrooms and 4.88 tons of medicinal plants annually from the forests of the study area. In each household, per year of extraction there were around 5.06 tons of fuel wood, 0.90 tons of fodder, 0.02 tons of wild fruits, 0.04 tons of wild vegetables, 0.01 tons of mushrooms and 0.04 tons of medicinal plants annually from the forests in the study area. Out of the total harvests, the NTFP collectors consumed a total of 100.77 tons of fuel wood, 78.85 tons of fodder, 2.44 tons of wild fruits, 4.23 tons of wild vegetables, 1.81 tons of mushrooms and 4.22 tons of medicinal plants annually. The average annual consumption rates in the sampled households were about 1.00 tons of fuel wood, 0.78 tons of fodder, 0.02 tons of wild fruits, 0.04 tons of wild vegetables, 0.01 tons of mushrooms and 0.04 tons of medicinal plants. The percentage involvement of households in NTFP collection ranged from 37% for mushrooms to 100% for fuel wood (Table 4).

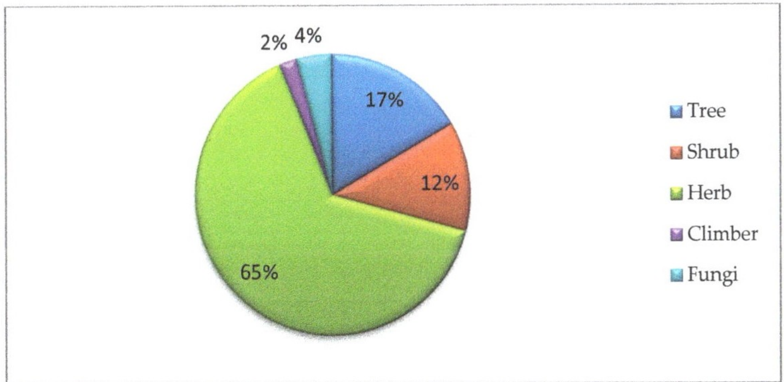

Figure 3. Proportion of NTFPs used by habit (N = 100).

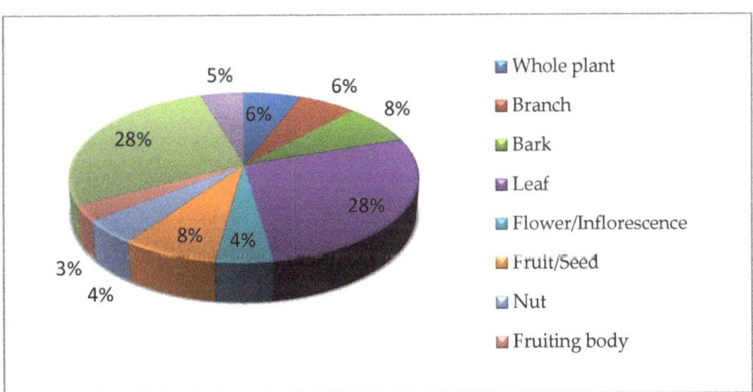

Figure 4. Proportion of NTFPs by percent parts used (N = 100).

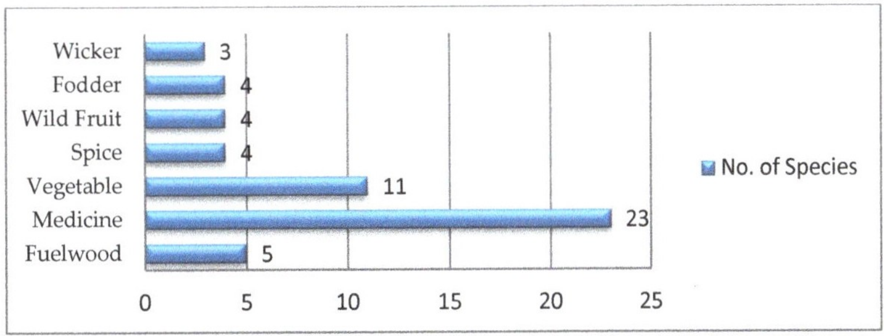

Figure 5. Proportion of NTFP Species by use category.

Table 3. Household descriptive variables determining NTFP-based bioeconomy (N = 100).

Variables (Code)	Explanation	Measurement Description	Minimum	Maximum	Mean	SD
Age (X_1)	Age of household head in years	Number of years lived by the respondent	20	84	48.84	14.57
Education level (X_2)	Household head undergone in education	0 = illiterate, 1 = < primary, 2 = primary, 3 = middle, 4 = high school, 5 = intermediate, 6 = graduate and over	0	6	1.55	1.74
Social membership (X_3)	Membership of household head in organisations	0 = no membership, 1 = membership of 1 organization, 2 = membership of >1 organization, 3 = office bearer, 4 = public leader	0	4	1.33	0.73
Household size (X_4)	No. of family members in a household	1 = ≤ 5 members, 2 = > 5 members	1	2	1.68	0.46
Household labour (X_5)	No. of workers in a household	1 = 1 worker, 2 = 2 workers, 3 = 3 workers, 4 = >3 workers	1	4	3.13	0.89
Farm size (X_6)	Land area under household management	0 = landless, 1 = marginal (up to 1.0 ha), 2 = small (1.1 to 2.0 ha), 3 = medium (2.1 to 4.0 ha), 4 = large (>4.0 ha)	1	2	1.68	0.46
Livestock ownership (X_7)	No. of livestock units owned by the household	0 = no livestock, 1 = ≤ 5 livestock, 2 = 6 to 10 livestock, 3 = > 10 livestock	0	3	2.29	1.14
Main occupation (X_8)	Occupation in which an individual was engaged for six months or more in a year	1 = wage labour, 2 = non-farm labour, 3 = cultivation, 4 = business, 5 = service, 6 = any other	1	6	3.27	1.29
Wealth status (X_9)	Relative position of households in the community in respect of wealth/physical assets	0 = poor, 1 = medium, 2 = rich	0	2	1.40	0.72
Gross annual income (X_{10})	Household income earned by all the on-farm and off-farm sources	INR/household/annum	5400.00	309,523.00	67,122.44	56,622.46
Proximity to forest (X_{11})	Distance between forests and house (km)	Distance of home to forests (km)	0.30	1.90	0.74	0.38
Forest visits (X_{12})	Frequency of forest visits in a year	0 = never, 1 = occasionally, 2 = frequently, 3 = very frequently	0	3	1.52	0.83

Table 4. Household extraction and consumption of NTFPs (N = 100).

NTFP	Involvement in Collection (%)	Total Extraction (ton/Year)	Average Extraction (ton/hh/Year)	Total Consumption (ton/Year)	Average Consumption (ton/hh/Year)
Fuel wood	100.00	506.43	5.06	100.77	1.00
Fodder	91.00	90.52	0.90	78.85	0.78
Fruits	34.00	2.62	0.02	2.44	0.02
Vegetables	64.00	4.42	0.04	4.23	0.04
Mushroom	37.00	1.99	0.01	1.81	0.01
Herbal medicine	88.00	4.88	0.04	4.22	0.04

3.4. Economic Valuation of NTFP Use

The diversity of NTFPs plays a crucial role in diversifying the household income; hence, significant proportions of the products extracted were marketed for income generation in the study area. The highest number of NTFPs marketed were medicinal plants (63%), followed by fuel wood (42%), vegetables (34%), mushrooms (32%), fruits (24%) and fodder (18%). Around 405.65 tons of fuel wood, 11.65 tons of fodder, 0.66 tons of medicinal plants, 0.18 tons of vegetables, 0.17 tons of mushrooms and 0.17 tons of fruits were sold in the market in a year. Fuel wood was the highest source of household income, at INR 3,124,839.00 (USD 41,623.73), with 69.88% income share to households; followed by medicinal plants, at INR 5,667,749.00 (USD 75,497.02), with 21.45% income share; and mushrooms, at INR 2,423,277.00 (USD 32,279.16), with 5.96% income share. Fodder, fruits and vegetables made comparatively low contributions to the household income, at 1.87%, 0.80% and 0.04%, respectively. NTFPs also contributed to the subsistence income of households, which is not usually accounted for. The monetary value of NTFPs used for subsistence consumption was estimated to be INR 8,611,614.31/year (USD 114,709.11) and INR 86,116.14/household/year (USD 1146.99). The economic valuation of NTFP extraction confirms that NTFPs generated a total income of INR 12,193,404.00/year (USD 162,421.75), which accounted for 29.38% of cash income and 70.62% of subsistence income in the study area (Table 5).

Table 5. Economic valuation of household NTFP use (N = 100).

NTFP	Involvement in Marketing (%)	Sale Price (INR/kg) USD *	Sale (Ton/ Year)	Subsistence Income (INR/Year) USD *	Cash Income (INR/Year) USD *	Total Income (INR/Year) USD *	Income Share (%)
Fuel wood	42.00	6.80 (0.091) *	405.65	621,839.10 (8283.07) *	2,503,000.00 (33,340.66) *	3,124,839.00 (41,623.73) *	69.88
Fodder	18.00	6.32 (0.83) *	11.65	452,157.90 (6022.87) *	66,940.00 (891.66) *	519,097.90 (6914.62) *	1.87
Fruits	24.00	180.37 (2.39) *	0.17	400,082.30 (5329.21) *	28,500.00 (379.63) *	428,582.30 (5708.91) *	0.80
Vegetables	34.00	7.44 (0.099) *	0.18	28,608.63 (381.07) *	1250.00 (16.65) *	29,858.63 (397.73) *	0.04
Mushroom	32.00	1344.02 (17.90) *	0.17	2,209,577.00 (29,432.18) *	213,700.00 (2846.54) *	2,423,277.00 (32,279.16) *	5.96
Herbal medicine	63.00	1278.53 (17.03) *	0.66	4,899,349.00 (65,260.70) *	768,400.00 (10,235.30) *	5,667,749.00 (75,497.02) *	21.45
Total	-		-	8,611,614.31 (114,709.11) *	3,581,790.00 (47,710.45) *	12,193,404.00 (162,421.75) *	100.00
Average	-		-	86,116.14 (1147.09) *	35,817.90 (477.10) *	121,934.00 (1624.22) *	-

* USD 1 = INR 75.08 as on 30 January 2022.

3.5. Contribution of NTFPs to Local Household Economy

The involvement of households in various economic activities is presented in Table 6. It was observed that NTFPs were the largest source of income across all the categories, with 53.33% contribution to the household income, followed by labour (15.27%), goat/sheep husbandry (11.46%), dairy (9.85%), and agricultural crops (6.80%) respectively. The art and crafts (1.49%), horticulture (1.46%) and service (0.34%) had significantly low contribution in the local economy.

Table 6. Contribution of NTFPs in household economy (N = 100).

Sources	Total Income (INR/Year) USD *	Average Income (INR/Year) USD *	Std. Dev.	Percentage
Agricultural crops	456,600.34 (6082.41) *	4566.00 (60.82) *	10,703.07	6.80
Horticulture	98,100.45 (1306.80) *	981.01 (13.07) *	2506.63	1.46
Dairy	661,200.28 (8807.90) *	6612.00 (88.08) *	13,315.71	9.85
Goat/sheep husbandry	770,000.42 (10,257.24) *	7700.00 (102.57) *	16,505.58	11.46
Labour	1,025,500.56 (13,660.77) *	10,255.01 (136.61) *	15,509.11	15.27
Art and craft	100,000.78 (1332.12) *	1000.01 (13.32) *	6590.47	1.49
NTFPs	3,581,790.56 (47,713.31) *	35,817.91 (477.10) *	37,310.56	53.33
Service	23,054.72 (307.11) *	230.55 (3.07) *	2019.32	0.34
Total	6,716,298.00 (89,468.34) *	67,122.44 (894.02) *	56,622.46	100.00

* USD 1 = INR 75.08 as on 30 January 2022.

The coefficient of correlation (r) was worked out to ascertain the relationship between the livelihood dependency on NTFPs and the socioeconomic characteristics of the sample households (Figure 6). Out of twelve socioeconomic characteristics of the people, eight characteristics—viz., education, social membership, household size, household labour, farm size, livestock ownership, age, proximity to forest and forest visits—exhibited positive and significant correlations with the livelihood dependency on NTFPs. By contrast, the characteristics of main occupation, wealth status and gross annual income had significant negative correlations with the livelihood dependency on NTFPs.

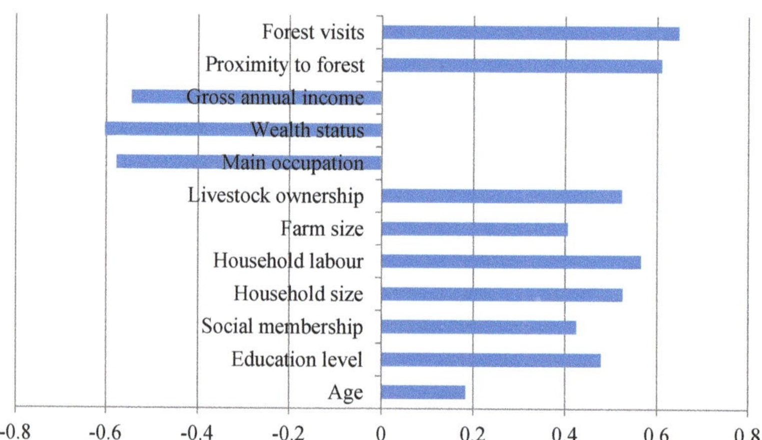

Figure 6. Correlation between household variables and NTFP income dependence (N = 100).

3.6. OLS Regression Model

The OLS regression analysis was carried out to determine the household dependence from NTFP income against household variables (Table 7). The coefficient values were determined for household variables including age (170.31), education level (3020.36), social membership (287.32), household size (2410.22), household labour (−195.03), farm size (10,287.28), livestock ownership (3057.57), main occupation (−4193.87), wealth status (−5895.07), gross annual income (0.00), proximity to forests (35,967.92) and forest visits (8835.33). The "*t*" values of the regression coefficients indicate that, out of twelve household variables, proximity to forests (5.25), forest visits (2.17), education level (0.14) and farm size (0.12) had significant influences on NTFP income levels. The coefficient of determination (R^2) 0.70 indicates that the explanatory variables contributed to 70.40% of the variation in household NTFP income. The degree of the F value (17.22) indicates that R^2 is statistically significant ($p < 0.05$), which establishes that the model is reliable and well prognostic. The OLS regression equation appropriated for the household NTFP income may be written as:

$$Y = 23{,}575.17 - 170.31\,X_1 + 3020.36\,X_2 + 287.32\,X_3 + 2410.22\,X_4 - 2410.22\,X_5 + 10{,}287.28\,X_6 + 3057.57\,X_7 - 4193.87\,X_8 - 5895.07\,X_9 + 0.000\,X_{10} + 35{,}967.92\,X_{11} + 8835.33\,X_{12}$$

where Y is household NTFP income (INR/year) and X_1–X_{12} are socioeconomic variables.

Table 7. OLS regression model of household NTFP income dependence against household variables.

Variables (Code)	Coefficient (b)	Standard Error of b	B	*t* Value	*p*
Age (X_1)	170.31	162.20	0.067	1.05	0.29
Education level (X_2)	3020.36	1695.03	0.14	1.78	0.07
Social membership (X_3)	287.32	3799.70	0.00	0.07	0.94

Table 7. *Cont.*

Variables (Code)	Coefficient (b)	Standard Error of b	B	*t* Value	*p*
Household size (X_4)	2410.22	16,275.12	0.03	0.14	0.88
Household labour (X_5)	−195.03	5124.99	0.00	−0.03	0.97
Farm size (X_6)	10,287.28	11,438.93	0.12	0.89	0.37
Livestock ownership (X_7)	3057.57	4195.01	0.09	0.72	0.46
Main occupation (X_8)	−4193.87	2502.01	−0.14	−1.67	0.09
Wealth status (X_9)	−5895.07	4744.92	−0.11	−1.24	0.21
Gross annual income (X_{10})	0.00	0.08	0.00	0.00	0.99
Proximity to forest (X_{11})	35,967.92	6841.32	0.36	5.25	0.00
Forest visits (X_{12})	8835.33	4064.83	0.19	2.17	0.03

A = −23,575.17; F = 17.22 *; R^2 =0.70; multiple R = 0.83; adjusted R^2 =0.66; * = significant at 5% level of probability.

The regression coefficient indicates that the proximity to forests, forest visits, education level and farm size of the NTFP collectors made a significant economic contribution to the households.

3.7. Income Inequality Mitigation by NTFPs

The income inequality mitigation potential of NTFPs was determined by the Lorenz curve. The study revealed that the household income without NTFPs deviated more from the line of equality than the Lorenz curve of the total household income (Figure 7). Similarly, the Ginicoefficient for the household income with NTFPs was 0.28, and 0.57 without the NTFP income. This means that NTFP income contributed to mitigating income inequalities among the households by 29.27%. Therefore, the values of Ginicoefficient and departure of the Lorenz curve from the line of equality clearly indicate that the NTFPs mitigated the income inequality significantly among the sampled households and had a substantial equalising effect on the total income distribution.

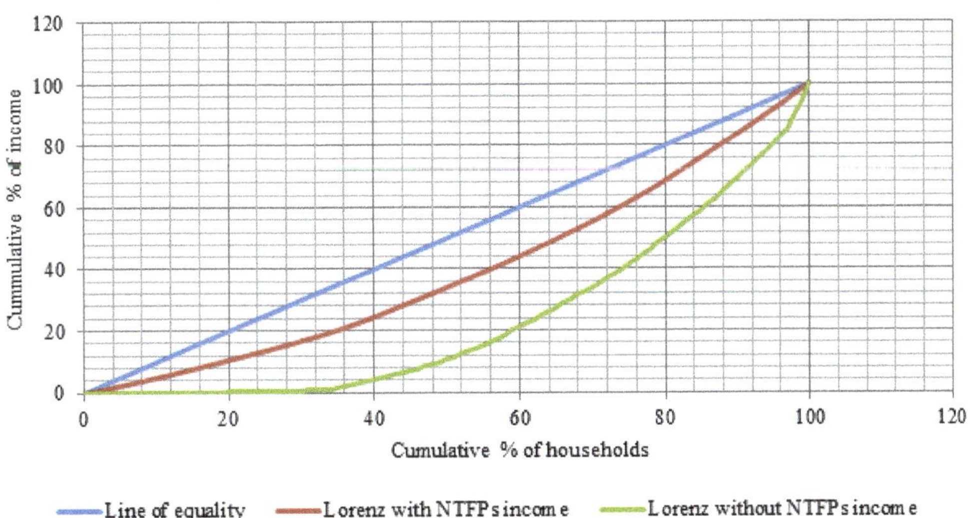

Figure 7. Lorenz curve of household income including and excluding NTFP income (N = 100).

4. Discussion

NTFP collection significantly contributed to the cash and subsistence income and the economic inequality mitigation among rural households. The entire sample of households derived substantial parts of their household annual income from NTFP-based activities. The absolute NTFP income was estimated to be INR 35,817.91 (USD 477.10), contributing

53.33% of the total annual household income and equalising economic inequalities by 29.27%. NTFP income constituted the first most viable income source because the alternative options were either scarce or even absent. The earnings from NTFPs enabled the people to purchase daily necessities, secure livelihood perspectives, create stock capital for income diversification and preserve that as savings to cope with adversity. Nonetheless, different NTFPs accrued different levels of income; the collectors reported fuel wood, medicinal plants, mushrooms, fodder, fruits and vegetables as the most important NTFPs for both subsistence and cash income. These NTFPs had more market demand and income generating opportunity than the others. The higher quantity of fuel wood collection and trade was only due to a lack of low-priced substitute energy sources. The higher collection of medicinal plants was due to its immense demand in traditional health-care systems and handsome sale return. The sale of fodder grasses was a forest-based self-employment of local people because livestock production was a major subsidiary occupation, and fodder security is a challenge due to a short growing season. Wild fruits and vegetables were prominent NTFPs consumed for food and nutritional security and sold by the primary collectors for revenue. The involvement of households in collection and marketing varied by NTFP type and availability, consumption requirement, market value and socioeconomic conditions of people. The cash incomes from NTFPs were variable across households and directly related to the degree of time and labour expended.

The correlation and OLS regression results indicate that household NTFP income was influenced by all explanatory variables, except the household head's age. The positive effect of education on household NTFP income is well articulated by the facts that, as low literacy prevailed among the NTFP collectors, ranging from illiterate to primary level, the more educated households had better awareness, skill bases and access to markets which accrued more NTFP return. Social participation facilitates information flow, sharing views and experiences, clarifying doubts, getting opinions and enriching knowledge among members in a social group; hence, this factor had a significant positive influence on household NTFP income. The findings also indicated that household size and labour significantly influenced the collection of NTFPs by the households. This is because the larger families had a greater labour force to support more NTFP extraction from the forests. The positive effect of farm size on NTFPs could be attributed to the fact that the households had limited farmland and were unable to produce sufficient food for their families; hence, they relied heavily on NTFPs for their food security, safety net and cash income. Similarly, livestock ownership had a positive effect on NTFP income. This is due to limited size of land holding, low fodder production, lack of grazing lands and heavy demand for fodder as safety nets. The economic attributes—viz., main occupation, wealth status and gross annual income— exhibited negative effects on the NTFP income. These factors were the major indicators of, and core contributors to, the household economic conditions that helped them to facilitate the other types of capital to be owned and traded. Thus, the households with higher occupation, wealth status and gross annual income had more financial opportunities, more earnings and less dependency on NTFP income. The involvement of household heads of different age groups in NTFP collection was more or less similar, indicating that the variation in age had no influence at all on the household NTFP income.

The descriptive analysis shows that the dependence of households on NTFPs was significant relative to other sources of livelihood such as agricultural crops, horticulture, dairy, goat/sheep husbandry, wage labour, arts, crafts and service. The households in the study area collected mostly medicinal plants, food and fuelwood. These results are consistent with other studies in the area [40], which highlighted the traditional use of plants for various purposes. However, the extent of NTFP collection and use differed widely across households. The possibility of household participation and intensity of involvement in NTFP collection was higher among households living near the forests. As the distance to the forest increased, the NTFP collection decreased. This is apparent due to the accessibility of the NTFP resource rich areas in the forests where households have to travel shorter distances. The proximity to forests, education level and farm size had positive effects on the household income compared to other activities, such as service, possibly due to multiple sources of income, such as forests, as well as farms, and higher levels of education gave them more power to negotiate NTFPs' prices in markets. Nearness to forests meant easy access to a number of NTFPs in a short span of time, and, hence, lower labour costs in the collection and transportation of the produce, and higher income. The results indicate that NTFP collectors lived in inaccessible areas, had poor socioeconomic conditions; and were mostly without access to services, provisions and government developmental schemes. Hence, they had inadequate access to employment, health and other welfare schemes of the government, resulting in low economic wellbeing. Therefore, the diversification of NTFP-based livelihoods has a great scope in these areas for improving the quality of life and human wellbeing from nature's contributions. This was also confirmed by the findings of other studies in other geographies [5,24,40–42].

The collection of NTFPs is a viable source of a subsistence livelihood, income and safety net option across the forest fringe communities of mountain areas, especially in Kashmir, due to a lack of any other sustainable income generation alternative. The income generated from NTFPs may not be the primary source of livelihood; however, domestic consumption in the form of fuel wood, fruits, vegetables and medicinal plants make a significant contribution to the subsistence of almost all households. Moreover, the income derived from NTFPs is a significant source of other domestic necessities, such as educating children, health, paying debt or providing a safety net against hardships during the severe winter months prominent in Kashmir. Our results are supported by other studies [5,11,43–46] as well, which show how people living in rural settings are dependent on NTFPs more for subsistence use than trade.

As indicated by the Lorenz curve, NTFPs play a significant role in income inequality mitigation and a safety net for underprivileged forest fringe communities. Therefore, transition to a NTFP-based bioeconomy has the potential to improve the local socioeconomic status, if recognised and managed properly. NTFPs consumed by households in forest fringe communities have not been fully accounted for. They have much greater worth than NTFPs traded in the local markets. Despite such an enormous contribution to local economy, the contributions of NTFPs in Kashmir have never been considered enough to be accounted for by the authorities. This has obstructed support for the potential NTFP-based bioeconomy in the region. Proper valuation of resources being extracted, consumed and traded at the local level must be included in the regional and national statistics for realising the actual potential of a NTFP-based bioeconomy. The results of this study are substantiated by the literature on the sustainable bioeconomy potential of NTFPs accumulated to date [47,48].

Indefinite harvesting of NTFPs from wild, without proper harvesting and management practices will negatively impact the sustainability and yield of the species. Therefore, the production of NTFPs through both *in-situ* and *ex-situ* mechanism is the only way forward. The management of NTFPs should be included in the forest working plans with amplified investments in the sector. The sustainable use of NTFPs has proven to be economical for local communities [49], and has the potential to enhance socio-ecological security in these multifunctional landscapes. On the other hand, the increasing industrial demand for NTFP-based diversified bio-products in emerging global markets can provide significant opportunities for NTFP-driven bioeconomies. The diversified bioactive compounds and genes of interest extracted from NTFPs have brought revolutions in pharmaceutical, nutraceutical, cosmeceutical, food and beverage industries [50]. Therefore, it is necessary to recognise the contribution of NTFPs in both the local as well as state economy. The diversification of raw materials, with an emphasis on production, processing and the establishment of an inclusive value chain, will significantly augment the livelihoods of, and mitigate income inequality for, forest fringe communities. It is equally important to have a shift from local to global value chains, in order to promote the economic value of NTFPs, from raw materials to end products, by adopting contemporary visions of bioeconomy [16]. The processing of NTFPs can positively influence the sustainable economic development of local-forest-dependent communities in J&K. However, despite local interest and the potential contributions of NTFPs as key sources of livelihood diversification and sustainable development, NTFP processing enterprises are still in the informal sector, and there is a tremendous lack of understanding of the underlying factors. Hence, transitioning from local NTFP commercialisation efforts to developing NTFP-based value chains that can help locals approach the export markets, and enhancing cooperation for a supportive institutional framework, are very much required. A largely successful NTFP-based bioeconomy can be supported by local socio-economics and ecological conditions that require more holistic approaches, which can address and support the local context and NTFP value chains. Hence, this can support the UT government to facilitate and accomplish its sustainable and equitable development goals by promoting an NTFP-based bioeconomy in J&K [16].

5. Conclusions

The findings of this study indicate that NTFPs make a significant contribution to supporting the subsistence use and income generation of households. People dwelling in remote and inaccessible areas of Kashmir, where market supplies are not organised, still extract fuelwood, fodder, vegetables, fruits, mushrooms and medicinal plants from the nearby forests for their consumption and income generation. The results presented in the study stress that NTFPs make a significant contribution to income inequality mitigation for the forest fringe communities. However, to realise the full potential of NTFPs, it is important that they are commercialised as a diversified product through a bioeconomy that ensures the sustainable use of wild species, following sustainable livelihoods, income generation and inclusive development. It is very important that NTFPs are managed properly and included in the forest working plans for commercial production of high value species. The contribution of NTFPs to local and regional economy must be considered in the state and national GDP and GNI calculations. The government can play an important support role in this context, where markets do not work inclusively. An inclusive economy includes resources which lack enough markets to manage supply and demand, and addresses the issues of under-delivery, non-reporting and overexploitation. Basically, these NTFPs have historically been an integral part of the day-to-day needs and traditional lifestyle of indigenous people and local communities. Progressive policies on forest resource management and trade must be interactive in nature and should acknowledge the local rights, knowledge and practices to ensure access and concessions for sustainable harvesting of NTFPs for socio-ecological and economic well-being. Livelihood promotion and income diversification for local communities may need sufficient support from the government to encourage a shift to an NTFP-based bioeconomy to keep pace with current and future development challenges in the region.

Author Contributions: Conceptualisation, I.A.P., S.S.; methodology, I.A.P. and M.A.I.; investigation, I.A.P.; data collection and interviewing, I.A.P.; writing—original draft preparation, I.A.P.; commenting, reviewing and suggesting improvements, M.A.I., J.C., S.D., S.S. and M.R.; revising and editing, J.C., S.D., S.S., M.R. and I.A.P.; visualisation and formatting, I.A.P., J.C., S.S. and M.A.I.; supervision and project administration, I.A.P. All authors have read and agreed to the published version of the manuscript.

Funding: This research was funded by the Scientific & Engineering Research Board, Government of India through its prestigious "Young Scientist Scheme, project ID SR/FT/LS-55/2012. We are thankful to the Open Access Publication Fund of the Karlsruhe Institute of Technology, Germany to cover the article processing charges.

Institutional Review Board Statement: Not applicable.

Informed Consent Statement: Written informed consent was obtained from all respondents involved in the study.

Data Availability Statement: The data presented in this study are available on request from the corresponding author. The data are not publicly available due to issues of respondent confidentiality and conditions of ethical approval.

Acknowledgments: The authors are grateful to the forest fringe communities, officials of the J&K Forest Department and traders of Langate Forest Division for sharing information during surveys. Thanks are also due to the 21 RR of the Indian Army for providing the necessary support during field visits. The constructive comments and suggestions of the anonymous learned reviewers for updating the manuscript are thankfully acknowledged.

Conflicts of Interest: The authors declare no conflict of interest.

References

1. de Beer, J.H.; McDermott, M. *The Economic Value of Non-Timber Forest Products in South–East Asia*; Committee for IUCN: Amsterdam, The Netherlands, 1989. Available online: https://portals.iucn.org/library/sites/library/files/documents/1996-020.pdf (accessed on 24 January 2020).
2. Unasylva. Towards a Harmonized Definition of Non-Wood Forest Products, FAO Forestry. 1999. Available online: https://www.fao.org/3/x2450e/x2450e0d.htm#fao%20forestry (accessed on 24 January 2020).
3. FAO. Forest Resource Assessment. 2015. Available online: https://www.fao.org/3/ap862e/ap862e00.pdf (accessed on 24 January 2020).
4. Shiva, M.P.; Verma, S.K. *Approaches to Sustainable Forest Management and Biodiversity Conservation: With Pivotal Role of Non-Timber Forest Products*; Centre for Minor Forest Products, Valley Offset Printers: DehraDun, India, 2002.
5. Talukdar, N.R.; Parthankar, C.; Barbhuiyaa, R.A.; Singh, B. Importance of Non-Timber Forest Products (NTFPs) in rural livelihood: A study in Patharia Hills Reserve Forest, northeast India. *Trees For. People* **2021**, *3*, 100042. [CrossRef]
6. Marko, L.; Riccardo, D.R.; Enrico, V.; Irina, P.; Jennifer, W.; Davide, P.; Pieter, J.V.; Robert, M. Collection and consumption of non-wood forest products in Europe. *For. Int. J. For. Res.* **2021**, *94*, 757–770. [CrossRef]
7. Shackleton, C.M.; Pandey, A.K. Positioning non-timber forest products on the development agenda. *For. Policy Econ.* **2014**, *38*, 1–7. [CrossRef]
8. Shanley, P.; Alan, R.P.; Laird, S.A.; Citlalli, L.B.; Manuel, R.G. From Lifelines to Livelihoods: Non-timber Forest Products into the Twenty-First Century. In *Tropical Forestry Handbook*; Pancel, L., Köhl, M., Eds.; Springer: Berlin/Heidelberg, Gemany, 2015. [CrossRef]
9. Dash, M.; Bhagirath, B.; Dil, B.R. Determinants of household collection of non-timber forest products (NTFPs) and alternative livelihood activities in Similipal Tiger Reserve, India. *For. Policy Econ.* **2016**, *73*, 215–228. [CrossRef]
10. FAO. *Assessing the Contribution of Bioeconomy to Countries' Economy: A Brief Review of National Frameworks*; Food and Agriculture Organisation of United Nations: Rome, Italy, 2018; 67p, Available online: http://www.fao.org/3/I9580EN/i9580en.pdf (accessed on 12 July 2020).
11. Mukul, S.A.; Manzoor, R.A.Z.M.; Mohammad, B.; Niaz, A.K. Role of non-timber forest products in sustaining forest-based livelihoods and rural households' resilience capacity in and around protected area: A Bangladesh study. *J. Environ. Plan. Manag.* **2016**, *59*, 628642. [CrossRef]
12. Arnold, J.E.M.; Ruiz-Pérez, M. Can non-timber forest products match tropical conservation and development objectives? *Ecol. Econ.* **2001**, *39*, 437–447. [CrossRef]
13. Shackleton, C.; Shackleton, S. The importance of non-timber forest products in rural livelihood security and as safety nets: A review of evidence from South Africa. *S. Afr. J. Sci.* **2004**, *100*, 658–664.
14. Nguyen, T.V.; Jie, H.L.; Thi, T.H.V.; Bin, Z. Determinants of Non-Timber Forest Product Planting, Development, and Trading: Case Study in Central Vietnam. *Forests* **2020**, *11*, 116. [CrossRef]
15. Pandey, A.K.; Tripathi, Y.C.; Kumar, A. Non timber forest products (NTFPs) for sustained livelihood: Challenges and strategies. *Res. J. For.* **2016**, *10*, 1–7. [CrossRef]
16. Peerzada, I.A.; Chamberlain, J.; Reddy, M.; Dhyani, S.; Saha, S. Policy and Governance Implications for Transition to NTFP-Based Bioeconomy in Kashmir Himalayas. *Sustainability* **2021**, *13*, 11811. [CrossRef]
17. Gangoo, S.A.; Islam, M.A.; Tahir, M. Wealth of Non Timber Forest Products and their Trade in Jammu and Kashmir. *Indian For.* **2017**, *143*, 827–833.
18. MEA. *Ecosystem and Human Well-being: A Framework for Assessment, Millennium Ecosystem Assessment Series*; Island Press: Washington, DC, USA, 2003; Available online: https://pdf.wri.org/ecosystems_human_wellbeing.pdf (accessed on 12 July 2020).
19. Muir, G.; Simona, S. Making NWFPS Visible: Disentangling Definitions and Refining Methodologies. In Proceedings of the HQ Agenda Item 10.2: Overview of Forestry and Environment Statistics in AP Region, 27th Session, Port Denarau, Fiji, 19–23 March 2018.
20. Haripriya, G.S. Integrated environmental and economic accounting: An application to the forest resources in India. *Environ. Res. Econ.* **2001**, *19*, 73–95. [CrossRef]
21. World Bank. *A Revised Forest Strategy for the World Bank Group*; World Bank: Washington, DC, USA, 2001.

22. CBD. *Linking Biodiversity Conservation and Poverty Alleviation: A State of Knowledge Review*; CBD Technical Series No: 55; Secretariat of the Convention on Biological Diversity: Montreal, QC, Canada, 2010.
23. Tugume, P.; Mukadasi, B.; Esezah, K.K. Non-Timber Forest Products Markets: Actors and Income Determinants. *Res. J. For.* **2019**, *6*, 1–19.
24. Sorrenti, S. *Non-Wood Forest Products in International Statistical Systems*; Non-wood Forest Products Series no. 22; FAO: Rome, Italy, 2017.
25. Arnold, J.E.M.; Ruiz-Perez, M. The Role of Non-Timber Forest Products in Conservation and Development. In *Incomes from the Forest: Methods for the Development and Conservation of Forest Products for Local Communities*; Wollenberg, E., Ingles, A., Eds.; CIFOR/IUCN: Bogor, Indonesia, 1998; pp. 17–42. Available online: https://www.cifor.org/knowledge/publication/479 (accessed on 12 July 2020).
26. FAO. *Contribution of the Forest Sector to National Economies, 1999–2011*; Lebedys, A., Li, T., Eds.; Forest Finance working paper; Food and Agriculture Organization of the United Nations: Rome, Italy, 2014.
27. Adamseged, M.E.; Philipp, G. Understanding Business Environments and Success Factors for Emerging Bioeconomy Enterprises through a Comprehensive Analytical Framework. *Sustainability* **2020**, *12*, 9018. [CrossRef]
28. Ros-Tonen, M.A.F.; Wiersum, K.F. The scope for improving rural livelihoods through non-timber forest products: An evolving research agenda. *For. Trees Livelihoods* **2005**, *15*, 129–148. [CrossRef]
29. Iponga, D.M.; Christian, M.Y.; Guillaume, L.; Fide'le, M.A.; Patrice, L.; Julius, C.T.; Alfred, N. The contribution of NTFP-gathering to rural people's livelihoods around two timber concessions in Gabon. *Agrofor. Syst.* **2018**, *92*, 157–168. [CrossRef]
30. ISFR. Indian State of Forest Report (ISFR) 2021, Forest Survey of India, Ministry of Environment, Forests and Climate Change, Government of India. Available online: https://fsi.nic.in/isfr-2021/chapter-13.pdf (accessed on 24 January 2020).
31. Damien, S.M.; Mark, F. MCD12Q1 MODIS/Terra+Aqua Land Cover Type Yearly L3 Global 500m SIN Grid V006 [Data set]. NASA EOSDIS Land Processes DAAC. 2019. Available online: https://doi.org/10.5067/MODIS/MCD12Q1.006 (accessed on 11 October 2021).
32. Ray, G.L.; Mondol, S. *Research Methods in Social Sciences and Extension Education*; Kalyani Publishers: New Delhi, India, 2004; pp. 66–76.
33. Mukherjee, N. *Participatory Rural Appraisal. Methodology and Applications*; Concept Publishing Company: Delhi, India, 1993.
34. Islam, M.A.; Wani, A.A.; Bhat, G.M.; Gatoo, A.A.; Shah, M.; Ummar, A.; Shah, S.S.G.S. Economic Contribution and Inequality Mitigation of Wicker Handicraft Entrepreneurship in Rural Kashmir, India. *Cur. J. Appl. Sci. Technol.* **2020**, *39*, 138–149. [CrossRef]
35. Snedecor, G.; Cochran, W.G. *Statistical Methods*; Iowa State University Press: Ames, IA, USA, 1967; pp. 17–36.
36. Lorenz, M.O. Methods of measuring the concentration of wealth. *Publ. Am. Stat. Assoc.* **1905**, *9*, 209–219. [CrossRef]
37. Gini, C. Measurement of inequality of incomes. *Econ. J.* **1921**, *31*, 124–126. [CrossRef]
38. Adeline, M.; Patrice, L.; Julius, C.T. The role of forest resources in income inequality in Cameroon. *For. Trees Livelihoods* **2017**, *26*, 271–285. [CrossRef]
39. Gujarati, D.N.; Porter, D.C. *Basic Econometrics*; Tata McGraw-Hill Publishing Company Limited: New Delhi, India, 2007.
40. Peerzada, I.A.; Sofi, A. Traditional use of medicinal plants among tribal communities of Bangus Valley, Kashmir Himalaya, India. *Stud. Ethno-Med.* **2017**, *11*, 318–331.
41. Fikir, D.; Tadesse, W.; Gure, A. Economic contribution to local livelihoods and households dependency on dry land forest products in Hammer District, Southeastern Ethiopia. *Int. J. For. Res.* **2016**, *2*, 11–22. [CrossRef]
42. Sharma, K.; Sharma, R.; Devi, N. Non-Timber Forest Products (NTFPs) and livelihood security: An economic study of high hill temperate zone households of Himachal Pradesh. *Econ. Aff.* **2019**, *64*, 305–315. [CrossRef]
43. Partakson, R.C.; Tomba, S.K.H. Forest based industry in Churachandpur district: From the entrepreneurial perspective. *J. Glob. Econ.* **2017**, *5*, 1–6. [CrossRef]
44. Baishya, R.A.; Begum, A. Promotion of rural livelihood through medicinal and aromatic plants based cottage industries for upliftment of rural economy of Assam. *Open Access Sci. Rep.* **2013**, *2*, 1–4. [CrossRef]
45. Ingram, V.; Ndoye, O.; Iponga, D.; Tieguhong, J.C.; Nasi, R. Non-timber forest products: Contribution to national economy and strategies for sustainable management. In *The Forests of the Congo Basin–State of the Forest 2010*; de Wasseige, C., de Marcken, P., Bayol, N., HiolHiol, F., Mayaux, P., Desclée, B., Nasi, R., Billand, A., Defourny, P., Eba'a, R., Eds.; Publications Office of the European Union: Luxembourg, 2012; pp. 137–154.
46. Banerjee, A.K.; Saha, S. Role of women in joint forest management at Arabari Forest (East Midnapore forest division) of South-West Bengal. *Int. J. For. Usufruct. Manag.* **2004**, *5*, 56–64.
47. Ahenkan, A.; Boon, E. Enhancing Food Security, Poverty Reduction and Sustainable Forest Management in Ghana through NonTimber Forest Products Farming: Case Study of SefwiWiawso District. 2008. Available online: www.grin.com/de/preview/ .html (accessed on 12 July 2020).
48. Bennett, B.C. Plants and people of the Amazonian rainforests. *BioScience* **1992**, *42*, 599–607. [CrossRef]
49. Mbuvi, D.; Boon, E. The livelihood potential of nonwood forest products: The case of Mbooni Division in Makueni District, Kenya. *Environ. Dev. Sustain.* **2008**, *11*, 989–1004. [CrossRef]
50. de Beer, J. On forest foods, a festival and community empowerment. *CFA NewsLett* **2011**, *54*, 1–5.

sustainability

Article

Versatile Green Processing for Recovery of Phenolic Compounds from Natural Product Extracts towards Bioeconomy and Cascade Utilization for Waste Valorization on the Example of Cocoa Bean Shell (CBS)

Christoph Jensch, Axel Schmidt and Jochen Strube *

Institute for Separation and Process Technology, Clausthal University of Technology, Leibnizstr. 15, 38678 Clausthal-Zellerfeld, Germany; jensch@itv.tu-clausthal.de (C.J.); schmidt@itv.tu-clausthal.de (A.S.)
* Correspondence: strube@itv.tu-clausthal.de; Tel.: +49-5323-72-2872

Abstract: In the context of bioeconomic research approaches, a cascade use of plant raw materials makes sense in many cases for waste valorization. This not only guarantees that the raw material is used as completely as possible, but also offers the possibility of using its by-products and residual flows profitably. To make such cascade uses as efficient as possible, efficient and environmentally friendly processes are needed. To exemplify the versatile method, e.g., every year 675,000 metric tons of cocoa bean shell (CBS) accrues as a waste stream in the food processing industry worldwide. A novel green process reaches very high yields of up to 100% in one extraction stage, ensures low consumption of organic solvents due to double usage of ethanol as the only organic solvent, is adaptable enough to capture all kinds of secondary metabolites from hot water extracts and ensures the usage of structural carbohydrates from precipitation. A Design of Experiments (DoE) was conducted to optimize the influence of pH value and phase ratio on the yield and purity of the integrated ethanol/water/salt aqueous-two-phase extraction (ATPS) system.

Keywords: aqueous two-phase systems (ATPS); liquid–liquid extraction (LLE); pressurized hot water extraction (PHWE); solid–liquid extraction (SLE); natural products; cocoa bean shell (CBS); precipitation; bioeconomy; total phenolic content (TPC)

Citation: Jensch, C.; Schmidt, A.; Strube, J. Versatile Green Processing for Recovery of Phenolic Compounds from Natural Product Extracts towards Bioeconomy and Cascade Utilization for Waste Valorization on the Example of Cocoa Bean Shell (CBS). *Sustainability* **2022**, *14*, 3126. https://doi.org/10.3390/su14053126

Academic Editors: Nallapaneni Manoj Kumar, Md Ariful Haque and Sarif Patwary

Received: 25 January 2022
Accepted: 3 March 2022
Published: 7 March 2022

Publisher's Note: MDPI stays neutral with regard to jurisdictional claims in published maps and institutional affiliations.

1. Introduction

Products based on renewable resources, such as plants, represent a growing market and the associated industry is an important supplier of versatile products. Applications include pharmaceutical products, the food, health and nutrition sectors, as well as plant protection for ecological farming or construction materials, basic chemicals and energy resources [1–7]. In the context of bioeconomic research approaches, a cascade use of plant raw materials makes sense in many cases. This not only guarantees that the raw material is used as completely as possible, but also offers the possibility of using its by-products and residual flows profitably. To make such cascade uses as efficient as possible, efficient and environmentally friendly processes are needed.

Pressurized hot water extraction (PHWE) has been studied in the solid–liquid extraction community for a while and is well-established [8–11]. One of the main advantages is the utilization of water as a solvent instead of organic solvents, which can help to reduce the cost of goods (COGs) and global warming potential (GWP) and therefore help to reach climate neutrality goals [12]. Additionally, PHWE extracts consist of the whole spectrum of components in the plant material. Besides secondary metabolites, such as polyphenols and flavonoids, matrix components such as lignin, cellulose and proteins are extracted [13]. However, if there are processing steps after the solid–liquid extraction, they are often associated with the usage of different organic solvents. Possible steps are precipitation, liquid–liquid extraction or chromatography, which usually come with organic solvents such as ethanol for precipitation, acetonitrile in chromatography or phase-forming solvents such as ethyl/butyl acetate in liquid–liquid extraction (LLE) [14,15].

Here, a promising approach is to reuse the ethanol, which is needed for the precipitation of glucans, in an ethanol/water/salt ATPS to recover the polyphenols from the precipitated extract [16,17]. The precipitate consists mainly of matrix components which can be further processed for usage in a variety of applications [18–22].

Every year, 675 kt of cocoa bean shell (CBS) accrues as a waste stream in the food processing industry worldwide [20] and has high research interest [20,23–31]. As a lignocellulosic biomass, CBS is a potential resource for β-glucans which can be utilized, for example, as bonding or binding agents. Besides β-glucans, CBS contains a high amount of polyphenols such as catechin and epicatechin [23,28] and methylxanthines such as theobromine and caffeine. The development of process routes for the use of co-products is essential for the economic viability of these processes [32,33]. To utilize the full potential of the CBS in the interest of bioeconomy, a process is developed using only minimal amounts ethanol as the only organic solvent. The process consists of an extraction with hot water, a precipitation with ethanol as an anti-solvent and a liquid–liquid extraction from the precipitation supernatant with salting-out of ethanol. The LLE is compared to a conventional LLE with ethyl acetate and butyl acetate. As the organic solvent LLE has to be conducted with an aqueous phase, the LLE has to be either conducted before precipitation or after precipitation and removal of the ethanol. This results in three possible process configurations, which are shown in Figure 1. The first process is the one with double utilization of ethanol from precipitation in an aqueous two-phase extraction (ATPE) with salt.

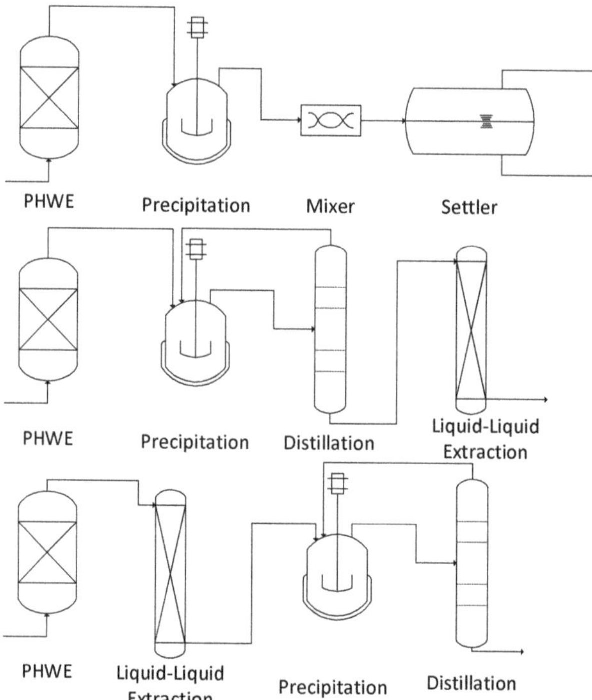

Figure 1. Overview of the three possible process configurations.

2. Materials and Methods

2.1. Extraction Setup

For temperature screening, 1 g of ground CBS was extracted in a 10 mL extraction column with a flow of 1 mL/min. The extraction plant consists of a pump, a GC oven for heating, a column for extraction and a water bath for cooling. For maximum yield, 120 mL of extract was collected as fractions.

For the solvent screening, 2 g of ground CBS is extracted with 40 mL of the respective solvent: 20–100% ethanol, methanol, iso-propanol, MTBE, butyl acetate, ethyl acetate, toluene and hexane. The vials were placed on a shaking device for 24 h to reach extraction equilibrium. All solvents were purchased at VWR International GmbH, 30163 Hannover, Germany.

The extraction was carried out as a pressurized hot water/liquid hot water/subcritical water extraction (PHWE). The extraction plant consists of an extraction column with a volume of 0.1 L, solvent vessel, heating unit with a heat exchanger, cooling unit with a heat exchanger, and an extract vessel. Extraction conditions were 140 °C at 1 L/h. The CBS was obtained as dried industrial waste from a project partner. The origin was not further specified. It was ground and sieved to 630–2000 µm, which is small enough to guarantee fast mass transfer but big enough to prevent blocking of the extraction column. The extraction column was filled with 20 g of ground plant material, and 500 mL of extract was collected. The extraction solvent was deionized water from an in-house deionization plant. The solvent ratio originates from the temperature screening, in which total yield was reached after a solvent ratio of 25.

2.2. Liquid–Liquid Extraction

The phase screenings for liquid–liquid extraction were conducted in 50 mL centrifugal vials, supplied by VWR International GmbH, 30163 Hannover, Germany. The extract/supernatant and solvent were measured into the centrifugal vials in the respective phase ratios. For the first LLE screenings for partition coefficients, the phase ratio was 50/50. The vials were placed on a shaking device for 2 h to reach extraction equilibrium.

The extracts used were collected as described in Section 2.1. For the ATPS, the extracts were precipitated with 60 wt.% ethanol, and supernatant was used for LLE. For the ATPS, a 40 wt.% citrate buffer and a 30 wt.% phosphate buffer were used at the respective pH values. Due to poor solubility properties at low pH values, phosphate buffers were only researched at pH 7 and pH 8. The salts were purchased at VWR International GmbH, 30163 Hannover, Germany.

2.3. Analytics

Offline analytics consists of three different analytic methods. For the target components theobromine, caffeine, catechin and epicatechin, reversed-phase high-performance liquid chromatography (RP-HPLC) analytics were conducted. The method is modified based on Rojo-Poveda et al. [28]. For detection, a diode array detector (DAD) is used, which detects theobromine and caffeine at 272 nm and catechin and epicatechin at 280 nm. For the separation, a Kinetex Phenyl-Hexyl C18 column (150 mm length × 4.6 mm internal diameter and 5 µm particle size; Phenomenex, Aschaffenburg, Germany) is used. Injection volume was 10 µL. The gradient consists of 0.1% formic acid as solvent A and methanol as solvent B with a flowrate of 1 mL/min. The elution program starts with 10% solvent B up to 12.5 min. The gradient reaches 80% solvent B at 37.5 min and a step up to 90% solvent B up to 42.5 min. For equilibration, the partition of solvent B is 10% from 42.5 to 45 min.

The column is heated to 35 °C. Methanol and formic acid in HPLC grade were bought from VWR International GmbH, 30163 Hannover, Germany. For calibration, theobromine, caffeine, catechin and epicatechin standards of the concentrations between 0.02 and 1 g/L from Sigma-Aldrich, St. Louis, MO, USA were used. The calibrations for HPLC analysis and Folin–Ciocalteu test are shown in Figure 2.

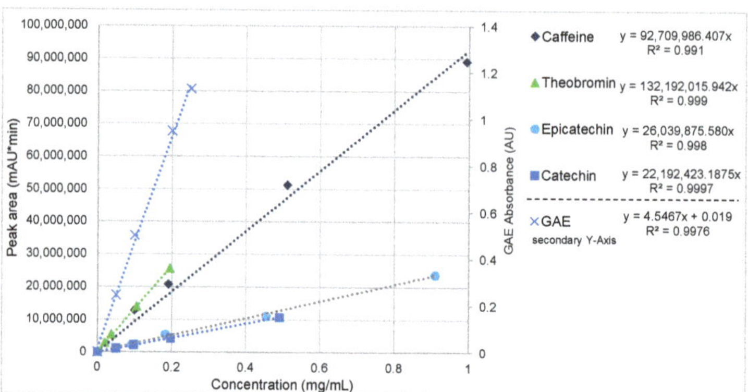

Figure 2. Calibration curves for caffeine, theobromine, epicatechin, catechin and gallic acid equivalents (GAE).

Determination of dry residue is conducted following the method described in the European pharmacopoeia (2.8.16 dry residue of extracts). An amount of 2 g per sample was dried in glass vials at 105 °C for 2 h, cooled down under a dry atmosphere and the residual mass was determined gravimetrically.

The total phenolic content of the samples is determined by UV/Vis spectroscopy using the Folin–Ciocalteu reagent. Gallic acid solutions of concentrations 0.05 g/L, 0.1 g/L, 0.2 g/L, and 0.25 g/L are utilized for the calibration. To prepare the calibration lines, 0.5 mL of gallic acid solution is mixed with 1.5 mL of Folin–Ciocalteu reagent, which is diluted to 10% of the original concentration beforehand and incubated for 5 min at room temperature. Then, 1.5 mL of 7% sodium carbonate solution is added, and everything is filled to 10 mL with HPLC grade water and incubated for 1.5 h. Measurement is conducted at 750 nm in triplicate. HPLC water is used as a blank sample instead of the gallic acid standards. The Folin–Ciocalteu reagent was supplied by VWR International, Hannover, Germany.

2.4. Statistical Analysis

The statistical analysis of the results from Design of Experiments (DoE) was conducted with JMP Statistical Discovery™ by SAS Institute, Cary, NC, USA.

2.5. Calculations

For the characterization of the solid–liquid extraction, the yield is calculated according to Equation (1), with m_{TC} as the extracted mass of the target component and m_{CBS} as the mass of ground CBS used in the extraction process.

$$\text{Yield} = \frac{m_{TC}}{m_{CBS}} \tag{1}$$

For characterization of the liquid–liquid extraction there are different target units. The partition coefficient describes the distribution of the target component in the two phases, the heavy phase and the light phase. The partition coefficient K is calculated according to Equation (2), with the concentration of the target component in the light phase $c_{TC,LP}$ and the concentration of the target component in the heavy phase $c_{TC,HP}$.

$$K = \frac{c_{TC,LP}}{c_{TC,HP}} \tag{2}$$

The yield in liquid–liquid extraction is calculated according to Equation (3), with the mass of the target component in the light phase $m_{TC,LP}$ and in the heavy phase $m_{TC,HP}$.

The purity of the liquid–liquid extraction for each component is calculated according to Equation (4), with the mass of the dry residue in the light phase $m_{DR,LP}$.

$$\text{Yield} = \frac{m_{TC,LP}}{m_{TC.HP} + m_{TC.LP}} \tag{3}$$

$$\text{Purity} = \frac{m_{TC,LP}}{m_{DR,LP}} \tag{4}$$

3. Results

3.1. Characterization of Solid–Liquid Extraction

The substance system of cocoa shells is first characterized according to a methodical procedure. Two different extraction methods are investigated. One is the pressurized hot water extraction (PHWE) and the other is a conventional extraction with organic or organic–aqueous extractants.

For the coupled glycan and polyphenol extraction based on prior knowledge, PHWE is the most suitable method, since the high temperature of the water and the associated slightly acidic properties of the water induce hydrolysis of the carbohydrate skeleton. The comparison with organic and organic–aqueous extraction agents provides, above all, a comparative overview of the substance properties of the polyphenols. Information on the solvents in which the polyphenols of the cocoa shells dissolve well can be informative with regard to the choice of extraction agent for the subsequent liquid–liquid extraction.

In the characterization, the experimental data are evaluated according to two target variables. The first is the dry residue. This describes the sum of all non-volatile components of the extract. In the present case, the main components of the dry residue are the glycans and the sum of the polyphenols.

The results for the solvent screening are shown in Figure 3. It can be seen that, in particular, ethanol/water mixtures between 20 and 80% ethanol show high solubilization properties for the polyphenols compared to the other extraction agents investigated, such as hexane, ethyl acetate or butyl acetate. All three would be suitable extractants for liquid–liquid extraction. The yield is calculated as the mass of the respective component in mg divided by the mass of CBS used in the extraction process.

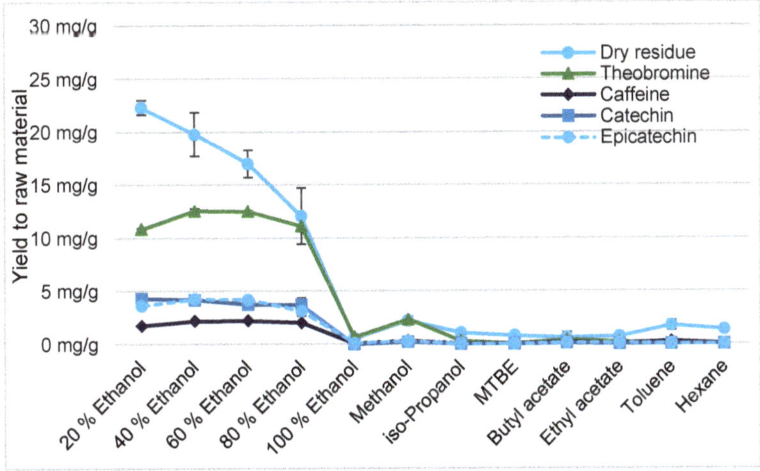

Figure 3. Yields for dry residue, methylxanthines and polyphenols in solvent screening with aqueous ethanol and organic solvents.

In addition to the extracted dry matter, the polyphenols of the cocoa shells are of interest. The yields achieved in milligrams of polyphenol per gram of extracted cocoa shell are shown in Figure 4. In addition to the investigated target components described at the beginning, where the two phenols catechin and epicatechin are considered, as well as the two methylxanthines theobromine and caffeine, a total phenol content appears here. These are remeasurements of old samples that were still available. The analysis by means of Folin–Ciocalteu was established additionally only in the later course of the project. Therefore, this analytical method has so far only been carried out in this series of measurements. It can be seen that in the temperature range between 120 °C and 160 °C, the yields obtained remain relatively constant. Accordingly, up to an extraction temperature of 160 °C, it can be assumed that no thermal decomposition processes of the investigated components take place. Based on the sum of extraction properties with respect to glycans and polyphenols, as well as operational and safety considerations, an extraction temperature of 140 °C is selected for the further extractions. Quercetin was not detected in any of the extracts from the present cocoa shells.

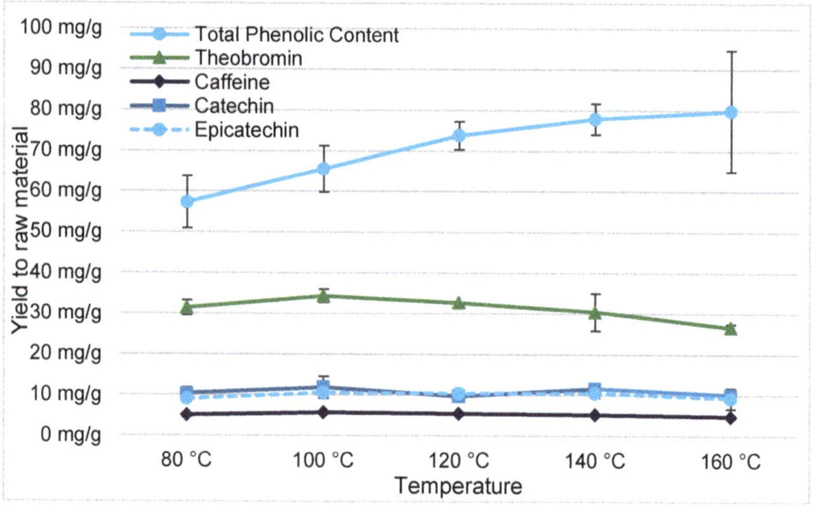

Figure 4. Yields for dry residue, methylxanthines and polyphenols in temperature screening.

3.2. Characterization of Liquid–Liquid Extraction

For the screening of suitable extraction solvents for liquid–liquid extraction, organic solvents that form a miscibility gap with water are required. Ethyl acetate, butyl acetate, methyl tert-butyl ether (MTBE), hexane and toluene are chosen for this purpose. The experiments are performed with volumetric phase ratios of 1:1 to determine the partition coefficients of the target components. Liquid–liquid extraction of the extracts with hexane and toluene resulted in emulsion formation in the organic phase with the fats from the cocoa shell. Accordingly, not only could these samples not be measured, but they also fall away for further processes considering that these processes could not be carried out. The further investigations are carried out with different starting extracts, which are based on considerations from the process synthesis. Here, there are different scenarios at which point in the process the LLE can be performed:

- Directly after extraction of the glycan–phenol mixture.
- After precipitation of the glycans with ethanol—an ethanol–water mixture is then present.
- After evaporation of the ethanol from the precipitation supernatant.

For LLE with ethanol–water mixtures, the above-mentioned organic extraction agents can only be used to a limited extent, if at all, because the ethanol content means that no more mixture gaps are formed, which is a prerequisite for LLE. However, it is possible to carry out an aqueous two-phase extraction in an ethanol/water/salt system. In this case, the addition of salt to an ethanol–water mixture displaces ethanol with a lower water content from the salt-rich phase. In principle, a wide range of different salts are suitable for this purpose. In view of environmental compatibility and green extraction processes, a citrate salt is used. Comparatively, but less green, a phosphate salt is used. The partition coefficients for the target components are shown in Figure 5. The partition coefficient is calculated with the concentration in the light phase divided by the concentration in the heavy phase. Here, large values represent a preferential distribution of the target components in the light phase, which in all cases is the organic phase, or the ethanolic, low-salt phase. Here, only very low partition coefficients are obtained for the organic solvents in question. In contrast, large partition coefficients are achieved for the ethanol/water/salt systems. Here, the citrate system shows the best results in comparison. The use of this system also has the advantage that no additional organic extractant needs to be added to the process. Only the ethanol is used, which is required for precipitation anyway.

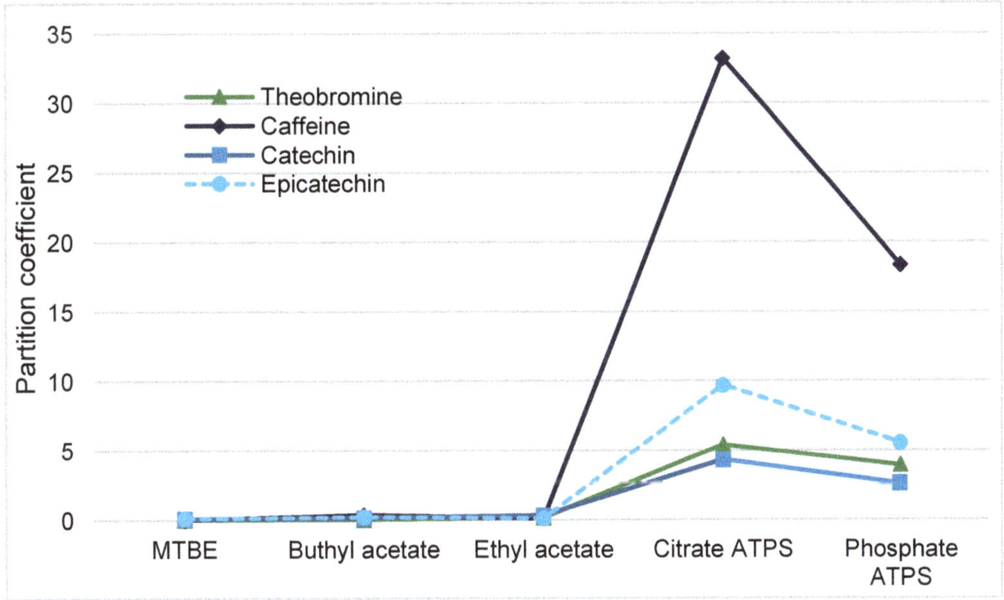

Figure 5. Partition coefficients for solvent screening for liquid–liquid extraction.

Therefore, the ethanol/water/salt systems will be characterized and investigated in more detail in further trials.

In Figure 6, the results of the phase screening with citrate ATPS at pH 6, 7 and 8, phosphate ATPS at pH 7 and 8, ethyl acetate and butyl acetate at phase ratios 30/70, 40/60, 50/50, 60/40 and 70/30 (m/m) expressed as feed/solvent are shown. The components researched are total phenolic content, theobromine, caffeine, catechin and epicatechin. The yield is calculated as the mass of the respective component in the light phase divided by the mass of the respective component which is brought into the system with the used CBS extract. The purity is calculated as the mass of the respective component divided by the mass of the dry residue within the light phase. The data show that, for all components, the aqueous two-phase systems reach exceptional high yields of up to 100% in one extraction stage, whereas the conventional extraction systems with ethyl acetate and butyl acetate reach only up to a maximum of 50% for caffeine and only up to 30% for the targeted polyphenols. The data show higher yields for the ATPS at a higher phase ratio, which represents a lower consumption of solvent, whereas the organic solvents show a contrary behavior. So, for better yields in the ATPS, less solvent is used, and for better yields with organic solvents, a higher amount of solvent is needed. The pH value seems to have a negligible influence on the yield. Regarding purity, organic solvents reach significantly higher values. This is because of rather high salt contents in the ATPS in the light, phenol-rich phase. This can be optimized with an adaption of the extraction system, e.g., higher ethanol content in the feed. This will be researched in a follow-up study.

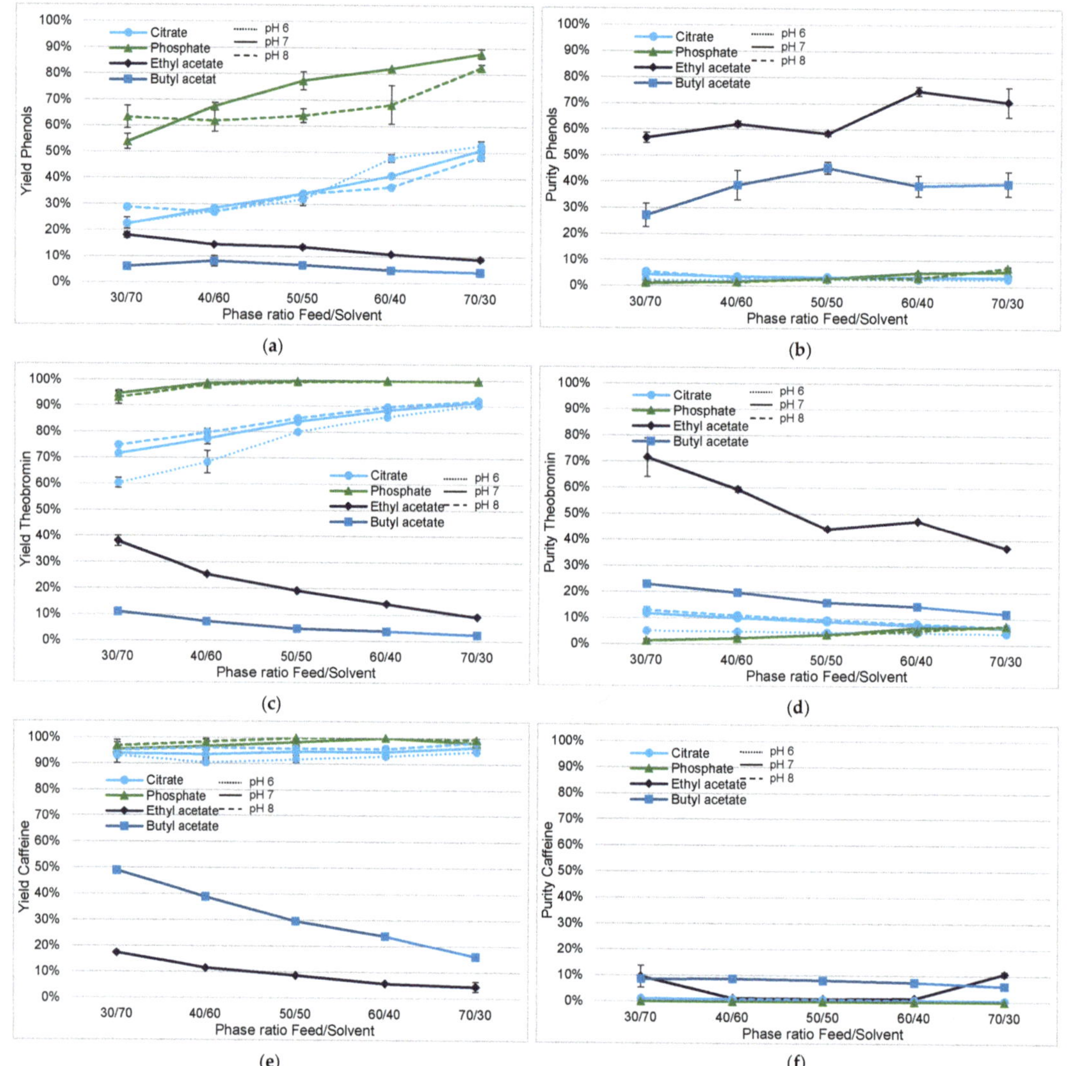

Figure 6. *Cont.*

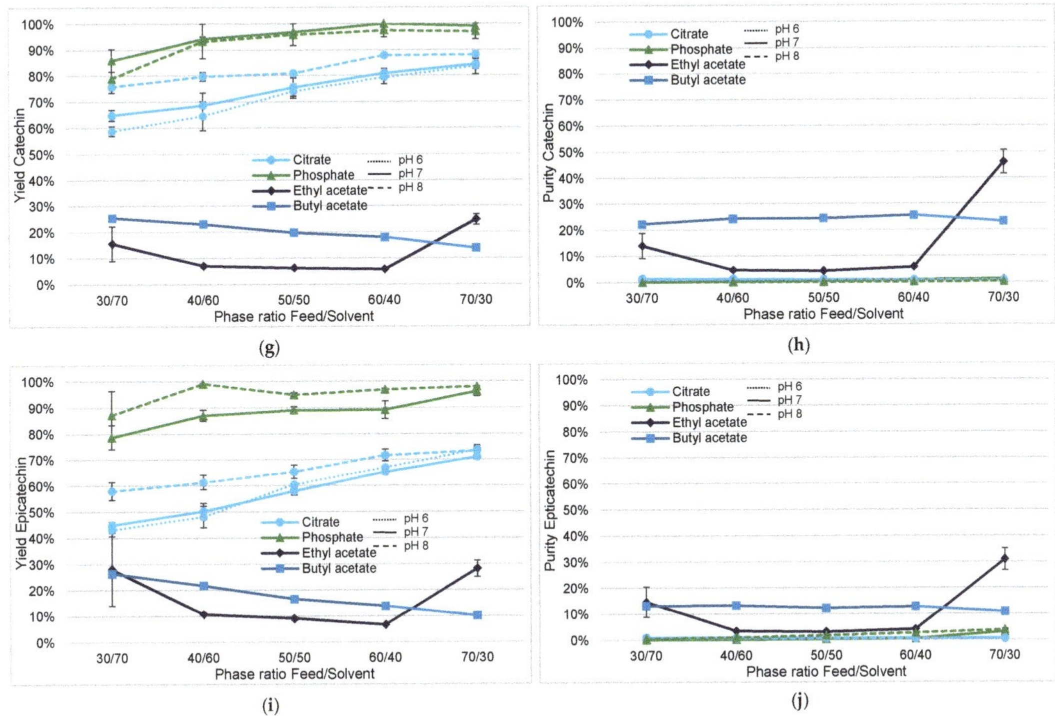

(g)

(h)

(i)

(j)

Figure 6. Yields and purities for total phenolic content (**a**,**b**), theobromine (**c**,**d**), caffeine (**e**,**f**), catechin (**g**,**h**) and epicatechin (**i**,**j**) in DoE with phase ratio and pH value including ethyl acetate and butyl acetate as reference.

In Figure 7, the statistical influence of pH value and phase ratio on the yield of phenol content, theobromine, caffeine, catechin and epicatechin are shown. The black squares in the plots represent the target values from the experimental data. These results are for the ATPS with a 40 wt.% citrate buffer. Due to low solubility of phosphate salts at pH 6, no full factorial DoE could be conducted.

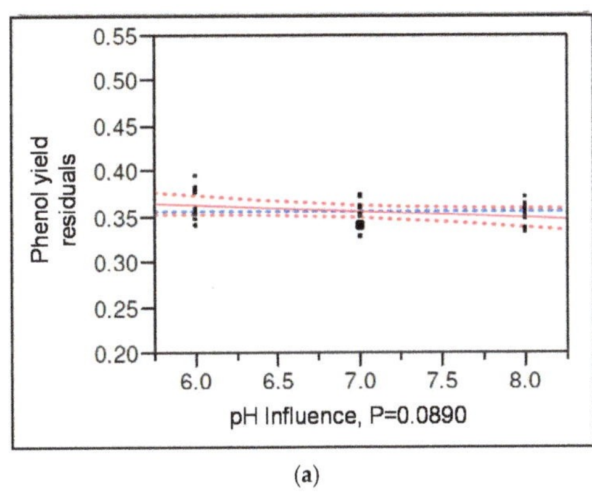

(a)

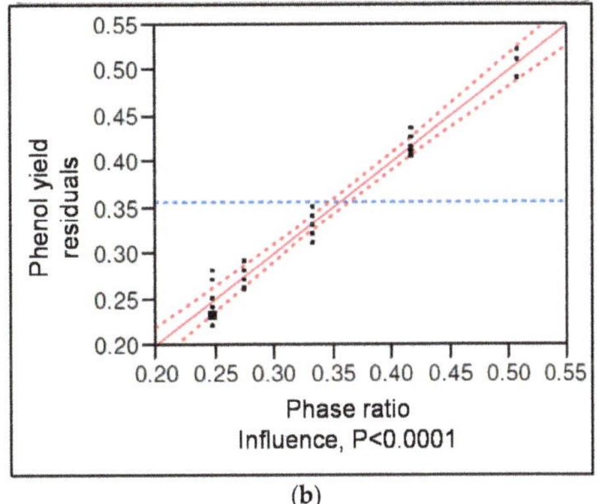

(b)

Figure 7. *Cont.*

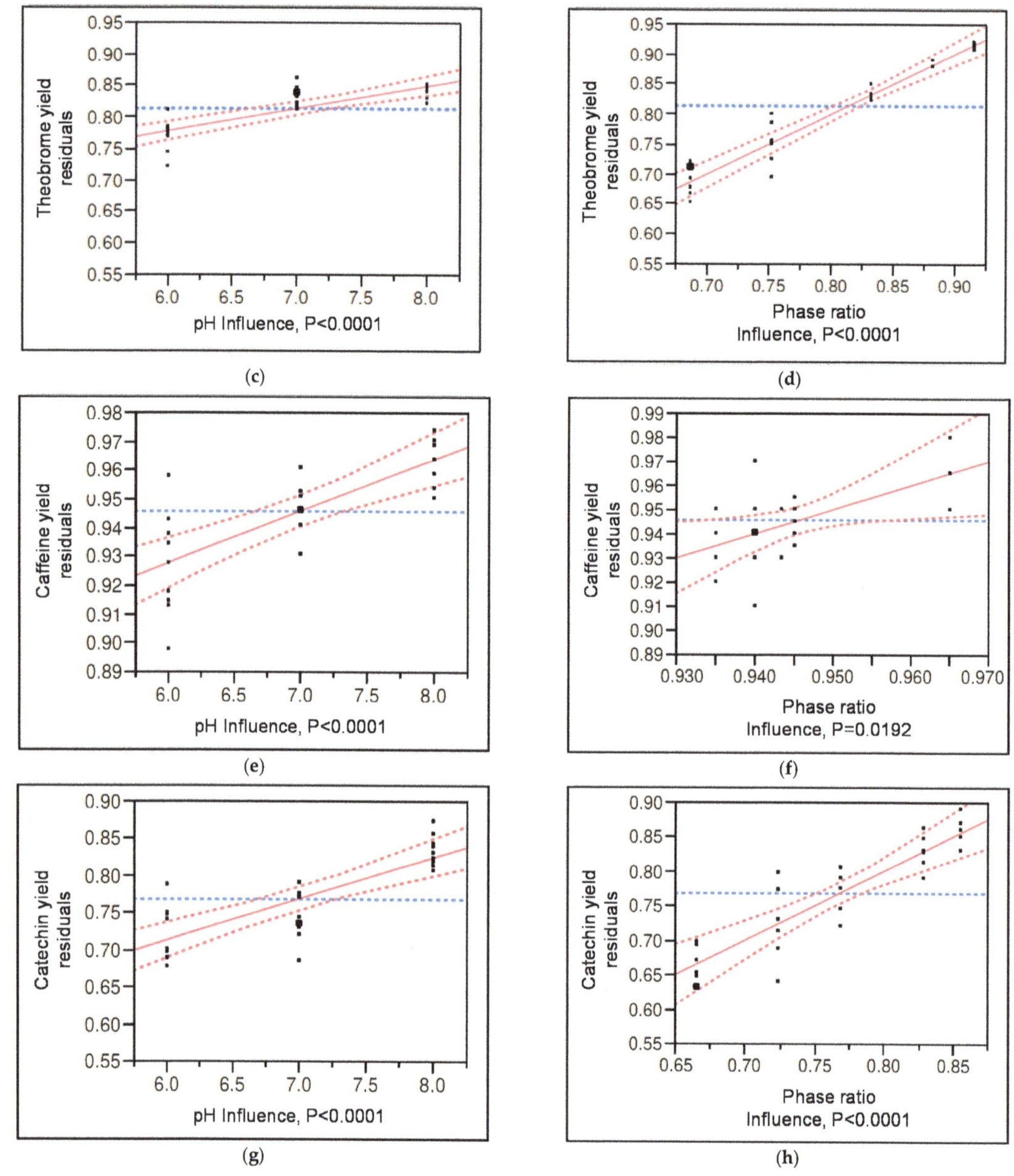

Figure 7. *Cont.*

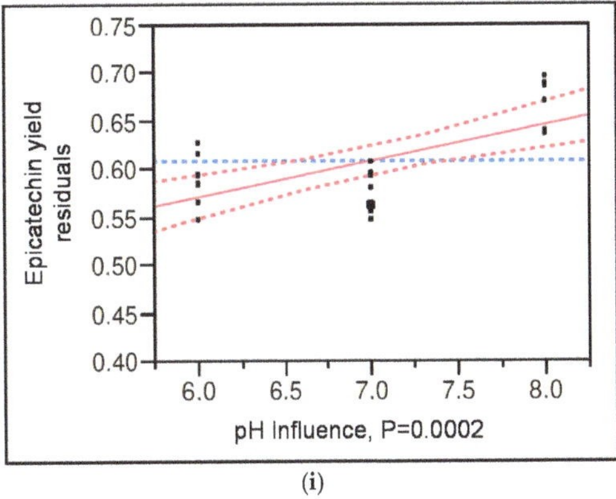

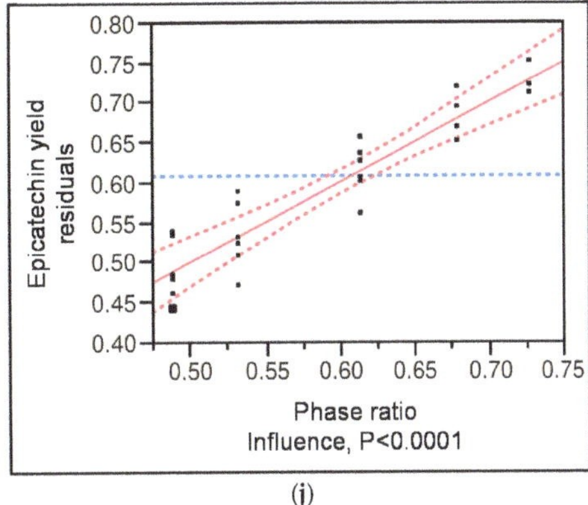

(i)   (j)

Figure 7. Influence of pH and phase ratio on yield for total phenolic content (**a**,**b**), theobromine (**c**,**d**), caffeine (**e**,**f**), catechin (**g**,**h**) and epicatechin (**i**,**j**).

For total phenolic content, the regression coefficient of the statistical model is 0.98. The pH value has no significant influence on the extraction yield, where the phase ratio has a high positive influence on the extraction yield. This behaviour matches with the observations in Figure 6. For theobromine, the regression coefficient of the statistical model is 0.96. Both pH value and phase ratio have a significant positive influence on the extraction yield. The regression coefficient of the statistical model for caffeine yield is 0.72. According to the P-values, pH value has a significant positive effect, while phase ratio has no significant effect. However, due to the low R^2 and the high scattering of the measured values in Figure 7e,f, this statement is questionable.

For catechin, the regression coefficient of the statistical model 0.88. Both pH value and phase ratio have a significant positive influence on the yield of catechin.

The regression coefficient of the statistical model for epicatechin is 0.91. The pH value has a medium significant positive influence on the yield, whereas phase ratio has a significant positive influence on the yield of epicatechin.

In Figure 8, the statistical influence of pH value and phase ratio on the purity of total phenolic content, theobromine, caffeine, catechin and epicatechin are shown.

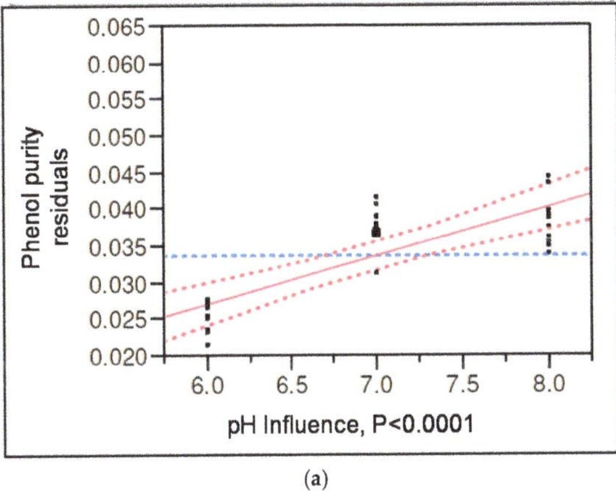

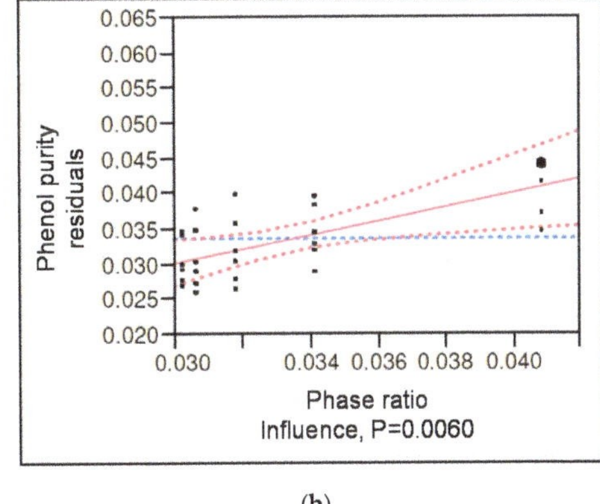

(a)   (b)

Figure 8. *Cont.*

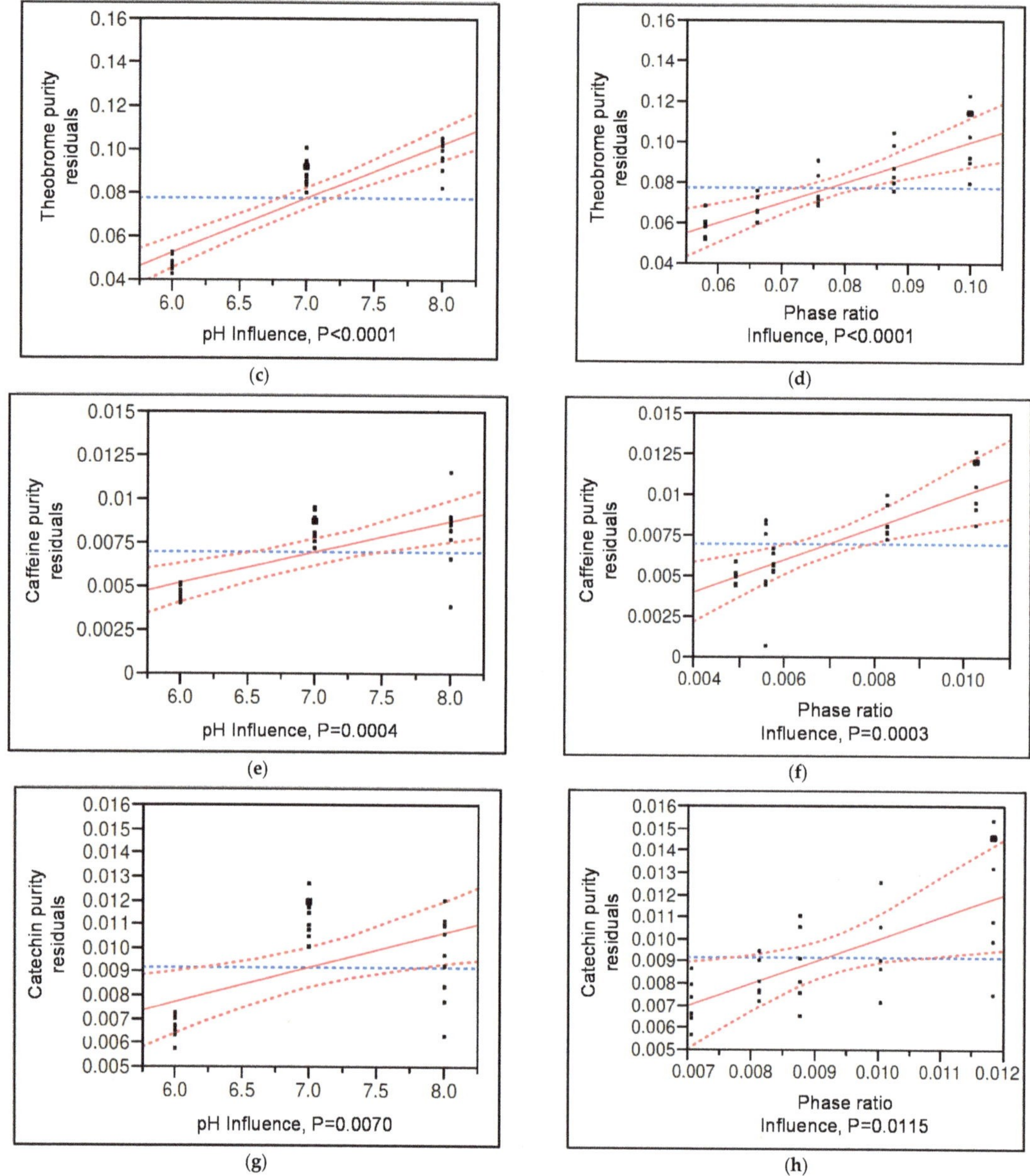

Figure 8. *Cont.*

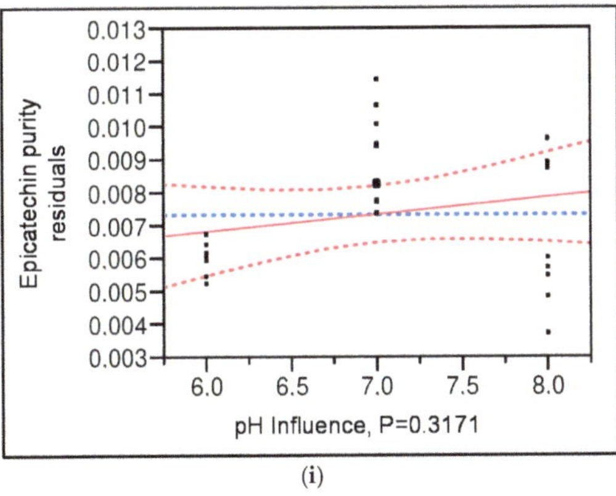

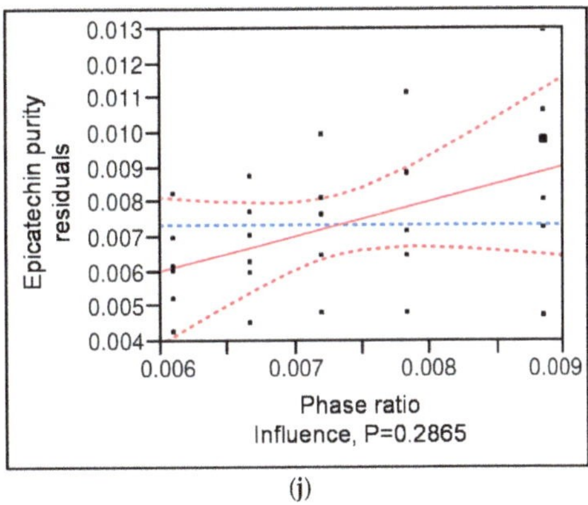

Figure 8. Influence of pH and phase ratio on purity for total phenolic content (**a,b**), theobromine (**c,d**), caffeine (**e,f**), catechin (**g,h**) and epicatechin (**i,j**).

For total phenolic content, the regression coefficient of the statistical model is 0.80. The pH value has a significant positive influence on the extraction yield, whereas the phase ratio has a low-to-no significant influence on the purity. This behaviour matches with the observations in Figure 6. For theobromine, the regression coefficient of the statistical model is 0.89. Both pH value and phase ratio have a significant positive influence on the purity of theobromine. The regression coefficient of the statistical model for caffeine purity is 0.76. According to the *p*-values, pH value and phase ratio have a medium significant positive effect on the purity of caffeine.

For catechin, the regression coefficient of the statistical model 0.58. Both pH value and phase ratio have no significant positive influence on the yield of catechin. However, due to the low R^2 and the high scattering of the measured values in Figure 8g,h, this statement has low confidence.

The regression coefficient of the statistical model for epicatechin is 0.28. Due to the low regression quality, there is no sophisticated statement to make.

In addition, for yield and purity, there is no significant influence of the interaction between pH value and phase ratio.

4. Discussion

In the present study, the extraction behaviour of the various target components in the CBS stock system was investigated. It was shown that the common organic solvents, with the exception of ethanol–water mixtures, give only poor extraction results. The comparatively good extraction properties of the ethanol–water mixtures already provided an indication, in the solvent screening of the SLE, that an ATPE with the ethanolic phase appears a promising method. Temperature screening showed very good extraction results for an extraction temperature of 140 °C, which is a common extraction temperature for the extraction of phenolic components.

In a first solvent screening for LLE, the three possible application points of LLE in the process alternatives are investigated. Whether LLE occurs before or after precipitation had no effect on the results in the present experiments. On the other hand, particularly strongly apolar solvents such as toluene or hexane are not suitable for fatty substance systems such as CBS. However, the potential of ATPE with ethanol, water and salt is confirmed, so that this process alternative is preferred. The only organic solvent required, ethanol, can even be used a second time for this purpose from the previous process step, precipitation. This saves resources and protects the climate; both points should be considered for processes in the bioeconomy. In the next step, the ATPE was investigated with two different salts. Here, pH of the salt buffer and the phase ratio were varied. From the statistical experimental design, there is a positive influence on the yield for higher phase ratios, i.e., for a low proportion of the salt buffer. The pH value has a low influence on the yield and only for some components. The purity also increases for higher phase ratios. Increasing pH also has a positive effect on purity. For the alternative studies with ethyl acetate and butyl acetate from the crude extract, it was shown that a high organic phase content increases both the yield and purity. For the ATPE systems, yields between 80 and 100% are achieved for the different target components. The organic comparative tests only deliver yields between 20 and 40% in one extraction stage. The comparatively low purities of the ATPE systems are due to a transition of salt into the light phase. The salt content in the light phase can be reduced by optimizing the system point. While high phase ratios should be considered to maximize the yield, pH value can be used to influence the purity of some components.

The precipitation with 60 wt.% ethanol, which was defined in the present study by a project framework, will be investigated in more detail in follow-up studies. Due to the double utilization of ethanol in the ATPE, precipitation and LLE are directly linked. However, it is also conceivable to adjust the ethanol content to the optimum of the LLE after precipitation with an ethanol content that is optimal for this process step.

5. Conclusions

Following this study, a process is available which integrates the PHWE into an overall process for cascade utilization for waste valorization in an environmentally friendly, green and efficient economic manner. The process consists of an extraction with hot water, a precipitation with ethanol as an anti-solvent and a liquid–liquid extraction from the precipitation supernatant with salting-out of ethanol. In this way, both the matrix components and the secondary plant compounds can be fully utilized by integrating unit operations appropriately. The versatile green process for waste valorization is shown in Figure 9.

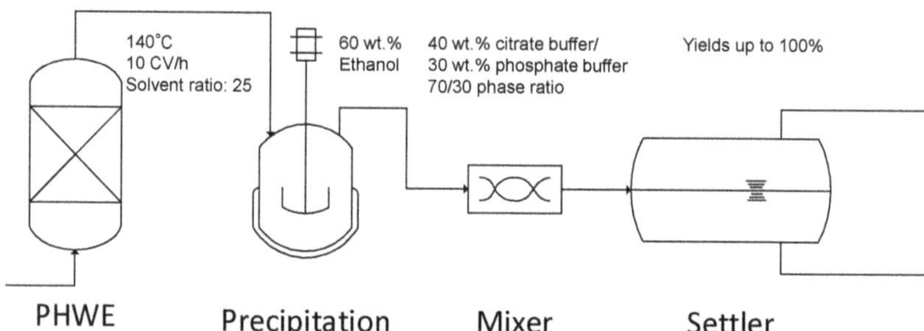

Figure 9. Overview of the novel process for the recovery of phenolic compounds from natural product extracts.

- The novel process reaches very high yields of up to 100% in one extraction stage.
- The novel process ensures low consumption of organic solvents due to double usage of ethanol as the only organic solvent.
- The process is adaptable enough to capture all kinds of secondary metabolites from hot water extracts and ensures usage of structural carbohydrates from precipitation. Ethanol is well-known as a precipitant for matrix components from hot water extracts. The ethanol content in the light phase is adaptable enough to match the solubility properties of the target component, usually between 50 and 80% ethanol [8,14].
- Follow-up studies will focus on process optimization, research on process analytical technology and complete dry residue characterization by component groups [13].

Author Contributions: Conceptualization, C.J. and J.S.; methodology, experimental design, and evaluation, C.J.; writing and editing, C.J.; reviewing and discussion, A.S. and J.S.; supervision, J.S. All authors have read and agreed to the published version of the manuscript.

Funding: This research origins from the GlyChem Project within the "BioBall Verbund" and was funded by the PtJ commissioned by the Bundesministerium für Bildung und Forschung (BMBF), grant number 031B0905C. We also kindly acknowledge the support by Open Access Publishing Fund of Clausthal University of Technology.

Institutional Review Board Statement: Not applicable.

Informed Consent Statement: Not applicable.

Data Availability Statement: Data cannot be made publicly available.

Acknowledgments: The authors would like to thank the ITVP lab-team, especially Frank Steinhäuser, Colin Herzberger, Volker Strohmeyer, and Thomas Knebel for their efforts and support. The authors would also like to thank Stefan Hanstein from Fraunhofer IWKS, Alzenau, Germany for the fruitful discussion.

Conflicts of Interest: The authors declare no conflict of interest.

References

1. Borges, A.; Abreu, A.; Dias, C.; Saavedra, M.; Borges, F.; Simões, M. New Perspectives on the Use of Phytochemicals as an Emergent Strategy to Control Bacterial Infections Including Biofilms. *Molecules* **2016**, *21*, 877. [CrossRef] [PubMed]
2. Leonel, M.; Sarmento, S.B.S.; Cereda, M.P. New starches for the food industry: *Curcuma longa* and *Curcuma zedoaria*. *Carbohydr. Polym.* **2003**, *54*, 385–388. [CrossRef]
3. Cardinali, A. Phytochemicals from artichoke by-product and their applications as natural ingredients for cosmetic industry. In *Green Extraction of Natural Products*; University of Bari: Bari, Italy, 2018.
4. Kleeberg, H. Method for the Production of Storage Stable Azadirachtin from Seed Kernels of the Neem Tree. U.S. Patent 5,695,763, 12 October 1997.
5. Fraunhofer-Projektgruppe für Wertstoffkreisläufe und Ressourcenstrategie. *Positionspapier zu Bioplastik*. 2018. Available online: https://www.iwks.fraunhofer.de/de/presse-und-medien/pressemeldungen-2018/positionspapier-zu-bioplastik.html (accessed on 3 December 2018).
6. Willner, T.; Sievers, A. *Fortschrittliche Alternative Flüssig Brenn- und Kraftstoffe: Für Klimaschutz im Globalen Rohstoffwandel*; HAW Hamburg: Hamburg, Germany, 2017.
7. Technische Universität Dresden. Bundesministerium für bildung und forschung. In *Nationale Forschungsstrategie BioÖkonomie 2030*; Technische Universität Dresden: Dresden, Germany, 2010.
8. Sixt, M. *Methoden zur Systematischen Gesamtprozessentwicklung und Prozessintensivierung von Extraktions- und Trennprozessen zur Gewinnung Pflanzlicher Wertkomponenten*; Shaker Verlag: Aachen, Germany, 2018; ISBN 978-3-8440-6169-7.
9. Plaza, M.; Turner, C. Pressurized hot water extraction of bioactives. *TrAC Trends Anal. Chem.* **2015**, *71*, 39–54. [CrossRef]
10. Jadeja, G.C.; Maheshwari, R.C.; Naik, S.N. Extraction of natural insecticide azadirachtin from neem (*Azadirachta indica* A. Juss) ceed kernels using pressurized hot solvent. *J. Supercrit. Fluids* **2011**, *56*, 253–258. [CrossRef]
11. Plaza, M.; Marina, M.L. Pressurized hot water extraction of bioactives. *TrAC Trends Anal. Chem.* **2019**, *116*, 236–247. [CrossRef]
12. Schmidt, A.; Uhlenbrock, L.; Strube, J. Technical Potential for Energy and GWP Reduction in Chemical–Pharmaceutical Industry in Germany and EU—Focused on Biologics and Botanicals Manufacturing. *Processes* **2020**, *8*, 818. [CrossRef]
13. Jensch, C.; Knierim, L.; Tegtmeier, M.; Strube, J. Development of a General PAT Strategy for Online Monitoring of Complex Mixtures—On the Example of Natural Product Extracts from Bearberry Leaf (*Arctostaphylos uva-ursi*). *Processes* **2021**, *9*, 2129. [CrossRef]
14. Sixt, M.; Schmidt, A.; Mestmäcker, F.; Huter, M.; Uhlenbrock, L.; Strube, J. Systematic and Model-Assisted Process Design for the Extraction and Purification of Artemisinin from *Artemisia annua* L.—Part I: Conceptual Process Design and Cost Estimation. *Processes* **2018**, *6*, 161. [CrossRef]
15. Schmidt, A.; Strube, J. Application and Fundamentals of Liquid-Liquid Extraction Processes: Purification of Biologicals, Botanicals, and Strategic Metals. In *Kirk-Othmer Encyclopedia of Chemical Technology*; Wiley & Sons: Hoboken, NJ, USA, 2018; Volume 16, pp. 1–52. [CrossRef]
16. Schmidt, A.; Strube, J. Distinct and Quantitative Validation Method for Predictive Process Modeling with Examples of Liquid-Liquid Extraction Processes of Complex Feed Mixtures. *Processes* **2019**, *7*, 298. [CrossRef]
17. Schmidt, A.; Richter, M.; Rudolph, F.; Strube, J. Integration of Aqueous Two-Phase Extraction as Cell Harvest and Capture Operation in the Manufacturing Process of Monoclonal Antibodies. *Antibodies* **2017**, *6*, 21. [CrossRef]
18. Santana, N.B.; Dias, J.C.T.; Rezende, R.P.; Franco, M.; Oliveira, L.K.S.; Souza, L.O. Production of xylitol and bio-detoxification of cocoa pod husk hemicellulose hydrolysate by Candida boidinii XM02G. *PLoS ONE* **2018**, *13*, e0195206. [CrossRef] [PubMed]
19. Rojo-Poveda, O.; Barbosa-Pereira, L.; Zeppa, G.; Stévigny, C. Cocoa Bean Shell-A By-Product with Nutritional Properties and Biofunctional Potential. *Nutrients* **2020**, *12*, 1123. [CrossRef] [PubMed]
20. Panak Balentić, J.; Ačkar, Đ.; Jokić, S.; Jozinović, A.; Babić, J.; Miličević, B.; Šubarić, D.; Pavlović, N. Cocoa Shell: A By-Product with Great Potential for Wide Application. *Molecules* **2018**, *23*, 1404. [CrossRef] [PubMed]

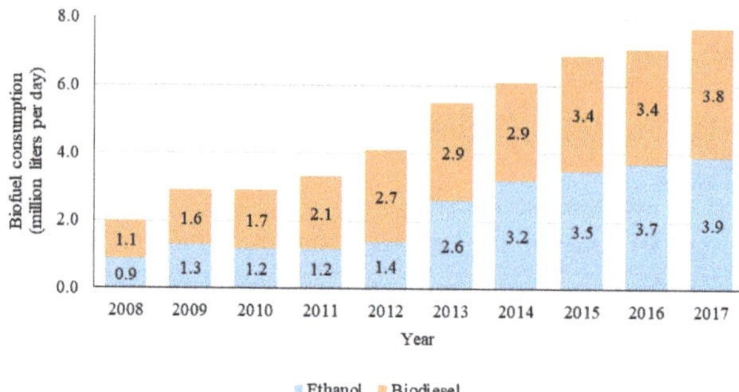

Figure 1. Thailand's biofuel consumption (million liters per day) [1–3].

However, the promotion of biofuels increases the demand for feedstock crops that in turn leads to the expansion of land dedicated to feedstock crops. Based on Thailand's Alternative Energy Development Plan (AEDP) 2015, lands for sugarcane and oil palm cultivation are being targeted to increase from 1.6 and 0.7 million ha in 2015 to 2.6 and 6.2 million ha in 2026, respectively (i.e., annually increased by 4.5 and 4.8 percent during 2015–2026) [5]. The expansion of land for feedstock crops can reduce the availability of land for other purposes, which then adversely affects economic opportunity for other activities. However, the economy-wide impact of land expansion induced by biofuel promotion is not found in any earlier studies, even though more than 40 percent of total land area is used by agriculture and the agricultural sector contributes approximately 10 percent to the Gross Domestic Product (GDP).

The measurement of the economy-wide effects of land expansion can be presented through GDP, as several earlier studies have shown. Despite ignoring the effects of land expansion, Silalertruksa and Gheewala [6] used GDP as an indicator to present the economic impact of bioethanol production in Thailand. Wianwiwat and Asafu-Adjaye [7], Kaenchan et al. [8], Phomsoda et al. [9], and Phomsoda et al. [10] revealed the dynamic effects of biofuel promotion on the economy through the intertemporal change in real GDP. However, although GDP is a standard measure for economic growth, it does not reflect actual human well-being as it does not account for social sustainability and future environmental consequences of present consumption [11,12]. Thus, Green GDP and other similar indices for sustainable development such as the Index of Sustainable Economic Welfare (ISWE) and the Genuine Progress Indicator (GPI) were developed to fill this lack [12,13].

Green GDP is an index of sustainable economic growth where the degradation and depletion of environmental and natural resources are subtracted from the conventional GDP. Since environmental and natural resources can be considered as the stocks of production factors used for generating the GDP of a country, their degradation and depletion should be deducted from the conventional GDP to derive the remaining stocks for the future. Green GDP has widely been adopted to promote more sustainable practices in several studies. For example, Li and Fang [14] presented Green GDP of all countries by integrating the total GDP with ecosystem services values obtained from spatial analysis based on Geographic Information System (GIS). Stjepanović et al. [15] measured Green GDP across countries by capturing emission, waste, and natural resource depletion. In addition, by incorporating greenhouse gas emissions, Kunanuntakij et al. [16] estimated Thailand's green GDP by using economic input–output life cycle assessment.

This study aimed to assess the effects of biofuel crop expansion on Thailand's Green GDP to address the lack of studies on the economy-wide effects of biofuel crop expansion that can in turn support policymakers in making decisions toward sustainable biofuel development in Thailand. The expansion of biofuel crops was incorporated relying on the targets officially published in AEDP 2015. Three scenarios of land expansion alternatives were considered in this study. In addition, the impacts of environmental interventions, i.e., air emissions, land transformation, water consumption, and fossil consumption, were captured.

2. Methods

Green GDP is defined as the *Conventional GDP* subtracted by the cost of environmental degradation and natural resource depletion, where environmental degradation refers to the effects of GHG emissions and land use and natural resource depletion denotes the depletion of water and fossil resources. The calculation of *Green GDP* is summarized in Equations (1) and (2), where *TEC* is the total environmental cost, *COP* is the cost of pollution (GHGs), *COL* is the cost of land degradation, *CWD* is the cost of water depletion, and *CFD* is the cost of fossil depletion.

$$Green\ GDP = Conventional\ GDP - TEC \tag{1}$$

$$TEC = COP + COL + CWD + CFD \tag{2}$$

The effects of biofuel crop expansion on *Green GDP* were estimated by comparing the business-as-usual (BAU) scenario with that in which biofuel crop expansion occurs. *Conventional GDP* was estimated using a static computable general equilibrium (CGE) model, a macroeconomic model for assessing the economy-wide impacts of policies that can also be modified to incorporate the environmental impacts of policies [8]. The procedure to formulate the CGE model used in this study is described in Section 2.1. The modification of the model to incorporate widespread environmental effects is presented in Section 2.2. The methods and equations for assessing the cost of environmental degradation and natural resource depletion are presented in Section 2.2.

2.1. CGE Model Setup

The standard CGE model developed by the Partnership for Economic Policy (PEP) research network [17] was used in the present study to estimate the effects of biofuel crop expansion. Model setup and simulation scenarios are detailed in the following subsections.

2.1.1. Model Description

Following the conventional structure of general equilibrium simulation, the model included four main economic agents: the production sectors, the aggregated household, the government, and the rest of the world. Main connectivities of transactions and activities are depicted in Figure 2. The consumption behavior of a household is governed by the Stone–Geary utility maximization framework, allowing for optimal adjustment of the consumption basket under the budget constraint. As illustrated in Figure 3, all production activities were structured based on the 4-level nested hierarchy, enabling the flexibility of selecting the optimal proportion of inputs and factors of production. In particular, the first level of this structure followed the Leontief production function, imposing the fixed ratio of value-added and total intermediate input. The second layer determined the distribution of value-added components and the selection of intermediate inputs. In the case of value-added allocation, a constant elasticity of substitution (CES) specification governed the optimal combination of labor and capital-land composite. For the total intermediate input, the selection was based on the Leontief production function, constantly demanding intermediate inputs by using a fixed proportion. In the third layer, the CES framework optimized the combination of land and capital. Considered one of the key features of this model, the last layer enriched the details of demand for land by specifically identifying the classification of land use into three categories: agricultural land, forest, and abandoned rice field.

Following the standard specification of CGE model, the CES mechanism determined the optimal composite of import and domestically produced goods. Similarly, a constant elasticity of transformation (CET) optimized the export decision, weighting the proportion of domestic sales and exports.

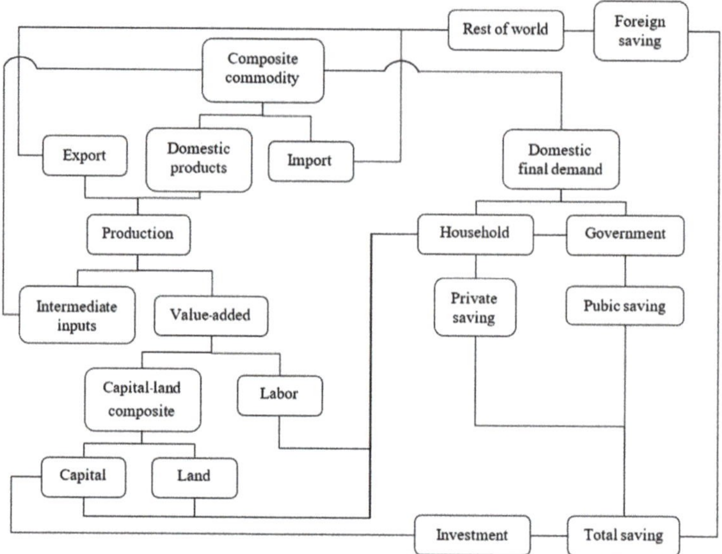

Figure 2. Main connectivities of economic transactions and activities within the CGE model.

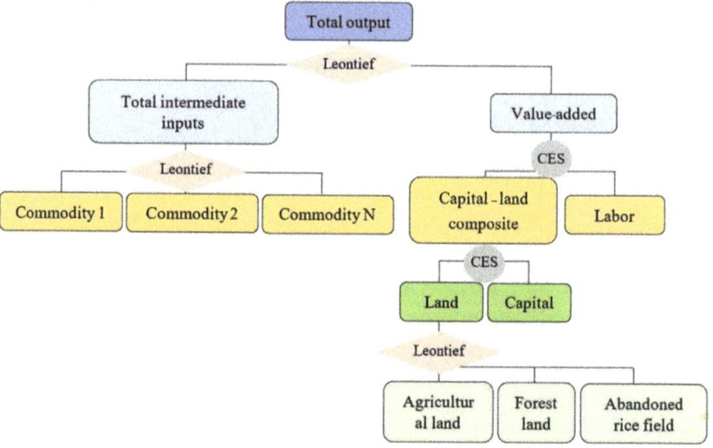

Figure 3. Structure of production.

2.1.2. Model Closure

To equalize the number of endogenous variables and equations, some variables were assigned to be exogenous. Following the conventional criteria introduced by Decaluwé et al. [17], variables influenced by the global economy and those determined by policymakers were specified as being exogenous. Thus, the international prices of imported and exported products, the current account balance, and the exchange rate were determined exogenously. Likewise, the policy-determined exogenous variables were government expenditure, domestic wage, capital demand, total investment, and the tax rate. Since the minimum requirements for foods and necessary goods are the primary demand for humans, the minimum consumption of a household was also specified exogenously.

Because the flexible adjustment of the biofuel crop sector was one of the main features of this model, the demand for the capital of biofuel crop plantation was endogenously determined to enable unconstrained variation in the nested structure of biofuel crop production. Also, this specification allowed the model to perform simulation scenarios by assigning the output of a specific biofuel crop exogenously.

2.1.3. Database

Similar to the conventional specification of the CGE model, the Social Accounting Matrix (SAM) was a primary source of data [18]. The SAM used in this research has been constructed based on the 2015 input–output (IO) table and officially produced and publicly distributed by Thailand's Office of the National Economic and Social Development Council [19].

The constructed SAM contained 39 production sectors, compromising between the mathematically solvable property of the model and obtaining sufficient detail for environmental and economic analysis. The SAM also included the main economic agents which are the government, the aggregated representative of households, and the rest of the world. The details of production sectors are exhibited in Appendix A, Table A1. To conform to the standard initialization of the model, elasticity parameters were obtained from Decaluwé et al. [17] and OECD/ILO [20] (Appendix A, Table A2). In accordance with the most recent published data of land use, pollution, and environmental indicators, the SAM and all variables of this CGE model were calibrated to the base year of 2017. Specifically, the calibration of SAM followed the steps introduced in Serag et al. [21]. The macroeconomic data were obtained from the official database of national income published by NESDC. Regularly produced and distributed by Thailand's National Statistical Office (NSO), details of production activities were obtained from the official industrial census and household consumption statistics were derived from the official socioeconomic survey. The compilation of data used the cross-entropy estimation technique as introduced by Robinson et al. [22].

2.1.4. Simulation Scenarios

Biofuel crop expansion has three simulation scenarios. Among them, the percent increment of biofuel crops was identically defined on the basis of the annual targets in AEDP 2015 [5]. That is, the output of cassava, sugarcane, and oil palm increased by 6.2, 4.5, and 3.7 percent, respectively. The output of sugarcane and oil palm production increased by expanding land while the output of cassava production increased due to productivity improvement in all scenarios

- S1: There is no transformation of forest area to cropland. Thus, the total dimension of agricultural land is constant, and the expansion of sugarcane and oil palm can diminish the size of other croplands.
- S2: Forest area (0.02 percent) is assumed to be transformed to agricultural land following the average annual decreasing rate of forest area during 2014–2016 [23]. Therefore, in this scenario, more agricultural land is available.
- S3: Abandoned rice fields (164,800 ha [24]) are utilized by transforming to agricultural land. Therefore, more agricultural land is available.

2.2. Expanding the Model to Capture Environmental Impacts

This study considered the environmental impacts caused by air emissions, land transformation, water consumption, and fossil resource consumption. Thus, the CGE model was expanded to capture these features and estimate the environmental impacts of scenarios S1–S3.

2.2.1. Air Emissions

This study focused on global warming [presented in a unit of kg carbon dioxide equivalent (kg CO_2 eq.)] caused by greenhouse gas (GHG) emissions. In particular, the standard CGE model was modified to incorporate the conversion factors, enabling the computation of CO_2 emissions from energy consumption and chemical fertilizer use.

The CO_2 conversion factors for energy consumption are shown in Table 1. As the sectoral production and commodity consumption in the CGE model are conventionally represented in monetary units, Table 1 exhibits all price factors (PEs) applied to convert the values of energy consumption into the physical base quantity unit. Conversion factors of CO_2 emissions are not applied in the use of crude oil and natural gas in the chemical industry and petroleum refineries and the use of petroleum products in the chemical industry because they are used as raw materials (feedstock) and not burned in these sectors. The conversion factors for CO_2 emissions from chemical fertilizer use (EFAG) were calculated by dividing the total CO_2 emissions from chemical fertilizer use of approximately 5547 million tonnes CO_2 eq. in 2015 by the total value of chemical fertilizer use of the whole nation in 2015 (derived from the SAM table). The 5547 million tonnes CO_2 eq. was derived based on the information on chemical fertilizer imports from the Office of Agricultural Economics [25] and the methods to calculate CO_2 emissions and emission factors of chemical fertilizer production and use given by the Thailand Greenhouse Gas Management Organization [26].

Table 1. CO_2 conversion factors for energy consumption.

Sources of Energy	EFEC (1000 Tonnes CO_2/ktoe) [a]	PE (1000 Million THB/ktoe) [b]
Coal and lignite	4.10533	0.004
Crude oil and natural gas	1.03978	0.016
Petroleum products	2.48847	0.053

Notes: [a] is CO_2 emission factors calculated from dividing the total emissions from each energy classification in 2015 [27] by its consumption amount in 2015 [28]; [b] is a ratio of the total emissions from each energy classification in 2015 per the total value of the corresponding energy consumption in 2015 (derived from the SAM table); ktoe is 1000 tonnes of oil equivalent; and THB is Thai baht.

The CO_2 conversion factors for energy consumption and chemical fertilizer use were attached to the database of the model. This study included Equations (3)–(7), modified from Kaenchan et al. [8], to compute the total CO_2 emissions.

The total amount of CO_2 emitted from each production sector can be estimated as shown in Equations (3)–(5), where $ECCO_{i,j}$ is the CO_2 emission caused by the consumption of energy product i by production sector j; $DI_{i,j}$ denotes the use of intermediate product i by production sector j; $EFEC_i$ is the emission coefficient corresponding to the consumption of product i; PE_i indicates the price of energy product i; $AGCO_{chem,jagri}$ represents the total amount of CO_2 emitted by the utilization of chemical fertilizer in farming activity j_{agri}; $DI_{chem,jagri}$ identifies the use of chemical fertilizer in farming activity j_{agri}; $EFAG_{chem,jagri}$ is the emission coefficient of using chemical fertilizer in farming activity j_{agri}, and $INTCO_j$ is the amount of CO_2 emitted by a production activity of sector j.

$$ECCO_{i,j} = \frac{DI_{i,j} \times EFEC_i}{PE_j} \tag{3}$$

$$AGCO_{chem,j_{agri}} = DI_{chem,j_{agri}} \times EFAG_{chem,j_{agri}} \tag{4}$$

$$INTCO_j = \sum_i ECCO_{i,j} + \sum_{chem} AGCO_{chem,j_{agri}} \tag{5}$$

Equation (6) specifies the computation of the total amount of CO_2 emitted by final consumption, where $FNCO_i$ is the emission caused by consumption of product i by household, government, and investment; $C_{i,h}$ denotes consumption made by household h of product i; I_i represents the investment-oriented deployment of goods i; and G_i indicates the governmental utilization of product i.

$$FNCO_i = \frac{\left(\sum_h C_{i,h} + I_i + G_i\right) \times EFEC_i}{PE_i} \tag{6}$$

Equation (7) mathematically identifies the total CO_2 emission, where TCO represents the sum of CO_2 emission constituted by intermediate utilization ($INTCO_j$) and final consumption ($FNCO_i$).

$$TCO = \sum_j INTCO_j + \sum_i FNCO_i \tag{7}$$

2.2.2. Land Transformation

Land is included in capital in the standard CGE model developed by the PEP research network [17]. As land plays an important role in the determined scenarios, it was separated from capital in this study, as shown in Figure 3. The land use of each agricultural subsector and its rental rate that is presented in Table 2 were employed to separate land from capital. The information on land use and the rental rate of land types in 2015 was mainly provided by OAE [29,30]. Only the area of livestock and forestry that are not provided by OAE were from Thailand's Land Development Department [24].

Table 2. Cropland and rental rate by agricultural subsector.

Land Use Types	2015 Land Use (ha) [a]	2015 Rental Rate (THB/ha) [b]
Paddy	10,643,878	5011
Maize	1,053,935	5072
Tapioca	1,491,155	6248
Sugarcane	1,534,632	8351
Oil palm	813,296	6031
Livestock	306,619	6222
Forestry	16,935,417	62
Other agricultures	7,606,344	6222

Notes: Rental rate of the land dedicated to livestock and other agricultures is assumed to be equivalent to the average rental rate of the first five land use types. As forest land has no rent, the rental rate for forestry is assumed to be 1 percent of the average rental rate of the first five land use types to enable model simulation. This assumption does not affect the results because relative prices are relied on in the model.

The effects of biofuel crop expansion (in each simulation scenario) on land transformation could be estimated by comparing the size of each land use type in the simulation scenario with that of the BAU. Equations (8) and (9) were used to calculate the size of land use types in the simulation scenarios. Mathematically, $Qland_j$ denotes the size of the land used by sector j (ha); a_qland_j is the coefficient for land use of sector j; $KNDC_j$ refers to the demand for land of sector j (Thai baht or THB); $QlandO_j$ is the initial size of the land used by sector j (ha; i.e., [a] in Table 2); and $KNDO_j$ is the initial value of the demand for land of sector j (THB; i.e., the product of [a] and [b] in Table 2).

$$Qland_j = a\_qland_j \times KNDC_j \tag{8}$$

$$a\_qland_j = \frac{QlandO_j}{KNDO_j} \tag{9}$$

Not only does land transformation decrease the number of species on land but it also contributes indirectly to global warming from burning and losing the ability to absorb carbon dioxide.

The impact of land transformation on the number of species could be estimated using the endpoint characterization factors for land transformation from Goedkoop et al. [31]. Following their computational technique, transforming one agricultural land to another one had no impact to the number of species, only the transformation of forest to agricultural land has. Further explanation on assessing the impact of land transformation on species loss can be found in Section 2.2.4.

Considering the impacts of land transformation on global warming, this study followed the method introduced by Silalertruksa and Gheewala [32] to compute GHG emissions that are caused by land transformation. The method is summarized in Equation (10), where *EFLUC* is the GHG emission factor for land transformation (tonne CO_2 eq./ha.yr); *BCL* stands for biomass carbon stock loss (the loss of the aboveground biomass carbon stock in the transformed land); *CSOC* is the change in soil carbon stock (i.e., the difference between soil organic carbon of the land before transformation (SOC_{before}) and soil organic carbon of the land after transformation (SOC_{after}), as shown in Equation (11)); *GHGLUC* is the amount of GHG emissions from land clearing (i.e., the sum of CO_2 *emissions* and *non-CO_2 GHG emissions* caused by burning biomass in the transformed land as presented in Equation (12)); and *T* refers to the time span of crop. The factor of 3.664 in Equation (10) was applied to convert carbon (12.01) to CO_2 (44.01). The information used for the calculation of Equations (10)–(12) is presented in Appendix A, Table A3.

$$EFLUC = \left(\frac{BCL \times 3.664}{T} \right) + \left(\frac{CSOC \times 3.664}{T} \right) + \left(\frac{GHGLUC}{T} \right) \tag{10}$$

$$CSOC = SOC_{before} - SOC_{after} \tag{11}$$

$$GHGLUC = CO_2 emissions + Non - CO_2 GHGemissions \tag{12}$$

Referring to Section 2.1.4, the two types of land being transformed were the forest (scenario S2) and abandoned rice field (scenario S3). The transformation of one type of agricultural land to another type of agricultural land in scenario S1 was considered to have no change in GHG emissions. The transformation of the forest in scenario S2 is based on the assumption that 50 percent of the 0.02 percent of Thailand's forest area in 2015 is transformed to crop fields and the remaining 50 percent is converted to perennial plants (using oil palm as a representative). Likewise, in scenario S3, 50 percent of the abandoned rice fields available in 2015 are assumed to be transformed to crop fields and another 50 percent to oil palm. Accordingly, the total amount of GHG emissions of scenarios S2 and S3 could be calculated using Equations (13) and (14), respectively, where $LMCO_{S2}$ and $LMCO_{S3}$ are the total GHG emissions from land transformation (tonne CO_2 eq.) under scenarios S2 and S3, respectively; and A_{S2} and A_{S3} are the size of the land transformed (ha) in scenarios S2 and S3, respectively.

$$LMCO_{S2} = \left(0.5 \times A_{S2} \times EFLUC_{Forest\ to\ crop} \right) + \left(0.5 \times A_{S2} \times EFLUC_{Forest\ to\ perennial\ plant} \right) \tag{13}$$

$$LMCO_{S3} = \left(0.5 \times A_{S3} \times EFLUC_{Abandoned\ land\ to\ crop} \right) + \left(0.5 \times A_{S3} \times EFLUC_{Abandonded\ land\ to\ perennial\ plant} \right) \tag{14}$$

2.2.3. Water Consumption

Irrigation water demand was considered in this study. The total irrigation water use of the country under each simulation scenario was computed using Equations (15)–(17), where *TQwater* is the total irrigation water use of the country; $Qwater_j$ denotes the total irrigation water used by sector *j*; a_water_j is the coefficient for irrigation water use of sector *j*; XST_j stands for the production output of sector *j*; and $QwaterO_j$ and $XSTO_j$ refer to the initial values of irrigation water used by sector *j* and the production output of sector *j*, respectively.

The total irrigation demand by agricultural subsectors ($Qwater_j$) are presented in Table 3. This study followed the method to derive the total irrigation demand of Kaenchan et al. [8] in which the amount of irrigation water required by the agricultural subsectors were calculated based on the actual amount of irrigation water used in the irrigated areas.

$$TQwater_t = \sum_j Qwater_j \tag{15}$$

$$Qwater_j = a\_water_j \times XST_j \tag{16}$$

$$a\_water_j = \frac{QwaterO_j}{XSTO_j} \tag{17}$$

Table 3. Irrigation demand by agricultural subsectors.

Agricultural Subsectors		Cultivated Area (ha) [a]	Irrigation Demand (m³/ha) [b]	Total Irrigation Demand (Million m³) [c]
Rice farming	Wet season rice	9,290,156	481	11,944
	Dry season rice	1,353,721	5526	
Maize cultivation		1,053,935	40	42
Tapioca cultivation		1,491,155	765	1140
Sugarcane cultivation		1,534,632	765	1173
Oil palm plantation		813,296	463	377

Notes: Cultivated area is the dimension of land use in 2015 from Table 2; [c] = [a] × [b].

2.2.4. Fossil Fuel Consumption

The effect of biofuel crop expansion on fossil resource depletion was estimated from the change in production outputs of the coal and lignite mining sector and the petroleum and natural gas drilling sector that could be directly obtained from the execution of the model.

After the environmental impacts of the simulation scenarios were derived, the impacts were characterized into damage categories, i.e., damage to human health, ecosystems, and resources, by using the endpoint characterization factors in the life cycle impact assessment (LCIA) method, as illustrated in Table 4. The damage to human health is represented in units of Disability Adjusted Life Year (DALY), the damage to ecosystems is presented in units of Potentially Disappeared Fraction of species (PDF.m².yr), and the damage to resources is quantified in monetary units. The damages could be converted into monetary units (THB) on the basis of the monetary conversion factors provided by Kaenchan and Gheewala [33]. However, before being utilized, the monetary conversion factors were adjusted for the time value of money following Haputta et al. [34] as explained in Equation (18) where MCF_{2017} indicates the value of monetary conversion factor in 2017; MCF_y denotes the value of monetary conversion factor in the year that it was initially calculated (year y); and r is an average inflation rate of Thailand over 2008–2017, i.e., approximately 0.02 [35]. The monetary conversion factors that were adjusted for the time value of money are shown in Table 5.

$$MCF_{2017} = MCF_y \times (1+r)^{(2017-y)} \tag{18}$$

Table 4. Endpoint characterization factors for the considered environmental impacts.

Midpoint Impact Category	Characterized Unit at Midpoint	Endpoint Characterization Factors		
		Human Health (DALY/Characterized Unit at Midpoint)	Ecosystems (PDF.m².yr/Characterized Unit at Midpoint)	Resources (USD₂₀₀₈/Characterized Unit at Midpoint)
Global warming potential	CO₂ eq.	1.40×10^{-6}	5.36×10^{-1}	-
Natural land transformation (from forest to agricultural land)	m²	-	7.90×10	-
Water depletion	m³	1.59×10^{-7}	1.32×10^{-1}	-
Fossil depletion	kg oil eq.	-	-	1.65×10^{-1}

Notes: Endpoint characterization factors for global warming potential, natural land transformation, and fossil depletion were based on Goedkoop et al. [31]; the endpoint characterization factors of ecosystems for global warming and natural land depletion, denoted as "species.yr" in Goedkoop et al. [31], were converted to the unit of PDF.m².yr by computing a ratio per the total number of species in a square meter (1,604,000 global species/1.08×10^{14} m² surface area); and the endpoint characterization factors for water depletion were obtained from Pfister et al. [36].

Table 5. Monetary conversion factors for endpoint damages.

	$THB_{2017}/DALY$	$THB_{2017}/PDF.m^2.yr$	THB_{2017}/kg Oil Eq. (THB_{2017}/USD_{2008})
Monetary Conversion Factor	576,595	1.00	6.70 (40.63)

All modifications incorporated in this extended CGE model enabled the in-depth investigation of simultaneous interactions between economic activities and environmental factors (e.g., GHG emission, land transformation, water demand and energy consumption). In particular, this framework provided the analytical foundation for circular economy analyses, allowing researchers and policymakers to conduct a cost–benefit assessment in order to achieve a sustainable growth path.

3. Results and Discussion

3.1. Conventional GDP and Other Economic Impacts

The change in conventional GDP and other macroeconomic impacts of the simulation scenarios are shown in Table 6. The direction of macroeconomic impacts among all scenarios were almost identical. Biofuel crop expansion and biofuel production could help generate more jobs, thus increasing employment. Such increase subsequently would raise household income and private consumption in the country. Concurrently, the government could earn more income taxes, leading to increased government income. Increasing domestic production and consumption simultaneously would encourage more exports, imports, and investment. As shown in Table 6, as the percent increase in exports was much higher than that of imports in all scenarios, biofuel promotion could bring about a trade surplus. The positive change in these macroeconomic indicators contributed to higher GDP at market price. The consumer price index, which is the representative price of all products purchased by households, of scenarios S1 and S2 was slightly higher due to the reduction in rice production (the explanation on the decrease in rice production is in the last paragraph of this section). By contrast, it was slightly lower in scenario S3 when the effect of biofuel crop expansion on rice production was eliminated. By considering GDP at market price along with the consumer price index (CPI), positive changes in real GDP in all scenarios were obtained.

Table 6. Macroeconomic impacts of biofuel crop expansion (% change from BAU).

Indicators	S1	S2	S3
GDP at market price	0.098	0.098	0.103
Consumer price index	0.006	0.006	−0.004
Real GDP	0.091	0.092	0.107
Employment	0.219	0.219	0.237
Export	0.112	0.112	0.120
Import	0.061	0.061	0.065
Private consumption	0.053	0.053	0.059
Government income	0.090	0.090	0.105
Household income	0.096	0.096	0.098
Gross fixed capital formation	0.154	0.155	0.172

Following Table 6, the economic impacts of scenarios S1 and S2 were mostly similar; however, the change in real GDP of scenario S2 was slightly higher than that of scenario S1. The change in real GDP was largest in scenario S3. This result showed that utilizing abandoned rice fields for agriculture is the best option for biofuel development from an economic point of view.

By multiplying the change in the real GDP of each scenario in Table 6 with the 2017 real GDP of 10,248 billion THB, the values of the change in the real GDP of scenarios S1–S3 of approximately 9, 9, and 12 billion THB, respectively, were derived. Accordingly, the values of conventional real GDP that could be used for Green GDP calculation (following Equation (1)) of scenarios S1–S3 were 10,257, 10,257, and 10,260 billion THB, respectively.

Table 7 shows the sectoral impacts of simulation scenarios in terms of percent change from BAU. The results demonstrated that biofuel promotion could reduce the production capability of several industries such as petroleum and natural gas, textile, rubber and plastic, iron and steel, engine, and electrical machinery and parts as shown in their lower output and employment in all biofuel promoting scenarios. The reason is to serve higher productions of biofuels. Simultaneously, labor mobility occured between these sectors to palm oil production, tapioca milling, and sugar milling.

Table 7. Sectoral impacts of biofuel crop expansion (% change from BAU).

Sector Number	Activities	S1		S2		S3	
		Output	Employment	Output	Employment	Output	Employment
1	Rice cultivation	−0.02	0.01	−0.01	0.01	0.20	0.12
2	Maize cultivation	0.03	0.07	0.03	0.07	0.17	0.01
3	Tapioca cultivation	6.20	3.98	6.20	3.98	6.20	3.98
4	Sugarcane cultivation	4.50	0.52	4.50	0.52	4.50	0.52
5	Oil palm plantation	3.70	0.22	3.70	0.22	3.70	0.22
6	Livestock	0.05	0.21	0.05	0.22	0.06	0.24
7	Forestry	0.00	0.00	0.00	0.00	0.00	0.01
8	Fishery	0.03	0.12	0.03	0.12	0.03	0.13
9	Other agricultural activities	0.01	0.03	0.01	0.03	0.09	0.08
10	Coal and lignite mining	0.01	0.02	0.01	0.02	0.01	0.03
11	Petroleum and natural gas	−0.09	−0.26	−0.09	−0.26	−0.09	−0.25
12	Other mining and quarrying	0.04	0.14	0.04	0.14	0.05	0.15
13	Other food manufacturing	0.13	0.44	0.13	0.44	0.15	0.48
14	Palm oil production	3.35	12.63	3.35	12.63	3.35	12.64
15	Rice milling	−0.02	−0.04	−0.01	−0.03	0.22	0.47
16	Tapioca milling	5.67	15.17	5.67	15.17	5.67	15.17
17	Maize drying and grinding	0.05	0.12	0.05	0.12	0.06	0.15
18	Sugar refinery	4.51	15.93	4.51	15.93	4.51	15.93
19	Textile production	−0.03	−0.09	−0.03	−0.09	−0.03	−0.09
20	Wood and furniture production	0.01	0.01	0.01	0.01	0.01	0.02
21	Paper production and printing	0.01	0.02	0.01	0.02	0.01	0.02
22	Chemical production	0.08	0.23	0.08	0.23	0.08	0.24
23	Petroleum refinery	0.02	0.10	0.02	0.10	0.03	0.12
24	Rubber and plastic production	−0.06	−0.17	−0.06	−0.17	−0.05	−0.14
25	Other non-metallic production	0.04	0.13	0.04	0.13	0.05	0.14
26	Iron and steel production	−0.04	−0.10	−0.04	−0.10	−0.04	−0.11
27	Fabricate metal production	0.00	0.00	0.00	0.00	0.00	0.00
28	Engine production	−0.01	−0.02	−0.01	−0.02	−0.01	−0.02
29	Electrical machinery production	−0.03	−0.13	−0.03	−0.13	−0.04	−0.14
30	Other manufacturing	−0.04	−0.09	−0.04	−0.09	−0.04	−0.09
31	Electricity production	0.07	0.18	0.07	0.18	0.08	0.19
32	Construction	0.08	0.25	0.08	0.25	0.09	0.28
33	Trade	0.11	0.46	0.11	0.46	0.12	0.51
34	Rail transportation	0.08	0.09	0.08	0.09	0.09	0.09
35	Road transportation	0.07	0.17	0.07	0.17	0.07	0.18
36	Water transportation	0.03	0.09	0.03	0.09	0.03	0.09
37	Air transportation	0.00	0.01	0.00	0.01	0.00	0.01
38	Other transportation	0.01	0.03	0.01	0.03	0.01	0.04
39	Services	0.05	0.09	0.05	0.09	0.05	0.10

Note: The impact on employment depends on the elasticity of substitution between production factors (Table A2 of Appendix A).

The expansion of biofuel crops led to a positive change in the production of all agricultural subsectors except for rice cultivation in scenarios S1 and S2. The enhancement of household income due to biofuel promoting policies drives the demand for agricultural products higher. Thus, the production of livestock, fishery, and other agricultural products increase. Based on scenario S1, the expansion of biofuel crops had a negative impact on the production capacity of rice cultivation and milling when the agricultural land was constant. A small negative effect on the production capacity of rice cultivation and rice milling was still found in scenario S2, where approximately 3270 ha of forest was transformed to agricultural land. More land would be required for agriculture to eliminate the negative change in output and employment in scenario S2. Utilizing abandoned rice fields (scenario S3) for agriculture could enhance the economic production of all agricultural subsectors, especially rice cultivation. Nevertheless, it brought about higher adverse impacts on the production capability and employment of iron and steel production and electrical machinery and parts industries than in scenarios S1 and S2. The reason is because workers of these sectors move to palm oil production, tapioca milling, and sugar milling to serve the increased productions of biofuels.

These obtained results are in accordance with those reported in previous publications using CGE models for examining the economy-wide impacts of biofuel policies in the case of Thailand [9,10]. Specifically, this study's simulation outcomes similarly showed that the expansion of biofuels could induce substitution effects on sectoral productions, leading to the manufacturing contraction of petroleum and natural gas. On the other hand, all participants in the biofuel supply chain (e.g., tapioca, sugarcane, and oil palm plantations) could benefit from this structural shift. Likewise, the macroeconomic indicators obtained from this study's simulations align with those shown in Phomsoda et al. [9] and Phomsoda et al. [10], indicating the same range of variation in real GDP and the essential role of productivity improvement on inflation (i.e., the percentage change of CPI).

3.2. Environmental Impacts

The change in environmental impacts (compared with the BAU) from the CGE model are exhibited in Table 8. The increase in global warming was highest in scenario S3, as greater the economic activity (real GDP in Table 6 and production output in Table 7), greater the consumption of energy and chemical fertilizers. The values of change in global warming outside the blanket in Table 8 was calculated only on the basis of the amount of GHG emissions from energy consumption and chemical fertilizer use. They were not combined with GHG emissions from land transformation. After combining with GHG emissions from land transformation, the increasing rate of global warming in scenarios S2 and S3 compared with the BAU changed to 0.241 percent and 0.004 percent, respectively, as shown in the parentheses. The high increasing rate of global warming in scenario S2 was contributed by GHG emissions from forest land clearing and the loss of carbon stock in biomass and soil (aboveground and belowground carbon stocks). By contrast, the low increasing rate of global warming in S3 was due to a small amount of GHG emissions from the abandoned land clearing and a slight loss of biomass carbon stocks. In addition, transforming the abandoned rice field to the agricultural land helped increase soil organic carbon (belowground carbon stock). Therefore, the reduction in GHG from increasing soil organic carbon was greater than the GHG emissions from land clearing and biomass carbon stock loss under the land transformation in scenario S3.

Table 8. Environmental impacts due to biofuel crop expansion (% change from BAU).

Impact Categories		S1	S2	S3
Global warming		0.075 (0.075)	0.075 (0.241)	0.083 (0.004)
Land transformation (from forest to agricultural land)		0.000	0.020	0.000
Water depletion		0.924	0.927	1.101
Fossil depletion	Coal and lignite	0.006	0.007	0.007
	Petroleum and natural gas	−0.087	−0.087	−0.085

Notes: Global warming of the BAU is 254 million tonnes CO_2 eq. (the amount of CO_2 emissions from energy use in 2015 was used as it is the most updated amount) [27]; the BAU values in 2017 for other environmental impacts are as follows: forest area = 16.35 million ha [23], irrigation water use = 14,676 million m^3 (i.e., the sum of the numbers in the column [c] of Table 3), coal and lignite consumption = 13,850 ktoe [37], and petroleum and natural gas consumption = 85,370 ktoe [37].

The effect of land transformation on ecosystem health was considered in the impact category of land transformation. In this case, only the transformation of forest to agricultural land was considered to have an effect on ecosystems. Therefore, only the transformation of forest to agricultural land was considered in Table 8, and thus, a 0.02 percent increase in land transformation was presented under scenario S2. Water depletion showed the volume of irrigation water demand in each scenario. More irrigation water was required in all scenarios, especially in scenario S3, implying that more biofuel crop cultivation could lead to increased demand for water and that the volume of water required is positively correlated to the area of agricultural land. As the dimension of agricultural area in scenario S3 was larger than that in the other scenarios after accounting for land transformation, scenario S3 required more water than scenarios S1 and S2. As for fossil depletion, a reduction in petroleum and natural gas use could be observed, while the consumption of coal and lignite was higher in all scenarios. The consumption of petroleum and natural gas was reduced as a result of the substitution of conventional fuels, i.e., gasoline and diesel, by biofuels. However, the increase in the production of electricity and chemical products led to more consumption of coal and lignite. Such increase in the production of electricity and chemical products was driven by more economic activities (as shown in Tables 6 and 7).

The value of environmental impacts after adjusting for the change in Table 8 could be obtained from combining the BAU of the impacts with the product of the BAU and the percent change of environmental impacts in Table 8. The obtained values were expressed in Table A4 of Appendix A. Then, the impacts in Table A4 were transformed into endpoint damages by using the characterization factors in Table 4. The endpoint damages of each scenario are shown in Table A5 of Appendix A.

3.3. Environmental Costs

The environmental costs are presented in Table 9. The costs were obtained by multiplying the environmental impacts in Table A5 with the monetary conversion factors in Table 5. In Table 9, the total environmental cost of each scenario was computed on the basis of Equation (2). The total environmental cost of scenario S2 was the highest among all scenarios due to the effects of forest transformation that induces CO_2 emissions higher and causes a loss of biodiversity on land. This finding also showed that converting a small piece of forest land (in this case, approximately 3300 ha) could lead to more environmental impacts than the transformation of large abandoned land (in this case, 164,800 ha). The lowest total environmental cost of BAU scenario implied that biofuel crop expansion could bring about adverse environmental impacts. However, the impacts could be alleviated by utilizing abandoned rice fields as the total environmental cost of scenario S3 was lower than that of the other biofuel crop expansion scenarios (scenarios S1 and S2).

Table 9. Environmental costs incurred by biofuel crop expansion.

Impact Categories	Environmental Costs (Billion THB$_{2017}$)			
	BAU	S1	S2	S3
Global warming (COP)	341.30	341.56	342.13	341.32
Land transformation (from forest to agricultural land) (COL)	0.00	0.00	2.57	0.00
Water depletion (CWD)	1.35	3.31	3.31	3.31
Fossil depletion (CFD)	0.67	0.66	0.66	0.66
Total (TEC)	343.31	345.53	348.67	345.30

3.4. Green GDP

The total environmental cost of each scenario in Table 9 was subtracted from its conventional GDP following Equation (1) to derive the Green GDP of each scenario. The Green GDP of each scenario is illustrated in Table 10. The highest Green GDP of all scenarios could be found in scenario S3, where biofuel crops were expanded along with the utilization of abandoned rice fields. Despite having higher environmental cost than the BAU, the Green GDP of all biofuel expansion scenarios were still higher than that of the BAU scenario. The increase in conventional GDP of all biofuel expansion scenarios could compensate for their higher environmental cost compared with the BAU scenario. Thus, considering Green GDP as an index for sustainable economic growth, biofuel crop expansion could be a policy leading towards sustainable development. However, as the Green GDP in scenario S2 was smaller than those of scenarios S1 and S3, expanding biofuel crops with forest transformation was considered to be less desirable. Policymakers should issue a law to prevent the transformation of forest to agricultural land, especially in remote areas.

Table 10. Conventional GDP, Green GDP, and GDP and environmental cost of the country.

Indicators	BAU	S1	S2	S3
Conventional GDP (real value) (billion THB)	10,248	10,257	10,257	10,260
Green GDP (real value) (billion THB)	9905	9912	9909	9914
GDP/monetary value of environmental damage	29.85	29.69	29.42	29.71

The GDP per unit of environmental cost in all scenarios showed that the value of economic production accounted for 29–30 times of the value of environmental damage. Furthermore, the GDP per unit of environmental cost was found to be the greatest in the BAU scenario, followed by scenarios S3, S1, and S2. As the GDP per unit of environmental cost implies how much value of economic production is contributed by one unit of environmental cost, an occurrence of environmental damage (derived from resource depletion and environmental degradation) in the BAU scenario was the most worthwhile. Therefore, where the efficiency of resource use and environmental degradation is considered, scenario S3, whose Green GDP is highest, may not be the best option. The policy under scenario S3 could maximize net social welfare, but it was not the most efficient scenario in terms of resource use and environmental degradation. Reduction in resource use, especially for water, and GHG emissions should be considered to achieve efficiency and welfare maximization. As biofuel crops expansion brings about larger water consumption and more GHG emissions (as a result of the enhancement of economic production), production technologies that can increase the productivity of sugarcane and oil palm cultivation and decrease GHG emissions should be applied. For example, green-cane cutting and mechanization should be utilized for sugarcane harvesting instead of burnt-cane cutting. Following Silalertruksa et al. [38] and Pongpat et al. [39], green-cane cutting and mechanization could provide more productivity for sugarcane cultivation than burnt-cane cutting while they generate less GHG emissions. Moreover, more serious regulations on industrial pollution control may help reduce GHG emissions.

Considering the pros and cons, as the efficiency of resource use and environmental degradation could be improved, the decision on biofuel crop expansion should be initially made based on economic welfare (in this case, Green GDP). Then, the policy to achieve increased efficiency can be improved. Therefore, this study showed that biofuel crop expansion could help enhance national economic welfare, and the most viable option for biofuel crop expansion is utilizing abandoned rice fields for agriculture. However, along with this policy, an improvement of production technologies and environmental mitigation measures to encourage more efficiency should be implemented.

4. Conclusions

In this study, the effects of biofuel crop expansion on Green GDP, the conventional GDP that is adjusted for environmental cost, were estimated. Three scenarios related to biofuel crop expansion policies were set to provide some policy implications towards sustainable biofuel development in Thailand. CGE modeling was used to estimate Green GDP of each scenario. Calculations based on LCIA were conducted, along with monetary conversion factors, to convert them into monetary units (environmental cost) to incorporate the environmental impacts (environmental degradation and resource depletion caused by GHG emissions, water resource use, land use, and fossil consumption) into the estimation. The results of the study could be concluded as follows:

- Biofuel crop expansion can help enhance economic growth and employment, but it can also lower the production of rice and some industrial outputs, which could be partially compensated by land expansion. As Green GDP, representing the net social welfare, for biofuel crop expansion policies was greatest when the abandoned rice fields are utilized for cultivation, this policy is recommended to be promoted.
- However, considering GDP per environmental cost, the policy of expanding biofuel crops along with utilizing abandoned rice fields for agriculture is still not the most efficient option. The efficiency of resource use and environmental degradation under this policy should be enhanced through technological improvements to achieve welfare maximization and efficiency. Furthermore, the government should support research on the productivity improvement of sugarcane and oil palm production and launch some environmental impact mitigating policies such as promoting green-cane cutting for sugarcane harvesting and supporting the utilization of alternative fuels in cultivation to encourage greater efficiency of natural resource use and environmental degradation.
- Increasing the cultivation of biofuel crops utilizing abandoned rice fields for agriculture may decrease the production capability and employment of iron and steel production and electrical machinery and parts industries. The reason is that the labor of these sectors moves to palm oil production, tapioca milling, and sugar milling to serve the increase in productions of biofuels. Increasing labor productivity by increasing the machinery to labor ratio, improving labor skill, and increasing working hours (overtime) can be considered to eliminate the labor shortage in iron and steel production and electrical machinery and parts industry.
- Expanding biofuel crop cultivation areas and utilizing forest areas provides even lower Green GDP than the scenario in which there is no land transformation, and its GDP per environmental cost is the lowest among all scenarios. This policy is thus considered inefficient. Therefore, strict laws and regulations must exist to prevent the illegal transformation of forest to agricultural land, especially in remote areas. Additionally, the governmental agency in charge should carefully make considerations on providing concessions for the regulated use of forest areas for other purposes, especially for oil palm plantation that has previously been mentioned.

The results of this study can support policymakers in making decisions on biofuel crop expansion. The provided information on environmental impacts can serve as a guideline for resource management and planning as well as environmental impact mitigating policies. The method to derive the effect of policy to Green GDP presented in this study is novel and can also be used for assessing the annual Green GDP of a country. Moreover, it can be applied to estimate the sustainability of public policies for which Green GDP is taken as an indicator.

For future policy formulations, the use of a dynamic CGE model would be preferable, especially for examining the dynamic adjustment and the long-term impact. Additionally, in this study, the rental rates of a few land use types were assumed. The actual rental rate of those land use types, if available, can instead be applied in future research.

Author Contributions: Conceptualization, P.H. and S.H.G.; data curation, P.H. and T.B.; formal analysis, P.H. and T.B.; funding acquisition, P.H.; investigation, P.H. and T.B.; methodology, P.H., T.B., N.P. and S.H.G.; project administration, P.H. and S.H.G.; resource, P.H. and S.H.G.; software, T.B. and N.P.; supervision, P.H. and S.H.G.; validation, T.B., N.P. and S.H.G.; visualization, P.H. and T.B.; writing—original draft preparation, P.H.; writing—review and editing, N.P. and S.H.G. All authors have read and agreed to the published version of the manuscript.

Funding: This research was funded by the National Science and Technology Development Agency (NSTDA) and King Mongkut's University of Technology Thonburi (KMUTT). The article processing charge was supported by Thammasat University.

Institutional Review Board Statement: Not applicable.

Informed Consent Statement: Not applicable.

Data Availability Statement: Not applicable.

Acknowledgments: This study was jointly carried out under the project "Network for Research and Innovation for Trade and Production of Sustainable Food and Bioenergy" supported by NSTDA and the post-doctoral fellowship program of KMUTT.

Conflicts of Interest: The authors declare no conflict of interest. The funders had no role in the design of the study; in the collection, analyses, or interpretation of data; in the writing of the manuscript, or in the decision to publish the results.

Abbreviations

AEDP	Alternative Energy Development Plan
BOT	Bank of Thailand
BAU	Business-as-usual
CGE	Computable general equilibrium
CPI	Consumer price index
DALY	Disability Adjusted Life Year
DEDE	Department of Alternative Energy Development and Efficiency
EPPO	Energy Policy and Planning Office
GIS	Geographic Information System
GPI	Genuine Progress Indicator
GHG	Greenhouse gas
GDP	Gross Domestic Product
ISWE	Index of Sustainable Economic Welfare
OAE	Office of Agricultural Economics
LDD	Land Development Department
LCIA	Life cycle impact assessment
NESDC	National Economic and Social Development Council
NSO	National Statistical Office
PEP	Partnership for Economic Policy
PDF	Potentially Disappeared Fraction of species
SAM	Social Accounting Matrix
THB	Thai baht
TGO	Thailand Greenhouse Gas Management Organization

Appendix A

Table A1. List of sectors and commodities in CGE model.

I-O Code [a]	Sector Number	Activities	Product Number	Products
001	1	Rice cultivation	1	Rice
002	2	Maize cultivation	2	Maize
004	3	Tapioca cultivation	3	Tapioca
009	4	Sugarcane cultivation	4	Sugarcane
011	5	Oil palm plantation	5	Oil palm
018-023	6	Livestock	6	Livestock
025-027	7	Forestry	7	Forest products
028-029	8	Fishery	8	Fish
003, 005-008, 010, 012-017, 024	9	Other agricultural activities	9	Other agricultural products
030	10	Coal and lignite mining	10	Coal and lignite
031	11	Petroleum and natural gas	11	Petroleum and natural gas
032-041	12	Other mining and quarrying	12	Mineral
042-046, 047-048, 052-054, 056-066	13	Other food manufacturing	13	Other food
047B	14	Palm oil production	14	Palm oil
049	15	Rice milling	15	Milled rice
050	16	Tapioca milling	16	Tapioca products
051	17	Maize drying and grinding	17	Grinded maize
055	18	Sugar refinery	18	Sugar
067-074	19	Textile production	19	Fabric
078-080	20	Wood and furniture production	20	Wooden products
081-083	21	Paper production and printing	21	Paper and printing products
084-092	22	Chemical production	22	Chemicals
093, 094, 136	23	Petroleum refinery	23	Petroleum products
095-098	24	Rubber and plastic production	24	Rubber and plastic
099-104	25	Other non-metallic production	25	Other non-metallic products
105-107	26	Iron and steel production	26	Iron and steel
108-111	27	Fabricate metal production	27	Fabricate metal
112-115, 123-128	28	Engine production	28	Engines
116-122	29	Electrical machinery production	29	Electrical machinery
075-077, 129-134	30	Other manufacturing	30	Products from other manufacturing
135	31	Electricity production	31	Electricity
138-144	32	Construction	32	Infrastructures
145-146	33	Trade	33	Trade
149	34	Rail transportation	34	Rail transportation
150-152	35	Road transportation	35	Road transportation
153-155	36	Water transportation	36	Water transportation
156	37	Air transportation	37	Air transportation
157	38	Other transportation	38	Other transportation
137, 147-148, 158-180	39	Services	39	Services

Note: [a] is based on NESDC [19].

Table A2. Parameters of elasticity of substitution.

Sector Number [Industry (j)]	Elasticity of Substitution between Capital–Land Composite and Labor [a]	Elasticity of Substitution between Capital and Land [b]	Sector Number [Industry (j)]	Elasticity of Substitution between Capital–Land Composite and Labor	Elasticity of Substitution between Capital and Land
1	0.20	0.20	21	1.50	0.50
2	0.20	0.20	22	1.50	0.50
3	0.20	0.43	23	1.50	0.50
4	0.20	0.20	24	1.50	0.50
5	0.20	0.20	25	1.50	0.50
6	0.20	0.20	26	1.50	0.50
7	0.20	0.20	27	1.50	0.50
8	0.20	0.20	28	1.50	0.50
9	0.20	0.20	29	1.50	0.50
10	1.50	0.50	30	1.50	0.50
11	1.50	0.50	31	1.50	0.50
12	1.50	0.50	32	1.50	0.50
13	1.50	0.50	33	1.50	0.50
14	1.50	0.50	34	1.50	0.50
15	1.50	0.50	35	1.50	0.50
16	1.50	0.50	36	1.50	0.50
17	1.50	0.50	37	1.50	0.50
18	1.50	0.50	38	1.50	0.50
19	1.50	0.50	39	1.50	0.50
20	1.50	0.50	-	-	-

Notes: Values in [a] were calculated following OECD/ILO [20]; values in [b] were determined based on the assumption that the elasticity of substitution between capital and land of agricultural subsectors are lower than that of other sectors because the agricultural subsectors are land intensive and the substitution of capital for land is rigid; the elasticity of substitution between capital and land of sector 3 (tapioca cultivation) is assumed to be higher than that of other agricultural subsectors, allowing more flexibility for the substitution of capital for land. Thus, this sector requires a lower marginal land for producing marginal output; for other types of elasticity, the standard elasticity parameters in Decaluwé et al. [17] were employed; the elasticity of transformation of sector j was set to 2.0; the elasticity of transformation between exports and domestic sales of product i of sector j was set to 2.0; the elasticity of substitution between imported and domestically produced commodity of product i was set to 2.0.

Table A3. Values of the parameters in Equations (8)−(10).

Parameters		Units	Values
Aboveground biomass of forest		tonne carbon C/ha	162.45
Aboveground biomass of set-aside land		tonne C/ha	7.58
Soil organic carbon (SOC) of forest land		tonne C/ha	47
SOC of cropland		tonne C/ha	45.34
SOC of oil palm		tonne C/ha	63.65
SOC of set-aside land		tonne C/ha	43.26
GHG emissions from forest land clearing	CO_2 emissions	tonne CO_2 eq./ha	261.42
	Non-CO_2 GHG emissions	tonne CO_2 eq./ha	37.99
GHG emissions from set-aside land clearing	CO_2 emissions	tonne CO_2 eq./ha	-
	Non-CO_2 GHG emissions	tonne CO_2 eq./ha	0.91
Time span of field crop		year (yr)	4
Time span of oil palm		yr	25

Notes: All values were derived based on the method of calculation introduced in Silalertruksa and Gheewala [32] and information from JGSEE [40] and IPCC [41].

Table A4. The environmental impacts after adjusting for the change in Table 8.

Impact Categories	BAU	S1	S2	S3
Global warming potential (million tonne CO_2 eq.)	254.43	254.62	255.04	254.44
Land transformation (from forest to agricultural land) (ha)	0.00	0.00	3,269.00	0.00
Water depletion (million m^3)	14,676.23	14,811.84	14,812.28	14,837.82
Fossil depletion (KTOE)	99,220.00	99,146.72	99,146.75	99,148.12

Table A5. Endpoint damages to the safeguard subjects, human health, ecosystem, and resources in each scenario.

Scenarios	Midpoint Impact Categories	Damage Categories		
		Human Health (DALY)	Ecosystems (PDF.m^2.yr)	Resources (USD$_{2008}$)
BAU	Global warming	3.6×10^5	1.4×10^{11}	0.0×10^0
	Land transformation	0.0×10^0	0.0×10^0	0.0×10^0
	Water depletion	2.3×10^3	0.0×10^0	0.0×10^0
	Fossil depletion	0.0×10^0	0.0×10^0	1.6×10^7
S1	Global warming	3.6×10^5	1.4×10^{11}	0.0×10^0
	Land transformation	0.0×10^0	0.0×10^0	0.0×10^0
	Water depletion	2.4×10^3	2.0×10^9	0.0×10^0
	Fossil depletion	0.0×10^0	0.0×10^0	1.6×10^7
S2	Global warming	3.6×10^5	1.4×10^{11}	0.0×10^0
	Land transformation	0.0×10^0	2.6×10^9	0.0×10^0
	Water depletion	2.4×10^3	2.0×10^9	0.0×10^0
	Fossil depletion	0.0×10^0	0.0×10^0	1.6×10^7
S3	Global warming	3.6×10^5	1.4×10^{11}	0.0×10^0
	Land transformation	0.0×10^0	0.0×10^0	0.0×10^0
	Water depletion	2.4×10^3	2.0×10^9	0.0×10^0
	Fossil depletion	0.0×10^0	0.0×10^0	1.6×10^7

References

1. DEDE. *Thailand Alternative Energy Situation 2008*; Department of Alternative Energy Development and Efficiency: Bangkok, Thailand, 2008.
2. DEDE. *Thailand Alternative Energy Situation 2013*; Department of Alternative Energy Development and Efficiency: Bangkok, Thailand, 2013.
3. DEDE. *Thailand Alternative Energy Situation 2017*; Department of Alternative Energy Development and Efficiency: Bangkok, Thailand, 2017.
4. Office of the Cane and Sugar Board (OCSB). Ethanol Situation 2018. 2018. Available online: http://www.ocsb.go.th/upload/bioindustry/fileupload/10208-8459.pdf (accessed on 6 February 2021).
5. DEDE. *Alternative Energy Development Plan: AEDP2015*; Department of Alternative Energy Development and Efficiency: Bangkok, Thailand, 2015.
6. Silalertruksa, T.; Gheewala, S.H. The environmental and socio-economic impacts of bio-ethanol production in Thailand. *Energy Procedia* **2011**, *9*, 35–43. [CrossRef]
7. Wianwiwat, S.; Asafu-Adjaye, J. Is there a role for biofuels in promoting energy self sufficiency and security? A CGE analysis of biofuel policy in Thailand. *Energy Policy* **2013**, *55*, 543–555. [CrossRef]
8. Kaenchan, P.; Puttanapong, N.; Bowonthumrongchai, T.; Limskul, K.; Gheewala, S.H. Macroeconomic modeling for assessing sustainability of bioethanol production in Thailand. *Energy Policy* **2019**, *127*, 361–373. [CrossRef]
9. Phomsoda, K.; Puttanapong, N.; Piantanakulchai, M. Economic Impacts of Thailand's Biofuel Subsidy Reallocation Using a Dynamic Computable General Equilibrium (CGE) Model. *Energies* **2021**, *14*, 2272. [CrossRef]
10. Phomsoda, K.; Puttanapong, N.; Piantanakulchai, M. Assessing Economic Impacts of Thailand's Fiscal Reallocation between Biofuel Subsidy and Transportation Investment: Application of Recursive Dynamic General Equilibrium Model. *Energies* **2021**, *14*, 4248. [CrossRef]
11. Giannetti, B.F.; Agostinho, F.; Almeida, C.M.V.B.; Huisingh, D. A review of limitations of GDP and alternative indices to monitor human wellbeing and to manage eco-system functionality. *J. Clean. Prod.* **2015**, *87*, 11–25. [CrossRef]
12. Wu, J.; Wu, T. Green GDP. In *Berkshire Encyclopedia of Sustainability, Vol. II—The Business of Sustainability*; Christensen, K., Fogel, D., Wagner, G., Whitehouse, P., Eds.; Berkshire Publishing: Great Barrington, MA, USA, 2010; pp. 248–250.
13. Kalimeris, P.; Bithas, K.; Richardson, C.; Nijkamp, P. Hidden linkages between resources and economy: A "Beyond-GDP" approach using alternative welfare indicators. *Ecol. Econ.* **2020**, *169*, 106508. [CrossRef]
14. Li, G.; Fang, C. Global mapping and estimation of ecosystem services values and gross domestic product: A spatially explicit integration of national 'green GDP' accounting. *Ecol. Indic.* **2014**, *46*, 293–314. [CrossRef]
15. Stjepanović, S.; Tomić, D.; Škare, M. A new approach to measuring green GDP: A cross-country analysis. *Entrep. Sustain. Issues* **2017**, *4*, 574. [CrossRef]
16. Kunanuntakij, K.; Varabuntoonvit, V.; Vorayos, N.; Panjapornpon, C.; Mungcharoen, T. Thailand Green GDP assessment based on environmentally extended input-output model. *J. Clean. Prod.* **2017**, *167*, 970–977. [CrossRef]

17. Decaluwé, B.; Lemelin, A.; Robichaud, V.; Maisonnave, H. *PEP-1-t: The PEP Standard Single-Country, Recursive Dynamic CGE Model, Version 2.1*; Partnership for Economic Policy: Kasarani, Kenya, 2013.
18. Mainar-Causapé, A.J.; Ferrari, E.; McDonald, S. *Social Accounting Matrices: Basic Aspects and Main Steps for Estimation*; Publications Office of the European Union: Luxembourg, 2018.
19. Office of National Economic and Social Development Council (NESDC). Input—Output Table of Thailand 2015 Preliminary. 2019. Available online: www.nesdc.go.th/main.php?filename=io_page (accessed on 27 February 2020).
20. OECD/ILO. *How Immigrants Contribute to Thailand's Economy*; OECD Publishing: Paris, France, 2017. [CrossRef]
21. Serag, E.; Ibrahim, F.; El Araby, Z.; Abd El Latif, M.; El Sarawy, M.; El Zaabalawy, D.; El Dib, S.A.; Salem, K.; Breisinger, C.; Raouf, M. *A 2019 Nexus Social Accounting Matrix for Egypt MENA RP Working Paper 35*; International Food Policy Research Institute (IFPRI): Washington, DC, USA, 2021; Volume 35. [CrossRef]
22. Robinson, S.; Cattaneo, A.; El-Said, M. Updating and estimating a social accounting matrix using cross entropy methods. *Econ. Syst. Res.* **2001**, *13*, 47–64. [CrossRef]
23. Office of the Forest Land Management, Royal Forest Department. Forest Area of Thailand Years 1973–2018. 2019. Available online: http://forestinfo.forest.go.th/Content.aspx?id=72 (accessed on 16 April 2020).
24. Land Development Department (LDD). Land Use of Thailand 2015/2016. 2016. Available online: www.ldd.go.th/www/lek_web/web.jsp?id=18671 (accessed on 16 April 2020).
25. Office of Agricultural Economics (OAE). Quantity and Value of Imported Chemical Fertilizer. 2020. Available online: www.oae.go.th/view/1/%E0%B8%9B%E0%B8%B1%E0%B8%88%E0%B8%88%E0%B8%B1%E0%B8%A2%E0%B8%81%E0%B8%B2%E0%B8%A3%E0%B8%9C%E0%B8%A5%E0%B8%B4%E0%B8%95/TH-TH (accessed on 11 June 2020).
26. Thailand Greenhouse Gas Management Organization (TGO). Emission Factors by Industries 2020. (Update: February 2020). 2020. Available online: www.tgo.or.th/ (accessed on 27 April 2020).
27. Energy Policy and Planning Office (EPPO). CO_2 Emission by Energy Type (Historical Statistics 1987–2018). 2020. Available online: www.eppo.go.th/index.php/en/en-energystatistics/co2-statistic?orders[publishUp]=publishUp&issearch=1 (accessed on 2 April 2020).
28. Energy Policy and Planning Office (EPPO). Energy Statistics of Thailand 2016. 2016. Available online: http://www.eppo.go.th/index.php/th/component/k2/item/11342-energy-statistics-2559 (accessed on 1 April 2020).
29. Office of Agricultural Economics (OAE). Agricultural Production Information. 2015. Available online: www.oae.go.th/view/1/%E0%B8%82%E0%B9%89%E0%B8%AD%E0%B8%A1%E0%B8%B9%E0%B8%A5%E0%B8%81%E0%B8%B2%E0%B8%A3%E0%B8%9C%E0%B8%A5%E0%B8%B4%E0%B8%95%E0%B8%AA%E0%B8%B4%E0%B8%99%E0%B8%84%E0%B9%89%E0%B8%B2%E0%B9%80%E0%B8%81%E0%B8%A9%E0%B8%95%E0%B8%A3/TH-TH (accessed on 24 October 2019).
30. OAE. *Production Cost of Agricultural Products 2009–2019 on Average*; Office of Agricultural Economics: Bangkok, Thailand, 2019.
31. Goedkoop, M.; Heijungs, R.; Huijbregts, M.; De Schryver, A.; Struijs, J.; Van Zelm, R. *ReCiPe 2008 Version 1.08 (Characterization Factors Spreadsheet Belonging to ReCiPe 2008: A Life Cycle Impact Assessment Method which Comprises Harmonised Category Indicators at the Midpoint and the Endpoint Level)*; Ministry of Housing, Spatial Planning and Environment: The Hague, The Netherlands, 2013.
32. Silalertruksa, T.; Gheewala, S.H. Food, Fuel, and Climate Change: Is Palm-Based Biodiesel a Sustainable Option for Thailand? *J. Ind. Ecol.* **2012**, *16*, 541–551. [CrossRef]
33. Kaenchan, P.; Gheewala, S.H. Budget constraint and the valuation of environmental impacts in Thailand. *Int. J. Life Cycle Assess.* **2017**, *22*, 1678–1691. [CrossRef]
34. Haputta, P.; Puttanapong, N.; Silalertruksa, T.; Bangviwat, A.; Prapaspongsa, T.; Gheewala, S.H. Sustainability analysis of bioethanol promotion in Thailand using a cost-benefit approach. *J. Clean. Prod.* **2020**, *251*, 119756. [CrossRef]
35. Bank of Thailand (BOT). EC_EI_027: Thailand Macroeconomics Indicators 1. 2019. Available online: www.bot.or.th/App/BTWS_STAT/statistics/BOTWEBSTAT.aspx?reportID=409&language=TH (accessed on 11 October 2019).
36. Pfister, S.; Koehler, A.; Hellweg, S. Assessing the environmental impacts of freshwater consumption in LCA. *Environ. Sci. Technol.* **2009**, *43*, 4098–4104. [CrossRef] [PubMed]
37. Energy Policy and Planning Office (EPPO). Energy statistics of Thailand 2018. 2018. Available online: http://www.eppo.go.th/index.php/th/component/k2/item/14166-energy-statistics-2561 (accessed on 16 April 2020).
38. Silalertruksa, T.; Gheewala, S.H.; Pongpat, P. Sustainability assessment of sugarcane biorefinery and molasses ethanol production in Thailand using eco-efficiency indicator. *Appl. Energy* **2015**, *160*, 603–609. [CrossRef]
39. Pongpat, P.; Gheewala, S.H.; Silalertruksa, T. An assessment of harvesting practices of sugarcane in the central region of Thailand. *J. Clean. Prod.* **2017**, *142*, 1138–1147. [CrossRef]
40. The Joint Graduate School of Energy and Environment (JGSEE). *Baseline Study for GHG Emissions in Palm Oil Production: Second Draft Final Report*; JGSEE: Bangkok, Thailand, 2010.
41. Intergovernmental Panel on Climate Change (IPCC). *2006 IPCC Guidelines for National Greenhouse Gas Inventories Volume 4—Agriculture, Forestry and Other Land Use (AFOLU)*; Institute for Global Environmental Strategies (IGES): Hayama, Japan, 2006.

Article

Sustainability-Focused Excellence: A Novel Model Integrating the Water–Energy–Food Nexus for Agro-Industrial Companies

Fernando Caixeta [1,2,*], André M. Carvalho [3], Pedro Saraiva [4,5] and Fausto Freire [6]

1   Federal Institute of Triângulo Mineiro, Campus Uberlândia, Fazenda Sobradinho, s/n, Uberlândia 38400-970, Brazil
2   Sustainable Energy Systems, University of Coimbra, 3000-456 Coimbra, Portugal
3   Polytechnic Institute of Cávado and Ave., 4750-810 Barcelos, Portugal
4   Department of Chemical Engineering, CIEPQPF, University of Coimbra, 3000-456 Coimbra, Portugal
5   Dean of NOVA IMS, NOVA University of Lisbon, 1070-312 Lisbon, Portugal
6   Department of Mechanical Engineering, ADAI, University of Coimbra, 3000-456 Coimbra, Portugal
*   Correspondence: fernandocaixeta@iftm.edu.br

Abstract: The water–energy–food (WEF) nexus approach is gaining attention due to the challenge of better managing natural elements. Agro-industrial companies, given their environmental impacts, need to take sustainability into proper account. However, this sector lacks the novel tools needed to integrate current methodologies with additional quality frameworks, such as business excellence models (BEMs). Therefore, the present research aims to propose a sustainability-focused excellence model by integrating the principles and objectives of the WEF nexus with existing BEM and proposing its application to agro-industrial companies. For that purpose, a new conceptual model to integrate sustainability and excellence was built. The proposed novel model can become a decision-support tool in helping agro-industrial companies transition toward improved sustainability while managing existing tradeoffs and synergies.

Keywords: Water-Energy-Food nexus; excellence models; agro-industrial companies; decision-support tool

Citation: Caixeta, F.; Carvalho, A.M.; Saraiva, P.; Freire, F. Sustainability-Focused Excellence: A Novel Model Integrating the Water–Energy–Food Nexus for Agro-Industrial Companies. *Sustainability* 2022, 14, 9678. https://doi.org/10.3390/su14159670

Academic Editor: Nallapaneni Manoj Kumar

Received: 22 February 2022
Accepted: 2 August 2022
Published: 5 August 2022

Publisher's Note: MDPI stays neutral with regard to jurisdictional claims in published maps and institutional affiliations.

1. Introduction

Sustainability actions are urgent in the face of the quick depletion of natural resources that are essential for humankind. There is a lack of adequate access to water, energy, and food resources for a substantial percentage of the global population: Despite efforts within the past 15 years, 800 million people are still considered food insecure, an equal number has no access to safe drinking water, and 1.2 billion people lack access to electricity [1]. In the face of this reality, new practices are needed transversally, and all industrial sectors benefit from focusing on impacts on environmental, social, and economic results. Many industries and sectors recognize the importance of better managing the aforementioned natural resources [2–4]. Nevertheless, the agro-industrial sector deserves to be highlighted, as it requires a huge contribution of those natural resources [5,6] and has high pollution emissions rates [7].

Given that, the good practices of natural resource management should be stimulated to overcome unsustainable practices in the agro-industrial area. In this sense, quality management (QM) could be used since it presents a framework that organizations are already familiar with and has proved to be valuable when adapted to new technologies [8]. Among them, one particular scope is business excellence models (BEMs), a framework that oversees all activities and tasks needed to maintain the desired level of performance in an organization and relate it to its external environment [9]. Given that, this approach provides the methods (tools, processes, metrics, and indicators) to do so [10]. Accordingly, and in principle, BEMs offer a useful toolset to also oversee the pursuit of organization-wide sustainability [8,11].

In this context, one important sustainability management concept used nowadays is the so-called water–energy–food (WEF) nexus. It means that the aforementioned elements are intrinsically managed together, and one action in one direction can affect both of the others [12,13]. As a result, these three issues should be considered in an integrated manner because they are connected, and their utilization may expose important tradeoffs [14]. Identifying and evaluating tradeoffs and synergies is essential for integrating other perspectives. Nevertheless, there is no clear framework to measure and assess the quality of processes and operations related to the deployment of the principles behind the WEF nexus [15]. The incorporation of the WEF nexus into an enterprise-level model is essential for integrating sustainability into efficient planning, development, and monitoring, but also for policymaking in key WEF-productive sectors of the economy [15,16].

In order to close this gap, our work highlights the use of BEMs as a possible solution for the integration of productivity and sustainability in such industries. BEMs best express the summary of different efforts and frameworks to promote the mentality of quality and continuous improvement in an organization, doing so by highlighting strategic opportunities and industry best practices [8,17]. Therefore, the objective of this work is to propose a sustainability-focused business excellence model, integrating the WEF nexus with BEM, exploring a bridge between sustainability and performance excellence, and allowing for a more complete management perspective to be made by agro-industrial companies operating according to WEF nexus principles.

2. Literature Review

2.1. Water, Food, and Energy Systems and the WEF Nexus

Concerns about water, food, and energy systems are rapidly growing due to differing regional availabilities and their impact on the interdependences amongst themselves [15,18]. WEF elements are essential for human life, sustainable development, and social equality [19]. In this sense, ensuring their security is a crucial activity concerning every individual worldwide.

Given that, the need for the integrated analysis of natural elements is not a novelty in the scientific literature. In this sense, some authors affirmed that WEF elements should be evaluated in an integrated manner. This concept is considered the first attempt at this new approach [20–23]. However, the WEF nexus concept was only launched in 2011, when it was incorporated into an international discussion on sustainable development by the Stockholm Environment Institute [12] and the World Economic Forum [13]. According to Hoff [12], the WEF nexus implies that water availability, energy production/consumption, and food security are inextricably linked. Consequently, actions in any one area have impacts on the others [19]. In addition, the World Economic Forum [13] affirms that focusing on the joint promotion of food security, in addition to water and energy accessibility, is the meaning of nexus thinking. These interlinkages, and the criticality of the scarcity of any of the three resources and its impact on the others, emphasize the need to manage them jointly and more efficiently [24].

Such a nexus approach can support the transition to implement the Sustainable Development Goals (SDGs)—Agenda 2030 [25]. In this sense, scientific literature has plenty of studies linking WEF nexus implementation with the SDGs, performing better in regional sustainability [26], for legal challenges [27], and for rural communities [28], among other uses. Furthermore, this addition should reduce environmental impacts and generate additional benefits that outweigh the integration across sectors, enhancing, for instance, circular economy implementation [29]. Such gains should appeal to national interest and encourage governments, private sectors, and civil society to engage [30]. In practical terms, this approach can represent the action required in bioeconomy to implement real solutions in sustainability actions [31].

Many scientific articles have tried to express some practical examples of how to implement this methodology. Saladini et al. [26] selected 12 indicators to monitor the Mediterranean area called the Partnership for Research and Innovation in the Mediterranean Area, based on the Sustainable Development Goals, which can be directly related to boosting sustainable business innovation [32]. Hussien et al. [33] assessed the impact of WEF elements using a risk-based method. Bijl et al. [34] showed differences in physical trade production, distance, and volume using indicators. Karan et al. [35] created indices based on the UN-Habitat's City Prosperity Index that specifically integrate the nexus-relevant indices into a weighted equity index. El-Gafy [36] proposed six indicators in order to quantify the nexus as a strategic tool applied to crop production. AbdelHady et al. [37] proposed three output indicators, agriculture, aquaculture, and net energy production, to assess the value of different ecosystem health conditions under three water management scenarios.

Additionally, managing water, energy, and food without efficient and synergistic actions may increase the risk of shortages. Consequently, one opportunity to improve the sustainable use of these sources is investigating the integration of the water–energy–food nexus with quality and business excellence initiatives.

2.2. Quality, Excellence, and Sustainability

Given the mentioned scientific gap in integrating the WEF nexus with the principles of quality and excellence, the use of BEMs has been considered. This decision is based on the understanding that these frameworks offer a clear opportunity for the development of a quality and continuous improvement mindset that can be aligned with needs and the best sustainability practices [8,38,39]. Regarding the agro-industrial sector, they are aligned with a major impact [40], and their actions can be responsible for proper natural resource management [41]. In this sense, excellence initiatives center on meeting stakeholders' needs and expectations, widening previous scopes of quality-oriented initiatives [42] and focusing on creating sustainable value for all "interested" parties. They have been used by organizations worldwide to improve their performance and achieve improved business results.

Excellence models are most widely used by organizations for self-assessment and improvement, including targeting sustainability [39], with companies opting to adapt and customize them in search of competitive advantages [17]. They promote a longitudinal management philosophy, highlighting a set of principles that orient managers' and associates' behaviors in the long-term, fostering continuous improvement [43]. BEMs focus on offering insights for organizations to manage processes, tools, or techniques, both old or new, with the goal of building value from new opportunities and achieving superior organizational results [44].

However, and regardless of their extensive use, there is still limited integration of the topic of environmental sustainability in excellence frameworks, in this particular case tackling the use of the WEF nexus approach. Despite other scientific authors cited in this research showing a growing pressure for the development of new models, the truth is that economic, social, and environmental sustainability are still often left outside of the scope of some major excellence models [45], although this situation is also changing. Thus, research on business and operational excellence does not usually include a clear focus on environmental sustainability.

Accordingly, there is a clear aim for promoting specific models for the deployment of sustainability-focused excellence in different industrial sectors. In the specific scope of this work—focusing on water, energy, and food as critical issues and looking specifically at the agro-industrial sector—this model proposes to integrate excellence and sustainability because of the WEF nexus.

Thus, in summary, the contributions of the present article are:

- Exploring possibilities and criteria for the use and selection of BEMs concerning WEF nexus elements.

- Proposing and showing the use of a theoretical and novel model to integrate the BEM and the WEF nexus approaches for agro-industrial companies.
- Stimulating and validating agro-industrial companies to perform more sustainable actions.

3. Research Methods

In the face of the interdisciplinary challenge to build a sustainability-focused excellence model, this article adopted the conceptual analysis methodology described by Jabareen [46]. This approach was extensively applied in the scientific literature, providing a comprehensive understanding of the researched phenomenon [47–51]. To this end, the criteria relationship interactions were highlighted throughout the following adapted steps: Selection and description of business excellence (BE) models (reading and categorizing of data); highlighting of the criteria used to evaluate them (identifying concepts); integration of the previous items, performed based on similarities that were found (deconstructivity, categorizing, synthesizing, and integrating concepts); and constructing a new conceptual model to integrate sustainability and excellence (presenting and validating the final framework) [46]. For that purpose, the researched keywords included synonyms dealing with combinations of the terms "water–energy–food nexus" or "water energy food nexus" ("WEF nexus"; "FEW nexus"), "WEF nexus analysis" ("WEF nexus", "WEF nexus model", "WEF nexus tools", "WEF nexus approaches"), "Agroindustrial WEF nexus", and "Agrifood industries WEF nexus", resulting in a database with 241 articles (with duplicates).

Therefore, we begin by describing and selecting the major BEMs, justifying their selection and relevance to provide context and reliability to the research process, as mentioned by Silverman [52], and Yin [53]. We also looked into criteria that may be used to evaluate the aforementioned BEMs. Finally, based on the former results, the integrated models were built and presented considering their applications in the agro-industrial area. For that purpose, indicators were grouped by using the similarity analysis method to form indices [54]. In the end, similar criteria were disposed into a conceptual model. A detailed explanation will be provided assessing sustainability standards that can be applied to WEF nexus approaches targeting agro-industrial production companies [55]. Additionally, we will highlight the differences between the proposed model and the existing ones found in the scientific literature.

4. Sustainability-Focused Model Development

4.1. Benchmarked Excellence Models

Our development of a sustainability-oriented excellence model followed the logic of aligning the different needs and concerns related to sustainability (in particular, those in the agro-industrial sector and pertaining to the WEF nexus) with the different enablers of excellence. Four different BEMs were considered for this purpose. The selection of these BEMs was based on their use, reach, and proximity to areas where agro-industrial companies have important production or sales volume. Additionally, the scientific literature connects them with sustainability [8,56–58].

Accordingly, the European Foundation for Quality Management (EFQM) excellence model was selected due to its widespread adoption by European companies, as well as for its results in improving organizational performance [44]. Similarly, the Malcolm Baldrige National Quality Award (MBNQA) excellence model was selected due to the fact that it is the most used in the USA and its proven advantages in promoting productivity in a large number of companies [59].

The Brazilian National Quality Award Model (BNQA) not only represents one of the largest and most important economies for the agro-industrial sector in the world but has also proven to be helpful in improving companies' performances by identifying their strengths and driving their improvement actions [60]. Likewise, the ZED Model—Zero Defect, Zero Effect scheme—from India [61] was selected. Not only is it promoted by the government, but it is also focused on the specific reality of micro, small, and medium enterprises in one of the most populated and largest economies of the world.

To better understand the contributions of each of the selected BEMs, further details are provided next.

4.1.1. EFQM—European Foundation for Quality Management (EFQM) Excellence Model

The European Foundation for Quality Management (EFQM) excellence model has been widely adopted in Europe and elsewhere. Efforts have been made to highlight the connection between the EFQM model and its impact from a management perspective [62]. Later in this article, we will focus on the importance of integration in the implementation of the EFQM model. Evidence shows that the more integrated this model is, the more effective and innovative it can be. However, we still lack mechanisms to achieve this integration [63]. One of the main impacts that is expected to demonstrate an innovation path for the purposes of its implementation.

There are eight fundamental enablers of excellence considered by the EFQM Model, which are the following: (1) achieving balanced results; (2) adding value for customers; (3) leading with vision, (4) inspiration and integrity; (5) managing by processes; (6) succeeding through people; (7) nurturing creativity and innovation; (8) building partnerships; and (9) taking responsibility for a sustainable future [44].

4.1.2. ZED—Zero Defect, Zero Effect scheme (Indian Program for MSMEs)

The "Zero Defect, Zero Effect" program was launched in 2014, as the Council of India established that companies should strive to manufacture their products with both zero defects and with zero effects on the environment. In this sense, a plan contemplating 50 parameters was designed to help micro, small, and medium enterprises pursue these objectives [61].

Its implementation includes process assessment and certification considering the corresponding maturity levels achieved. For that purpose, it includes a maturity model designed for different organizational areas. Among its specific parameters, the following can be found [61]: (1) manufacturing capabilities; (2) design capabilities; (3) quality/environment/safety assurance systems; (4) people development and engagement systems for quality and environment; (5) learning and improvement Systems; and (6) legal compliances (hygiene factors).

4.1.3. Malcolm Baldrige National Quality Award (MBNQA) Excellence Model

The MBNQA and its associated excellence model were launched in 1987 in the United States of America. It is aimed at increasing the productivity of national and international companies through the implementation of quality criteria. Indeed, many companies have improved their quality performance after applying the MBNQA [64]. A significant difference in terms of quality performance is evident when comparing organizations using the excellence model and winning the MBNQA with those that do not [65].

Therefore, the excellence model of the MBNQA is built on the following set of interrelated enablers: (1) visionary leadership; (2) customer-driven excellence; (3) organizational and personal learning; (4) valuing workforce members and partners; (5) agility; (6) focus on the future; (7) managing for innovation; (8) managing by the fact; (9) societal responsibility; (10) focus on results and creating value; and (11) a systems perspective [59].

4.1.4. Brazilian National Quality Award (BNQA)

Similar to its worldwide counterparts, the Brazilian National Quality Award is aimed at enhancing the national quality performance of industrial companies [66]. It was launched by the Brazilian National Quality Foundation (FNQ) around 1992. For the present research, we considered the following enablers of the BNQA criteria: (1) leadership; (2) strategies and plans; (3) clients; (4) society; (5) information and knowledge; (6) people; (7) processes; and, finally, (8) results [60].

4.2. Sustainability-Focused Model Presentation and Discussion

To build a sustainability-focused excellence model, several evaluation-related items have been considered. Thus, isolated efforts are no longer effective given the complexity of such an interdisciplinary research effort. In this sense, quality management systems have proved to provide a multiscale and holistic view, ranging from product production to societal challenges defined on a global level [67]. Based on this perspective, and on the previous excellence model principles explicated in Table 1, the following dimensions were considered: drivers, agents, process, and outcomes. Based on the previous discussion, our sustainability-focused model targeting agro-industrial companies is proposed based on the WEF nexus approach, as illustrated in Figure 1.

Table 1. Enablers used by the EFQM [44], ZED [61], MBNQA [59], and BNQA [60] models with their similar correspondents on the proposed sustainability-focused model.

Sustainability-Focused Model	EFQM	ZED	MBNQA	BNQA
DRIVERS				
Legislation		Legal compliances		
Requirements		Manufacturing capabilities	Managing by the fact	
Operational	Managing by process			Processes
ESG (Environmental, Social, and Governance)	Taking responsibility for a sustainable future	Quality/environment/safety assurance systems People development and engagement systems for quality and environment	Focus on the future Societal responsibility	Society
Optimization		Learning and improvement systems	Agility	
Financial	Achieving balanced results		Focus on results and creating value Systems perspective	Results
AGENTS				
Leaders	Leading with vision, inspiration, and integrity		Visionary Leadership	Leadership
Workforce	Succeeding through people	People development and engagement systems for quality and environment	Valuing workforce members and partners	People
Partners	Building partnerships		Valuing workforce members and partners	
PROCESS				
Learning		Design capabilities	Organizational and personal learning	
Innovation	Nurturing creativity and innovation		Managing for innovation	Strategies and plans
Valorization	Adding value for customers		Customer-driven excellence	Clients Information and knowledge
OUTCOME		Quality assurance diagnosis sustainability recommendations		

Source: Own elaboration.

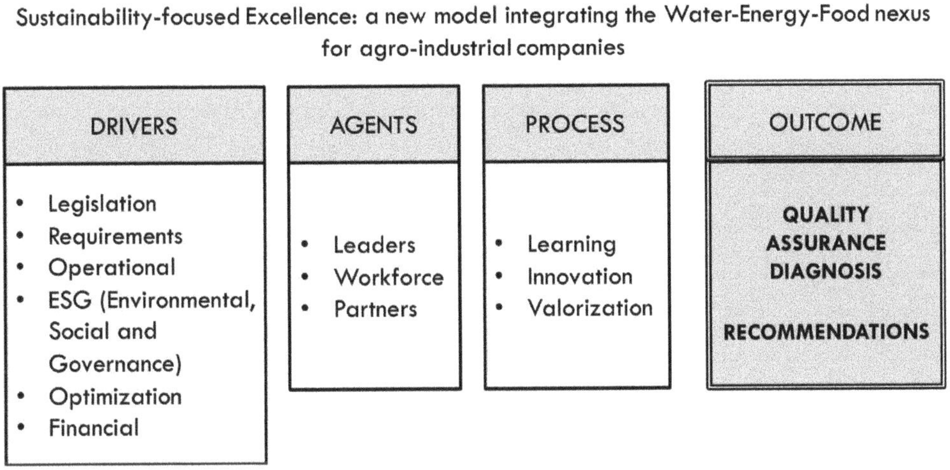

Sustainability-focused Excellence: a new model integrating the Water-Energy-Food nexus for agro-industrial companies

DRIVERS	AGENTS	PROCESS	OUTCOME
• Legislation • Requirements • Operational • ESG (Environmental, Social and Governance) • Optimization • Financial	• Leaders • Workforce • Partners	• Learning • Innovation • Valorization	**QUALITY ASSURANCE DIAGNOSIS** **RECOMMENDATIONS**

Figure 1. A novel model for sustainability-focused excellence based on the WEF nexus approach for agro-industrial companies. Source: Own elaboration.

A comprehensive description of these dimensions is presented in the following topics, and the similarities between the models can be seen in Table 1.

For the fulfillment of the proposed objective, this article presents this overall conceptual model at the level of its main dimensions and criteria, but in future publications, additional details about its indicators and guidelines on how it can be used will be provided.

4.2.1. DRIVERS

Drivers guide the actions of organizations. They are the reason behind the need to change and encourage companies to transition to a more sustainable perspective. Based on the reality of the agro-industrial sector, the present research proposes a discussion that includes the following constituents: (1) legislation; (2) requirements; (3) operational drivers; (4) environmental, social, and governance drivers; (5) optimization; and (6) financial drivers.

Legislation

When it comes to environmental protection, legislation is required in order to control and limit the use and exploration of resources. In this sense, governments and national institutions play a significant role in the stimulation of better operational practices. In fact, legislation has been responsible for a positive impact on organizations regarding sustainability changes, innovation processes, and even safety and consumer protection [68]. From another perspective, inefficient legislation can cause damage to sustainable operations, for instance, when low policy measures permit non-circularity actions [69]. In this sense, many initiatives must be proposed to overcome these problems.

The previous concepts are consonant with what the WEF nexus approach targets when it calls attention to the need to reduce inequalities in natural resource uses and generate additional cost–benefit outputs. To illustrate this item, previous examples of legislative drivers being considered in excellence models can be found in the ZED model under the *"Legal Compliance"* criterion.

Requirements

In an agro-industrial activity, requirements rely on many factors, such as finality, food specificity, and consumer needs, among others. In this sense, for inclusion in our model, requirements are considered from two perspectives: manufacturing vision and customer behavior. Regarding the selected BE models, the following were chosen to represent this perspective: manufacturing capabilities from the ZED program, *managing by the fact*, and *society* from the MBNQA excellence model

The first consideration is directly connected to the production site, and the second one is related to the customer relationship management department. Therefore, considering the aim of the present research, those actions relate to the WEF nexus approach. When specifying the connection between quality and the management of requirements, quality management systems (QMS) and the ISO 9001 standard can be useful. ISO 9001 is a valuable tool to determine, validate, and implement the quality policy, which includes the management of information and knowledge, thus supporting the management and development of production capabilities [70].

Regarding customer behavior, it is important to highlight that this may require a transformation in the organization's policy. In this sense, companies need to be concerned with multiple issues, not only from the environmental perspective but also economic and social ones [71]. Thus, as mentioned before, this article aims to fulfill the knowledge gap to transform the WEF nexus approach—generally used for macroanalysis—into a micro-evaluation for, in this specific case, agro-industrial companies.

Operational Drivers

Operational performance is also important when reflected in the company's competitiveness [72]. It can be measured by many indicators, such as cost and quality, among others. This research highlights operational drivers as a result of the following excellence criteria: *managing by process* from the EFQM model and *process* from the BNQA model. Therefore, a major focus on processing activity is highlighted, as well as concerning the overall quality that will be delivered to the consumers.

We look at operational performance in this model from the point of view of external processes and operations related to supply chain management (SCM). SCM deals with all the processes, including how to go from raw material acquisition all the way to packaging disposal after consumer usage [73]. In this sense, throughout the various food processing stages, SCM-controlling tools become critical. Therefore, water, energy, and food should be included in the mentioned tools but, as stated in this research, in an integrated way. On this subject, it is important to innovate and find new indicators that can quantify the relationship between these elements [12].

Environmental, Social, and Governance (ESG)

Over the last few decades, environmental, social, and governance aspects have received increasing interest from a company perspective. Elements such as consumer behavior, policy, and emerging technologies are driving these concerns worldwide. In this scope, and in our proposition, this reflects a novel sustainability-focused excellence model; many ESG perspectives must be considered. To that end, the following excellence criteria are emphasized: *taking responsibility* for a sustainable future from the EFQM excellence model; *quality/environment/safety assurance, people development*, and *engagement systems* for quality and environment from the ZED program; *focus on the future* and *societal responsibility* from the MBNQA excellence model; and *society* from the BNQA model.

Most excellence models already have a perspective that considers environmental, social, and governance concerns. This is a consequence of the increasing interest in corporate social responsibility [74]. Agro-industrial companies are increasingly exposed to social responsibility concerns, as their operations contribute to natural resource depletion and face huge challenges to avoid and reduce waste [75]. This exposure is reinforced by changes in consumer behavior, with growing expectations for companies to adopt alternatives with less environmental damage, as well as to create the possibility for social inclusion along the production chain [76].

In the scope of the agro-industrial sector, the role of the environmental, social, and governance concerns promoted in driving the transition to a greener and more sustainable state can be easily connected to the United Nation's Sustainable Development Goals (SDGs), whose use presents great potential. The SDGs adopted by the United Nations, and propelled by the United Nations Sustainable Development Solution Network [25], offer a suitable framework for concerns about ESGs.

It is strongly recommended that agro-industrial companies relate all of this evidence to daily production, building, for instance, indicators that are consolidated and included in their quality management systems.

Optimization

The scientific literature has a wide variety of articles tackling optimization in the agro-industrial chain. They concern important challenges, such as aiming to monitor food production [77] or the development of new products [78,79]. Nevertheless, optimization efforts are mostly concerned with a productivity perspective. In this sense, implementing integrated quality systems helps to better standardize the operational activities, with recognized positive effects on financial gains and product quality [80,81].

In our sustainability-focused excellence model, the inclusion of the optimization perspective follows criteria from the ZED program—learning and improvement systems—and the MBNQA excellence model—agility.

Financial Drivers

Financial performance is, obviously, a driver beyond any agro-industrial company. For that reason, in the present research, the inclusion of financial drivers is supported by the following criteria: *Achieving balanced results* from the EFQM model; *focus on results and creating value* and the *systems perspective* from the MBNQA excellence model; and, finally, *results* from the BNQA model. Under this scope, for the present research, two concepts will be highlighted: the production chain cost analysis and the green finance perspective.

Financial analysis is a well-known challenge in the agro-industrial production chain due to the need for a proper definition of the cost method, investments, and other economic indicators [82]. In this sense, the WEF nexus approach aims to quantify the relationship among natural elements, and one of those numbers can be expressed using cost analysis [83]. Thus, this point is a driver of sustainable change because it can stimulate more agro-industrial companies to make their own investments, considering the mentioned costs. Consequently, this action needs to be considered in the definition of a company's subsequent steps, guiding the best ecofriendly decision and balancing it with financial imperatives.

4.2.2. AGENTS

Agents are change promoters that significantly impact the transition to sustainability. In the scope of this work, they are the stakeholders of agro-industrial companies. Agents will be divided here into the following categories: leaders, the workforce, and partners.

Leaders

Undoubtedly, sustainability driving change is a challenge typically faced by leaders, especially targeting social aspects [84]. Hence, they are responsible for promoting good conditions from which the whole organization can advance. In the present research, support for the attention given to leaders is based on the criteria *leading with vision, inspiration, and integrity* from the EFQM excellence model; *visionary leadership* from the MBNQA excellence model; and *leadership* from the model of the BNQA.

Organizations rely on the responsibility and impact that the leader's vision has on the transition path. For the proposed model, this research highlights the importance of leadership in two different perspectives: climate governance leadership and leadership under the quality management systems approach.

According to Carvalho et al. [85], leaders need to make decisions based on climate governance. This means that the sustainability implications of each decision should be considered. To that end, leaders need to use diverse information and knowledge to decide what is best in the context of transitioning to sustainability. Recent studies on the WEF nexus indicate that internal policies should include, among other issues, how to deal with wastage in the production chain, as well as guidance to implement a circular economy perspective in agro-industrial companies [86].

Workforce

As in the previous agent, the workforce is also an important driver to change behavior. Efforts should be made to include more productivity by enhancing people's knowledge. For that purpose, in the present research, this topic will be represented by the following concerns: *succeeding through People* (EFQM excellence model); *people development* and *engagement systems for quality and environment* (ZED program); *valuing workforce members and partners* (MBNQA excellence model); and *people* (BNQA). In practice, the proposed model will focus on evaluating the promotion of increased workforce skills, tools, and capabilities that can support the pursuit of both increased productivity and sustainability. According to Ingram [87], these skills, tools, and capabilities are aimed at improving workforce knowledge of the food processing chain's impact on topics such as food security and safety, sustainability, and health. WEF nexus approaches should be presented from a comprehensive and educational perspective. Workforce knowledge proved to be beneficial, with evidence that it impacts the collaborators' environmental mindset, as well as consumer behavior [88]. In this sense, continued training constitutes a critical action to help agro-industrial companies become more sustainable.

Partners

Building partnerships is required to control the agro-industrial production chain. One of the main reasons for that is the need for traceability, which is essential to ensure final product quality. Thus, it needs to be constructed with records from all processes, including origin and production processes, but also from product characteristics such as temperature, size, and color, among others. These can have a positive impact on standards audits, and, finally, they can also potentially avoid sanitary barriers [89]. According to Tomich et al. [90], partnerships are more effective when they have multiple stakeholders. In the case of the agro-industrial sector, this means that the production chain should be built upon partnerships. Specific to the present research, in order to lead to a successful implementation, all food transformation paths should be aware of the WEF nexus approach information. Quality control should become the responsibility of every actor, and all partners can become agents of sustainable change by using audits, requirements, and other requests in agro-industrial companies.

The inclusion of partners in this model is inspired by the EFQM excellence criterion *building partnerships* and the MBNQA excellence criterion *valuing workforce members and partners*.

PROCESS

In the context of sustainable excellence, continuous improvement should be a highlighted aim of any process. Agro-industrial companies should look for process improvements that can be translated into more agility, fewer costs, and, finally, more sustainability. Under this scope, the process improvement presented is based on three main objectives in the present research: learning, innovation, and valorization.

Learning

Organizational learning is required to transform the process, both in terms of productivity and sustainability. It can be defined as the recognition that society changes and that organizations should change accordingly [91]. It can also lead to more sustainable practices. The development of learning process classes is critical. In the case of the proposed model, continuous organizational learning regarding both the WEF nexus and quality and excellence are required. According to Moore et al. [92] learning those issues can avoid biased decisions, and pieces of evidence indicate that the results positively impact organizational improvement.

The focus on organizational learning is helpful for the following uses: *Design capabilities* in the ZED program and *organizational and personal learning* in the MBNQA model.

Innovation

Innovation has a strong link to excellent performance and can also support the pursuit of greener, more sustainable solutions. According to Kafetzopoulos and Gotzamani [93] enablers from some excellence models are intrinsically linked to innovation. In this sense, it is essential to provide conditions to develop such activities. In the specific case of agro-industrial companies, all actors are expected to be able to contribute to innovation, regardless of their jobs. For that purpose, it is important to highlight that there is a huge challenge in transforming the WEF nexus perspective into a feasible methodology that can be used by a company since this concept has mainly been applied at the macro level, for instance, to cities, regions, and countries, among other similar analyses [94]. Therefore, to address this concern, it is strongly recommended that the companies should include in their innovative actions the three nexus elements combined together.

In this present research, the innovation issue mainly has to do with joining the following criteria: *nurturing creativity and innovation* (from the excellence model of the EFQM); *managing for innovation* (MBNQA excellence model); and *strategies and plans* (from the model of the BNQA).

Valorization

Increasing the valorization of resources, as a social aspect, is a growing concern for organizations. Customers are willing to pay for differentiation in some products and then valorize them. In the specific case of agro-industrial companies, sustainability valorization is highlighted [95]. Changing customer behavior is responsible for several companies promoting a transition to greener products. Likewise, the perception that the quality of a product depends on how it is produced is changing.

In the present research, this topic is supported by covering the following topics: *adding value for customers* (EFQM excellence model); *customer-driven excellence* (MBNQA); and *clients* (BNQA).

4.2.3. OUTCOMES

In the context of sustainable excellence, continuous improvement should be highlighted as a major aim for any process. In particular, agro-industrial companies should look for process improvements that can be translated into more agility, fewer costs, and, finally, more sustainability. In this scope, process improvement in our present research can be related to the following main objectives: quality assurance diagnosis and sustainability recommendations.

Quality Assurance Diagnosis

One of the outcomes that is expected from the proposed model is a quality assurance diagnosis (QAD). In practical terms, companies using the model should be able to describe actual conformities and nonconformities according to an evaluation made according to an integrated vision analysis. As stated by Thrän et al. [96], environmentally integrated analysis can positively impact final products. In this case, the result will be translated into a more cost-competitive final product. Therefore, the mentioned QAD can be used as a control tool to establish indicators that should be followed. In the specific case of agro-industrial companies, it should help in the identification of impacts that deal directly with WEF nexus elements.

Sustainability Recommendations

Based on the previous QAD, recommendations should guide an action plan regarding the definition and implementation of proper corrective actions. According to Sharma et al. [73] an action plan is one of the required tools to implement sustainable performance in agro-industrial companies. The mentioned action plan should capture the experienced model results and transform the suggestions into practical terms.

Additionally, those recommendations should reflect the WEF nexus concept, which is to manage the elements in a joint manner. Likewise, those actions should enhance efforts to overcome the tradeoff in order to take the concept from a macro-perspective to a micro-perspective applied, in this case, to agro-industrial companies. These recommendations are expected to result in actions such as increases in productivity, decreases in costs, and, most critically, greener, more effective, and more sustainable production.

4.3. How Agro-Industrial Companies and WEF Nexus Aspects Can Be Targeted in the Proposed Model

The agro-industrial sector plays a significant role in feeding mankind in relation to the WEF nexus approach. However, this sector is very energy intensive, with important negative environmental impacts. In this sense, for the purpose of reducing GHG emissions, alternative energy sources (solar and wind energy for instance) and water resources (sustainable water production) should be employed [97].

Additionally, water pollution owing to agro-industrial activities is a crucial problem to human beings and the ecosystem [98]. Approximately 85% of global water consumption is used for irrigation, and the demand is still increasing [99]. In order to avoid wastage, adequate water treatment and reuse through non-pollutant methodologies are suggested [100].

The food scenario is no different. One out of nine people in the world go hungry every day according to the "State of Food Insecurity" section in the World Report [101]. Annually, one-third of food produced for human consumption is wasted in the food production chain [102]. Thus, it is a major challenge for all agro-industrial chains to produce food while avoiding food wastage.

Therefore, to implement the WEF nexus in the agro-industrial sector, these aspects should be evaluated during the model's implementation. It means, that during its implementation, these aspects should be evaluated in each of the following aspects: drivers, agents, processes, and outcomes. In this sense, problems such as the implementation of renewable energy sources [29], the treatment of water sources [103], and actions to avoid food wastage [104] should be ranked in an evaluation model. Regarding the particular aspects of this sector, it is good to integrate with quality global standards such as the Hazard Analysis Critical Control Point (HACCP) [105], ISO 22.000 [106], and good manufacturing practices [107], among others.

5. Model Verification and Validation

To verify and validate the model, the proposed results were validated by experts. Thus, based on the citation of Carvalho et al. [108], the presented model was discussed by researchers and agro-industrial producers from Portugal, the United States of America, Italy, Brazil, and Kenya. Following that, to clarify possible uncertainties, this article highlights how to use the proposed model and a benchmark study on the diversity of evaluation models.

5.1. How to Use the Proposed Sustainability-Focused Model

Integrated management approaches have proved that organizations can become more effective, more efficient, more responsive, and enjoy better performance outcomes [109]. In this scope, the integration of productivity and sustainability also manages the perceived tradeoff that is felt by organizations regarding the need to have a superior level of business performance and the need to promote social, economic, and environmental sustainability [110]. Additionally, the proposed model has synergies spawning from this integration, and it shows that a proper balance is essential to achieve the mentioned goals. Thus, the practical deployment of this integrative vision's model represents a well-structured approach, promoting two main visions: a framework promoting performance assessment and improvement while also ensuring an increased focus on the sustainability demanded by multiple stakeholders.

To practically apply the proposed model, it is important to highlight its constituents and interdependencies. Beginning with the drivers, these elements should be discussed after the evaluation with respect to the reasons that connections between the WEF elements are not occurring. After that, based on the previous results, agents must show how committed they are to leading these organizational changes. Following this path, the process must be evaluated from a sustainability perspective; in this sense, it could bring some ideas on how to promote innovation in agro-industrial companies by solving problems with the underscored evaluation. Finally, summarizing all the insights, the outcomes should bring, firstly, a diagnosis of the real situation; after that, it should find suggestions on how to perform better, with indications for an action plan (ideally with changing actions, employee responsibility, and dates to implement these suggestions). Accordingly, it is possible to see the interdependence between the elements completing the proposed analysis made by this model.

Following the previous discussion, with respect to many quality management models (ISO standards, Food Safety System Certification (FSSC), and national legislations, among others), the presented novel model can be integrated within the sequence of the Plan, Do, Check, and Act (PDCA) cycle program. The first part is represented by the driver and agent dimensions so that every analyzed item should be included in the planning stage. The application part of the program—relative to the "do part"—focuses on the process. In this case, every issue discussed before should be implemented. The check components will be credited to the quality assurance diagnosis from the outcome, highlighting the non-conformities that should be targeted. Finally, the Act part should try to overcome the previous diagnosis, thus finishing the quality cycle and restarting the evaluation. Therefore, the proposed model should be responsible for the sustainability transition that was discussed in this article. An illustrated overview can be seen in Figure 2.

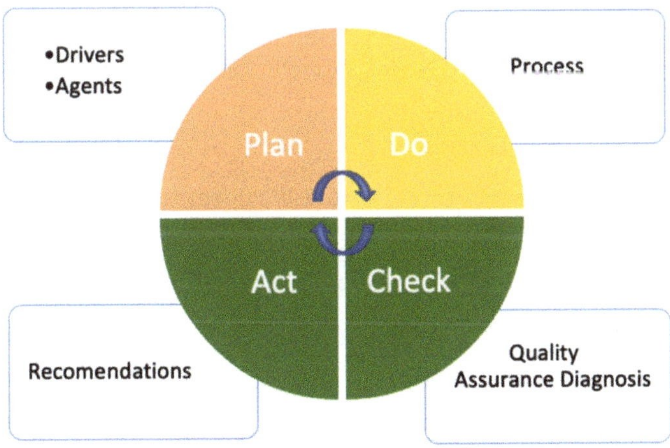

Figure 2. Illustrated overview of the sustainability-focused excellence model integrated with the PDCA cycle. Source: Own elaboration.

5.2. Benchmarking with other Evaluation Models

The scientific literature covers plenty of articles relying on the connection between quality management systems and sustainability concerns [111–117]. Additionally, according to Zink [113], the concept of quality management implicitly brings a sustainability perspective, especially when considering total quality management principles. In this sense, some studies rely on the efficiency of indicators, expressing items relative to these concerns [112,115]. However, a major novelty in our proposed sustainability-focused model is the fact that it aims to unite the WEF nexus approach with quality management systems models, specifically BE models. Thus, BE models can be integrated with an increasing variety of other subsystems implemented according to other standards and subsystems raised from specific standards designed for specific activity sectors [70,118]. Finally, it is important to consider that, in 2020, the EFQM model adopted a new version that reinforces sustainability indicators [92], although it is not customized to agro-industrial companies, as is the case of our suggested model.

Following the previous discussion, agro-industrial companies have been implementing many greener actions, not just because of environmental concerns, but also because of the benefits deriving from sustainable initiatives [119–123]. In this case, it is important to highlight that the mentioned actions are independent of company size and are trying to overcome situations such as the provision of natural resources [119]. Thus, demands for WEF nexus resources, in this sector, are increasing rapidly [124]. In this context, it is a major challenge for the overall agro-industrial chain to produce food while avoiding the excessive use of energy and water and preventing food waste. For that purpose, Table 2 lists five articles that involve WEF nexus approaches for agro-industrial companies, highlighting their objectives and evaluation methods, as well as the major differences and advantages connected with our suggested conceptual model.

Table 2. Articles discussing the implementation of the WEF nexus approach in the agro-industrial production chain, highlighting their objectives and evaluation methods, as well as their main contributions to our sustainability-focused excellence model.

Reference	Objective	Evaluation Method	Sustainability-Focused Excellence (SFE) Model
[124]	This article aims to mitigate climate change in the agri-food sector.	Construction of a method to evaluate the WEF nexus and energy use through the agri-food production chain.	SFE model aims to build a model based on BE model approaches and, additionally, focuses on a specific company, not in a sector.
[125]	Analyzes the WEF nexus approach in rice production on a watershed scale.	Material flow analysis using a Sankey diagram.	SFE model applies QMS approaches and is focused on process operational phases.
[126]	Improving water governance in a farm operation in South Africa.	Use of the WEF nexus approach integrated to financial indicators, enhancing the farm's profitability.	SFE model uses the WEF nexus approach for food supply chain transformation, not only in the farm component. Additionally, the impact can be viewed in three aspects, not just with respect to water concerns.
[127]	Improving the offer of WEF resources in beef cattle production.	Model development based on the WEF nexus that focuses on water reuse, greener energy resources, and improving food production.	SFE model focuses on a food transformation process and is based on the QMS models, specifically benchmarked with BE models.
[128]	Improving biorefinery processes according to the WEF nexus approach.	Integration between the WEF nexus approach and sustainable biorefinery processes.	SFE model focuses on agro-industrial processes, also dealing with productivity indicators using BE models as benchmarks.

Source: Own elaboration.

Table 2 shows the important contribution of the sustainability-focused excellence (SFE) model to the present discussion. Thus, existent WEF nexus models focus on a macro-perspective, and, in this case, we propose to focus on a microanalysis, namely, with respect to agro-industrial companies. The contents of Table 2 underscore that the use of BE models should bring other evident novelties to the system. We believe that the use of this model will be an important tool for leaders, academics, decisionmakers, and lawmakers, as well as for guiding new policies.

6. Conclusions and Future Work

This article presents a sustainability-focused excellence model by integrating the principles and objectives of the water–energy–food (WEF) nexus with existing excellence models and proposing its application for agro-industrial companies. The development of this model was based on the conceptual links between productivity and sustainability and the need to transition from one perspective together with the other. Accordingly, the novel model is presented with a focus on change, being constituted by the following dimensions (and respective evaluation criteria): drivers (legislation, requirements, operational drivers, ESG drivers, optimization, and financial drivers); agents (leaders, workforce, and partners); process (learning, innovation, and valorization); and outcome (quality assurance diagnosis, and recommendations).

We believe that this novel model can become a decision-support tool to help agro-industrial companies make the transition toward improved sustainability while managing existing tradeoffs and synergies. From this perspective, the sustainability-excellence model can be used in a dynamic and integrated way with the WEF nexus.

In terms of limitations and future research work, it is important to note that this model has not yet been tested in practice, an essential step toward complete validation. However, our aim with this publication is to perform a broader validation. Model validation can be performed in several ways. Although practical validation is often used, the theoretical validation of models through publication is notably prevalent as a first step in the validation process, with reviewers' inputs and insights serving as a basis for improvement and scientific validation. Some authors, such as Maqsood Ahmad Sandhu and Ahm Shamsuzzoha [129], and Carvalho et. al. [108], have published their conceptual models before advancing to empirical validation. Accordingly, our aim is to use this publication to make the model available to the scientific community, which should be considered the first step in this validation process, gaining broader acceptance from academics and practitioners before advancing the model's applicability validation in practice. Furthermore, during this time, the model can serve as a basis for the development of other models in this context of action.

As such, once the evaluation criteria and methods are defined, the next step will be to consider additional model validation with actual agro-industrial company case studies. From this perspective, the use of companies in different countries is desired since every country has different legislation and may provide different perspectives.

Author Contributions: Conceptualization, F.C., A.M.C. and P.S.; writing—original draft preparation, F.C., A.M.C. and P.S.; writing—review and editing, F.C., A.M.C., P.S. and F.F. All authors have read and agreed to the published version of the manuscript.

Funding: This research was funded by the Brazilian National Council for Scientific and Technological Development—CNPq, Ministry of Science, Technology, Innovations, and Communications, Brazil—201490/2018-8.

Institutional Review Board Statement: Not applicable.

Informed Consent Statement: Not applicable.

Data Availability Statement: Not applicable.

Acknowledgments: We acknowledge the Energy for Sustainability Initiative from the University of Coimbra for academic support.

Conflicts of Interest: The authors declare no conflict of interest.

References

1. Scanlon, B.R.; Ruddell, B.L.; Reed, P.M.; Hook, R.I.; Zheng, C.; Tidwell, V.C.; Siebert, S. The food-energy-water nexus: Transforming science for society. *Water Resour. Res.* **2017**, *53*, 3550–3556. [CrossRef]
2. Clift, R. Climate change and energy policy: The importance of sustainability arguments. *Energy* **2007**, *32*, 262–268. [CrossRef]
3. Schoenherr, T.; Talluri, S. Environmental sustainability initiatives: A comparative analysis of plant efficiencies in Europe and the US. *IEEE Trans. Eng. Manag.* **2013**, *60*, 353–365. [CrossRef]
4. Yeung, S.M. UNSDGs and future quality management—Social policy for developing sustainable development mindset. *Corp. Gov. Sustain. Rev.* **2019**, *3*, 26–34. [CrossRef]
5. Viles, E.; Santos, J.; Muñoz-Villamizar, A.; Grau, P.; Fernández-Arévalo, T. Lean-green improvement opportunities for sustainable manufacturing using water telemetry in agri-food industry. *Sustainability* **2021**, *13*, 2240. [CrossRef]
6. Adelodun, B.; Kareem, K.Y.; Kumar, P.; Kumar, V.; Choi, K.S.; Yadav, K.K.; Yadav, A.; El-Denglawey, A.; Cabral-Pinto, M.; Son, C.T.; et al. Understanding the impacts of the COVID-19 pandemic on sustainable agri-food system and agroecosystem decarbonization nexus: A review. *J. Clean. Prod.* **2021**, *318*, 128451. [CrossRef]
7. Karwacka, M.; Ciurzyńska, A.; Lenart, A.; Janowicz, M. Sustainable Development in the Agri-Food Sector in Terms of the Carbon Footprint: A Review. *Sustainability* **2020**, *12*, 6463. [CrossRef]
8. Fonseca, L.; Amaral, A.; Oliveira, J. Quality 4.0: The efqm 2020 model and industry 4.0 relationships and implications. *Sustainability* **2021**, *13*, 3107. [CrossRef]
9. Carvalho, A.M.; Sampaio, P.; Rebentisch, E.; Carvalho, J.A.; Saraiva, P. Operational excellence, organisational culture and agility: The missing link? *Total Qual. Manag. Bus. Excell.* **2019**, *30*, 1495–1514. [CrossRef]
10. Isaksson, R. Excellence for sustainability–maintaining the license to operate. *Total Qual. Manag. Bus. Excell.* **2019**, *32*, 489–500. [CrossRef]
11. Siltori, P.F.S.; Simon Rampasso, I.; Martins, V.W.B.; Anholon, R.; Silva, D.; Souza Pinto, J. Analysis of ISO 9001 certification benefits in Brazilian companies. *Total Qual. Manag. Bus. Excell.* **2020**, *32*, 1614–1632. [CrossRef]
12. Hoff, H. *Understanding the Nexus: Background Paper for the Bonn2011 Nexus Conference*; Stockholm Environment Institute: Stockholm, Sweden, 2011; pp. 1–52.
13. World Economic Forum. *Water Security: The Water-Food-Energy-Climate Nexus: The World Economic Forum Water Initiative*; Island Press: Washington, DC, USA, 2011.
14. Taniguchi, M.; Endo, A.; Gurdak, J.J.; Swarzenski, P. Water-Energy-Food Nexus in the Asia-Pacific Region. *J. Hydrol. Reg. Stud.* **2017**, *11*, 1–8. [CrossRef]
15. Purwanto, A.; Sušnik, J.; Suryadi, F.X.; de Fraiture, C. Water-energy-food nexus: Critical review, practical applications, and prospects for future research. *Sustainability* **2021**, *13*, 1919. [CrossRef]
16. Botai, J.O.; Botai, C.M.; Ncongwane, K.P.; Mpandeli, S.; Nhamo, L.; Masinde, M.; Adeola, A.M.; Mengistu, M.G.; Tazvinga, H.; Murambadoro, M.D.; et al. A review of the water-energy-food nexus research in Africa. *Sustainability* **2021**, *13*, 1762. [CrossRef]
17. Carvalho, A.M.; Sampaio, P.; Rebentisch, E.; Saraiva, P. Operational excellence as a means to achieve an enduring capacity to change—Revision and evolution of a conceptual model. *Procedia Manuf.* **2017**, *13*, 1328–1335. [CrossRef]
18. Leung Pah Hang, M.Y.; Martinez-Hernandez, E.; Leach, M.; Yang, A. Designing integrated local production systems: A study on the food-energy-water nexus. *J. Clean. Prod.* **2016**, *135*, 1065–1084. [CrossRef]
19. Daher, B.T.; Mohtar, R.H. Water-energy-food (WEF) Nexus Tool 2.0: Guiding integrative resource planning and decision-making. *Water Int.* **2015**, *40*, 748–771. [CrossRef]
20. Koutsoyiannis, D.; Makropoulos, C.; Langousis, A.; Baki, S.; Efstratiadis, A.; Christofides, A.; Karavokiros, G.; Mamassis, N. HESS opinions: "Climate, hydrology, energy, water: Recognizing uncertainty and seeking sustainability". *Hydrol. Earth Syst. Sci.* **2009**, *13*, 247–257. [CrossRef]
21. Hülsmann, S.; Sušnik, J.; Rinke, K.; Langan, S.; van Wijk, D.; Janssen, A.B.; Mooij, W.M. Integrated modelling and management of water resources: The ecosystem perspective on the nexus approach. *Curr. Opin. Environ. Sustain.* **2019**, *40*, 14–20. [CrossRef]
22. Benson, D.; Gain, A.K.; Rouillard, J.J. Water governance in a comparative perspective: From IWRM to a "nexus" approach? *Water Altern.* **2015**, *8*, 756–773.
23. Roidt, M.; Avellán, T. Learning from integrated management approaches to implement the Nexus. *J. Environ. Manag.* **2019**, *237*, 609–616. [CrossRef]
24. Weitz, N.; Strambo, C.; Kemp-Benedict, E.; Nilsson, M. Closing the governance gaps in the water-energy-food nexus: Insights from integrative governance. *Glob. Environ. Chang.* **2017**, *45*, 165–173. [CrossRef]
25. UN. *2030 Agenda for Sustainable Development*; United Nations: New York, NY, USA, 2015; Volume 16301.
26. Saladini, F.; Betti, G.; Ferragina, E.; Bouraoui, F.; Cupertino, S.; Canitano, G.; Gigliotti, M.; Autino, A.; Pulselli, F.M.; Riccaboni, A.; et al. Linking the water-energy-food nexus and sustainable development indicators for the Mediterranean region. *Ecol. Indic.* **2018**, *91*, 689–697. [CrossRef]
27. Olawuyi, D. Sustainable development and the water-energy-food nexus: Legal challenges and emerging solutions. *Environ. Sci. Policy* **2020**, *103*, 1–9. [CrossRef]
28. Cansino-Loeza, B.; Tovar-Facio, J.; Ponce-Ortega, J.M. Stochastic optimization of the water-energy-food nexus in disadvantaged rural communities to achieve the sustainable development goals. *Sustain. Prod. Consum.* **2021**, *28*, 1249–1261. [CrossRef]

29. Parsa, A.; Van De Wiel, M.J.; Schmutz, U. Intersection, interrelation or interdependence? The relationship between circular economy and nexus approach. *J. Clean. Prod.* **2021**, *313*, 127794. [CrossRef]
30. Garcia, D.J.; You, F. The water-energy-food nexus and process systems engineering: A new focus. *Comput. Chem. Eng.* **2016**, *91*, 49–67. [CrossRef]
31. D'Adamo, I.; Gastaldi, M.; Morone, P.; Rosa, P.; Sassanelli, C.; Settembre-Blundo, D.; Shen, Y. Bioeconomy of Sustainability: Drivers, Opportunities and Policy Implications. *Sustainability* **2022**, *14*, 200. [CrossRef]
32. Merino-Saum, A.; Baldi, M.G.; Gunderson, I.; Oberle, B. Articulating natural resources and sustainable development goals through green economy indicators: A systematic analysis. *Resour. Conserv. Recyc.* **2018**, *139*, 90–103. [CrossRef]
33. Hussien, W.A.; Memon, F.A.; Savic, D.A. A risk-based assessment of the household water-energy-food nexus under the impact of seasonal variability. *J. Clean. Prod.* **2018**, *171*, 1275–1289. [CrossRef]
34. Bijl, D.L.; Bogaart, P.W.; Dekker, S.C.; van Vuuren, D.P. Unpacking the nexus: Different spatial scales for water, food and energy. *Glob. Environ. Chang.* **2018**, *48*, 22–31. [CrossRef]
35. Karan, E.; Asadi, S.; Mohtar, R.; Baawain, M. Towards the optimization of sustainable food-energy-water systems: A stochastic approach. *J. Clean. Prod.* **2018**, *171*, 662–674. [CrossRef]
36. El-Gafy, I. Water-food-energy nexus index: Analysis of water-energy-food nexus of crop's production system applying the indicators approach. *Appl. Water Sci.* **2017**, *7*, 2857–2868. [CrossRef]
37. AbdelHady, R.S.; Fahmy, H.S.; Pacini, N. Valuing of Wadi El-Rayan ecosystem through water-food-energy nexus approach. *Ecohydrol. Hydrobiol.* **2017**, *17*, 247–253. [CrossRef]
38. Rivera, D.E.; Piferrer, M.R.T.; Mundet, M.H.B. Measuring territorial social responsibility and sustainability using the EFQM excellence model. *Sustainability* **2021**, *13*, 2153. [CrossRef]
39. Henríquez-Machado, R.; Muñoz-Villamizar, A.; Santos, J. Sustainability through operational excellence: An emerging country perspective. *Sustainability* **2021**, *13*, 3165. [CrossRef]
40. Rajakal, J.P.; Ng, D.K.S.; Tan, R.R.; Andiappan, V.; Wan, Y.K. Multi-objective expansion analysis for sustainable agro-industrial value chains based on profit, carbon and water footprint. *J. Clean. Prod.* **2021**, *288*, 125117. [CrossRef]
41. Amoriello, T. Sustainability: Recovery and Reuse of Brewing-Derived. *Sustainability* **2021**, *13*, 2355. [CrossRef]
42. Klefsjö, B.; Bergquist, B.; Garvare, R. Quality management and business excellence, customers and stakeholders: Do we agree on what we are talking about, and does it matter? *TQM J.* **2008**, *20*, 120–129. [CrossRef]
43. Miller, R.D.; Raymer, J.; Cook, R.; Barker, S. *The Shingo Model for Operational Excellence*; Logan, U., Ed.; Utah State University: Logan, UT, USA, 2013.
44. EFQM. EFQM Model for Business Excellence. 2003. Available online: https://efqm.org/ (accessed on 2 December 2021).
45. Asif, M.; Searcy, C.; Garvare, R.; Ahmad, N. Including sustainability in business excellence models. *Total Qual. Manag. Bus. Excell.* **2011**, *22*, 773–786. [CrossRef]
46. Jabareen, Y. Building a Conceptual Framework: Philosophy, Definitions, and Procedure. *Int. J. Qual. Methods* **2009**, *8*, 49–62. [CrossRef]
47. Eizenberg, E.; Jabareen, Y. Social sustainability: A new conceptual framework. *Sustainability* **2017**, *9*, 68. [CrossRef]
48. Hong, Q.N.; Pluye, P. A Conceptual Framework for Critical Appraisal in Systematic Mixed Studies Reviews. *J. Mix. Methods Res.* **2019**, *13*, 446–460. [CrossRef]
49. Khandpur, N.; Zatz, L.Y.; Bleich, S.N.; Taillie, L.S.; Orr, J.A.; Rimm, E.B.; Moran, A.J. Supermarkets in cyberspace: A conceptual framework to capture the influence of online food retail environments on consumer behavior. *Int. J. Environ. Res. Public Health* **2020**, *17*, 8639. [CrossRef]
50. Kørnøv, L.; Lyhne, I.; Davila, J.G. Linking the UN SDGs and environmental assessment: Towards a conceptual framework. *Environ. Impact Assess. Rev.* **2020**, *85*, 106463. [CrossRef]
51. Hidayati, A.N.; Riadi, I.; Ramadhani, E.; Amany, S.U. AI Development of conceptual framework for cyber fraud investigation. *Regist. J. Ilm. Teknol. Sist. Inf.* **2021**, *7*, 125–135. [CrossRef]
52. Silverman, D. *Doing Qualitative Research: A Practical Guide*; Sage: London, UK, 2000.
53. Yin, R.K. *Case Study Research: Design and Methods*, 4th ed.; Sage: Thousand Oaks, CA, USA, 2009.
54. Nardi, P.C.C.; da Silva, R.L.M.; Ribeiro, E.M.S.; de Oliveira, S.V.W.B. Proposal for a methodology to monitor sustainability in the production of soft drinks in Ref PET. *J. Clean. Prod.* **2017**, *151*, 218–234. [CrossRef]
55. Mohr, M.; Renz, P. Naga City: Septage treatment and wastewater concept for Del Rosario. Fraunhofer IGB Report. 2014. Available online: https://www.unescap.org/sites/default/files/Report_PH_Naga_SeptageAndWastewater_2014.pdf (accessed on 15 January 2022).
56. Khalil, M.K.; Muneenam, U. Total quality management practices and corporate green performance: Does organizational culture matter? *Sustainability* **2021**, *13*, 21. [CrossRef]
57. Adapa, S. Factors influencing consumption and anti-consumption of recycled water: Evidence from Australia. *J. Clean. Prod.* **2018**, *201*, 624–635. [CrossRef]
58. Jamal, C.M.C.; Albrecht Anversa, M.V.; De Souza Chacon, P.A. A conexão do Sistema de Gestão da Qualidade Total (SGQ) com a Gestão da Inovação (GI). *Sist. Gestão* **2021**, *16*, 3–10. [CrossRef]
59. Baldrige Performance Excellence Program Criteria for Performance Excellence, US, NIST, Gaithersburg, MD. Available online: https://www.nist.gov/baldrige (accessed on 15 January 2022).

60. Brazilian National Quality Award Foundation FPNQ—Performance Excellence Criteria. Available online: https://fnq.org.br/empresas-reconhecidas-em-gestao-meg-e-pnq-de-1992-a-2020/ (accessed on 3 November 2021).
61. Ministry of Micro Small and Medium Enterprises. ZED Zero Deffect Zero Effect Statement. Available online: https://zed.msme.gov.in/ (accessed on 2 December 2021).
62. Criado-García, F.; Calvo-Mora, A.; Martelo-Landroguez, S. Knowledge management issues in the EFQM excellence model framework. *Int. J. Qual. Reliab. Manag.* **2019**, *37*, 781–800. [CrossRef]
63. Davies, J. Integration: Is it the key to effective implementation of the EFQM Excellence Model? *Int. J. Qual. Reliab. Manag.* **2008**, *25*, 383–399. [CrossRef]
64. Miller, J.; Parast, M.M. Learning by Applying: The Case of the Malcolm Baldrige National Quality Award. *IEEE Trans. Eng. Manag.* **2019**, *66*, 337–353. [CrossRef]
65. Tettey, A.; Sampson, G.; Mesmer, B. Exploratory Analysis of the Malcolm Baldrige National Quality Award Model. *J. Manag. Eng. Integr.* **2019**, *12*, 2019.
66. Cauchick Miguel, P.A. Comparing the Brazilian national quality award with some of the major prizes. *TQM Mag.* **2001**, *13*, 260–272. [CrossRef]
67. Sampaio, P.; Saraiva, P.; Rodrigues, A.G. ISO 9001 certification research: Questions, answers and approaches. *Int. J. Qual. Reliab. Manag.* **2009**, *26*, 38–58. [CrossRef]
68. Leialohilani, A.; de Boer, A. EU food legislation impacts innovation in the area of plant-based dairy alternatives. *Trends Food Sci. Technol.* **2020**, *104*, 262–267. [CrossRef]
69. Canali, M.; Amani, P.; Aramyan, L.; Gheoldus, M.; Moates, G.; Östergren, K.; Silvennoinen, K.; Waldron, K.; Vittuari, M. Food waste drivers in Europe, from identification to possible interventions. *Sustainability* **2017**, *9*, 37. [CrossRef]
70. Sampaio, P.; Saraiva, P.; Rodrigues, A.G. The economic impact of quality management systems in Portuguese certified companies: Empirical evidence. *Int. J. Qual. Reliab. Manag.* **2011**, *28*, 929–950. [CrossRef]
71. Nguyen, N.; Johnson, L.W. Consumer behaviour and environmental sustainability. *J. Consum. Behav.* **2020**, *19*, 2020–2022. [CrossRef]
72. Gupta, S.; Dangayach, G.S.; Singh, A.K.; Meena, M.L.; Rao, P.N. Adoption of sustainable supply operation quality practices and their impact on stakeholder's performance and sustainable performance for sustainable competitiveness in Indian manufacturing companies. *Int. J. Intell. Enterp.* **2018**, *5*, 125–140. [CrossRef]
73. Sharma, V.K.; Chandna, P.; Bhardwaj, A. Green supply chain management related performance indicators in agro industry: A review. *J. Clean. Prod.* **2017**, *141*, 1194–1208. [CrossRef]
74. Govindan, K.; Khodaverdi, R.; Jafarian, A. A fuzzy multi criteria approach for measuring sustainability performance of a supplier based on triple bottom line approach. *J. Clean. Prod.* **2013**, *47*, 345–354. [CrossRef]
75. Dania, W.A.P.; Xing, K.; Amer, Y. Collaboration behavioural factors for sustainable agri-food supply chains: A systematic review. *J. Clean. Prod.* **2018**, *186*, 851–864. [CrossRef]
76. Doukidis, G.I.; Matopoulos, A.; Vlachopoulou, M.; Manthou, V.; Manos, B. A conceptual framework for supply chain collaboration: Empirical evidence from the agri-food industry. *Supply Chain Manag. Int. J.* **2007**, *12*, 177–186. [CrossRef]
77. Gao, T.; Tian, Y.; Zhu, Z.; Sun, D.W. Modelling, responses and applications of time-temperature indicators (TTIs) in monitoring fresh food quality. *Trends Food Sci. Technol.* **2020**, *99*, 311–322. [CrossRef]
78. Estevão, S.T.; Batista de Almeida e Silva, J.; Lourenço, F.R. Development and optimization of beer containing malted and non-malted substitutes using quality by design (QbD) approach. *J. Food Eng.* **2020**, *289*, 110182. [CrossRef]
79. Zayed, A.; Farag, M.A. Valorization, extraction optimization and technology advancements of artichoke biowastes: Food and non-food applications. *Lwt* **2020**, *132*, 109883. [CrossRef]
80. Costa, L.B.M.; Godinho Filho, M.; Fredendall, L.D.; Ganga, G.M.D. The effect of Lean Six Sigma practices on food industry performance: Implications of the Sector's experience and typical characteristics. *Food Control* **2020**, *112*, 107110. [CrossRef]
81. Purwanto, A.; Asbari, M.; Santoso, P.B. Does Culture, Motivation, Competence, Leadership, Commitment Influence Quality Performance? *Inovbiz J. Inov. Bisnis* **2019**, *7*, 201. [CrossRef]
82. Villalva-Catano, A.; Ramos-Palomino, E.; Provost, K.; Casal, E. A Model in Agri-Food Supply Chain Costing Using ABC Costing: An Empirical Research for Peruvian Coffee Supply Chain. In Proceedings of the 7th International Engineering, Sciences and Technology Conference (IESTEC), Panama City, Panama, 9–11 October 2019; pp. 1–6. [CrossRef]
83. McCarl, B.A.; Yang, Y.; Schwabe, K.; Engel, B.A.; Mondal, A.H.; Ringler, C.; Pistikopoulos, E.N. Model Use in WEF Nexus Analysis: A Review of Issues. *Curr. Sustain. Energy Rep.* **2017**, *4*, 144–152. [CrossRef]
84. Blanco-Portela, N.; R-Pertierra, L.; Benayas, J.; Lozano, R. Sustainability leaders' perceptions on the drivers for and the barriers to the integration of sustainability in Latin American Higher Education Institutions. *Sustainability* **2018**, *10*, 2954. [CrossRef]
85. Carvalho, A.M.; Sampaio, P.; Rebentisch, E.; Carvalho, J.Á.; Saraiva, P. The influence of operational excellence on the culture and agility of organizations: Evidence from industry. *Int. J. Qual. Reliab. Manag.* **2020**, *38*, 1520–1549. [CrossRef]
86. Del Borghi, A.; Moreschi, L.; Gallo, M. Circular economy approach to reduce water-energy-food nexus. *Curr. Opin. Environ. Sci. Health* **2020**, *13*, 23–28. [CrossRef]
87. Ingram, J.; Ajates, R.; Arnall, A.; Blake, L.; Borrelli, R.; Collier, R.; de Frece, A.; Häsler, B.; Lang, T.; Pope, H.; et al. A future workforce of food-system analysts. *Nat. Food* **2020**, *1*, 9–10. [CrossRef]

88. Uhrin, A.; Bruque-Camara, S.; Moyano-Fuentes, J. Lean production, workforce development and operational performance. *Manag. Decis.* **2017**, *55*, 103–118. [CrossRef]

89. Shuvo, S.D. Assessing food safety and associated food hygiene and sanitary practices in food industries: A cross-sectional study on biscuit industry of Bangladesh. *Nutr. Food Sci.* **2015**, *45*, 39–53. [CrossRef]

90. Tomich, T.P.; Lidder, P.; Dijkman, J.; Coley, M.; Webb, P.; Gill, M. Agri-food systems in international research for development: Ten theses regarding impact pathways, partnerships, program design, and priority-setting for rural prosperity. *Agric. Syst.* **2019**, *172*, 101–109. [CrossRef]

91. Argote, L.; Miron-Spektor, E. Organizational learning: From experience to knowledge. *Organ. Sci.* **2011**, *22*, 1123–1137. [CrossRef]

92. Moore, A.A.; Weckauf, R.; Accouche, W.F.; Black, S.A. The value of consensus in rapid organisation assessment: Wildlife programmes and the Conservation Excellence Model. *Total Qual. Manag. Bus. Excell.* **2020**, *31*, 666–680. [CrossRef]

93. Kafetzopoulos, D.; Gotzamani, K. Investigating the role of EFQM enablers in innovation performance. *TQM J.* **2019**, *31*, 239–256. [CrossRef]

94. Kurian, M. Monitoring versus modelling of water–energy–food interactions: How place-based observatories can inform research for sustainable development. *Curr. Opin. Environ. Sustain.* **2020**, *44*, 35–41. [CrossRef]

95. Aschemann-Witzel, J.; Giménez, A.; Ares, G. Convenience or price orientation? Consumer characteristics influencing food waste behaviour in the context of an emerging country and the impact on future sustainability of the global food sector. *Glob. Environ. Chang.* **2018**, *49*, 85–94. [CrossRef]

96. Thrän, D.; Schaldach, R.; Millinger, M.; Wolf, V.; Arendt, O.; Ponitka, J.; Gärtner, S.; Rettenmaier, N.; Hennenberg, K.; Schüngel, J. The Milestones modeling framework: An integrated analysis of national bioenergy strategies and their global environmental impacts. *Environ. Model. Softw.* **2016**, *86*, 14–29. [CrossRef]

97. Pashangpour, R.; Faghihi, F.; Soleymani, S. Optimized scheduling for electric lift trucks in a sugarcane agro-industry based on thermal, biomass and solar resources. *Int. J. Environ. Sci. Technol.* **2018**, *15*, 2349–2358. [CrossRef]

98. Chen, Y.; Wang, F.; Duan, L.; Yang, H.; Gao, J. Tetracycline adsorption onto rice husk ash, an agricultural waste: Its kinetic and thermodynamic studies. *J. Mol. Liq.* **2016**, *222*, 487–494. [CrossRef]

99. Zou, Y.; Duan, X.; Xue, Z.; Mingju, E.; Sun, M.; Lu, X.; Jiang, M.; Yu, X. Water use conflict between wetland and agriculture. *J. Environ. Manag.* **2018**, *224*, 140–146. [CrossRef]

100. Mo, J.; Yang, Q.; Zhang, N.; Zhang, W.; Zheng, Y.; Zhang, Z. A review on agro-industrial waste (AIW) derived adsorbents for water and wastewater treatment. *J. Environ. Manag.* **2018**, *227*, 395–405. [CrossRef] [PubMed]

101. FAO; IFAD; WFP. *Strengthening the Enabling Environment for Food Security and Nutrition*; FAO: Rome, Italy, 2014.

102. Gustavsson, J.; Cederberg, C.; Sonesson, U.; Emanuelsson, A. *The Methodology of the FAO Study: Global Food Losses and Food Waste—Extent, Causes and Prevention*; SIK Report; FAO: Rome, Italy, 2011; Volume 2013.

103. Sagbansua, L.; Balo, F. Ecological impact & financial feasibility of Energy Recovery (EIFFER) Model for natural insulation material optimization. *Energy Build.* **2017**, *148*, 1–14. [CrossRef]

104. Ghedini, G.; Loreau, M.; White, C.R.; Marshall, D.J. Testing MacArthur's minimisation principle: Do communities minimise energy wastage during succession? *Ecol. Lett.* **2018**, *21*, 1182–1190. [CrossRef]

105. Pardo, J.E.; Zied, D.C.; Alvarez-Ortí, M.; Peñaranda, J.Á.; Gómez-Cantó, C.; Pardo-Giménez, A. Application of hazard analysis and critical control points (HACCP) to the processing of compost used in the cultivation of button mushroom. *Int. J. Recycl. Org. Waste Agric.* **2017**, *6*, 179–188. [CrossRef]

106. Purwanto, A.; Putri, R.S.; Ahmad, A.H.; Asbari, M.; Bernarto, I.; Santoso, P.B.; Sihite, O.B. The effect of implementation integrated management system ISO 9001, ISO 14001, ISO 22000 and ISO 45001 on Indonesian food industries performance. *Test Eng. Manag.* **2020**, *82*, 14054–14069.

107. Malavi, D.N.; Abong, G.O.; Muzhingi, T. Effect of food safety training on behavior change of food handlers: A case of orange-fleshed sweetpotato purée processing in Kenya. *Food Control* **2021**, *119*, 107500. [CrossRef]

108. Rebentisch, E.; Prusak, L. *Integrating Program Management and Systems Engineering: Methods, Tools, and Organizational Systems for Improving Performance*; John Wiley & Sons: Hoboken, NJ, USA, 2017.

109. Melnyk, S.A.; Sroufe, R.P.; Calantone, R. Assessing the impact of environmental management systems on corporate and environmental performance. *J. Oper. Manag.* **2003**, *21*, 329–351. [CrossRef]

110. Reed, R.; Lemak, D.J.; Mero, N.P. Total quality management and sustainable competitive advantage. *J. Qual. Manag.* **2000**, *5*, 5–26. [CrossRef]

111. Isaksson, R. Total quality management for sustainable development: Process based system models. *Bus. Process Manag. J.* **2006**, *12*, 632–645. [CrossRef]

112. Klaus, J. Zink from total quality management to corporate sustainability based on a stakeholder management. *J. Manag. Hist.* **2007**, *13*, 394–401.

113. Bastas, A.; Liyanage, K. Sustainable supply chain quality management: A systematic review. *J. Clean. Prod.* **2018**, *181*, 726–744. [CrossRef]

114. Souza, J.P.E.; Alves, J.M. Lean-integrated management system: A model for sustainability improvement. *J. Clean. Prod.* **2018**, *172*, 2667–2682. [CrossRef]

115. Chen, R.; Lee, Y.D.; Wang, C.H. Total quality management and sustainable competitive advantage: Serial mediation of transformational leadership and executive ability. *Total Qual. Manag. Bus. Excell.* **2020**, *31*, 451–468. [CrossRef]

116. Abbas, J. Impact of total quality management on corporate sustainability through the mediating effect of knowledge management. *J. Clean. Prod.* **2020**, *244*, 118806. [CrossRef]

117. Sampaio, P.; Saraiva, P.; Domingues, P. Management systems: Integration or addition? *Int. J. Qual. Reliab. Manag.* **2012**, *29*, 402–424. [CrossRef]

118. Namagembe, S. Enhancing environmentally friendly practices in SME agri-food upstream chains. *Int. J. Qual. Reliab. Manag.* **2020**, *38*, 505–527. [CrossRef]

119. Serhan, H.; Yannou-Lebris, G. The engineering of food with sustainable development goals: Policies, curriculums, business models, and practices. *Int. J. Sustain. Eng.* **2020**, *14*, 12–25. [CrossRef]

120. Gaddis, J.E.; Jeon, J. Sustainability transitions in agri-food systems: Insights from South Korea's universal free, eco-friendly school lunch program. *Agric. Hum. Values* **2020**, *37*, 1055–1071. [CrossRef]

121. Cao, Y.; Tao, L.; Wu, K.; Wan, G. Coordinating joint greening efforts in an agri-food supply chain with environmentally sensitive demand. *J. Clean. Prod.* **2020**, *277*, 123883. [CrossRef]

122. Fortunati, S.; Morea, D.; Mosconi, E.M. Circular economy and corporate social responsibility in the agricultural system: Cases study of the Italian agri-food industry. *Agric. Econ.* **2020**, *66*, 489–498. [CrossRef]

123. Tortorella, M.M.; Di Leo, S.; Cosmi, C.; Fortes, P.; Viccaro, M.; Cozzi, M.; Pietrapertosa, F.; Salvia, M.; Romano, S. A methodological integrated approach to analyse climate change effects in agri-food sector: The TIMES water-energy-food module. *Int. J. Environ. Res. Public Health* **2020**, *17*, 7703. [CrossRef]

124. Ngammuangtueng, P.; Jakrawatana, N.; Nilsalab, P.; Gheewala, S.H. Water, energy and food nexus in rice production in Thailand. *Sustainability* **2019**, *11*, 5852. [CrossRef]

125. Seeliger, L.; De Clercq, W.P.; Hoffmann, W.; Cullis, J.D.S.; Horn, A.M.; De Witt, M. Applying the water-energy-food nexus to farm profitability in the middle breede catchment, South Africa. *S. Afr. J. Sci.* **2018**, *114*, 1–10. [CrossRef]

126. De Castro Sobrosa Neto, R.; Berchin, I.I.; Magtoto, M.; Berchin, S.; Xavier, W.G.; de Andrade Guerra, J.B.S.O. An integrative approach for the water-energy-food nexus in beef cattle production: A simulation of the proposed model to Brazil. *J. Clean. Prod.* **2018**, *204*, 1108–1123. [CrossRef]

127. Martinez-hernandez, E.; Samsatli, S. ScienceDirect Biorefineries and the food, energy, water nexus—Towards a whole systems approach to design and planning. *Curr. Opin. Chem. Eng.* **2017**, *18*, 16–22. [CrossRef]

128. Sandhu, M.A.; Ahm Shamsuzzoha, P.H. Business excellence in a volatile, uncertain, complex and ambiguous environment (BEVUCA). *TQM J.* **2016**, *34*, 1–5. [CrossRef]

129. De Carolis, A.; Macchi, M.; Negri, E.; Terzi, S. A maturity model for assessing the digital readiness of manufacturing companies. *IFIP Adv. Inf. Commun. Technol.* **2017**, *513*, 13–20. [CrossRef]

sustainability

Article

Searching for Novel Sustainability Initiatives in Amazonia

Gabriel Medina [1,*], Cassio Pereira [2], Joice Ferreira [3], Erika Berenguer [4] and Jos Barlow [5]

[1] Faculty of Agronomy and Veterinary Medicine, University of Brasilia, Brasília 70910-900, Brazil
[2] Iniama, Belém 66095-105, Brazil
[3] Embrapa, Belém 66095-903, Brazil
[4] Ecosystems Lab, University of Oxford, Oxford OX1 2JD, UK
[5] Lancaster Environment Centre, Lancaster University, Lancaster LA1 4YW, UK
* Correspondence: gabriel.medina@unb.br

Abstract: Amazonia is facing growing environmental pressures and deep social injustices that prompt questions about how sustainable development may emerge. This study sought novel sustainability initiatives in the Brazilian Amazon based on interviews conducted with diverse practitioners in 2021 using a horizon-scanning approach and snowball sampling for selecting interviewees, who then described the initiative most familiar to them. The interviews resulted in 50 described initiatives and 101 similar initiatives that were listed but not described. The results reveal the emergence of a range of sustainability initiatives, which we classify into seven types of new seeds of change ranging from eco-business opportunities, territorial protection by grassroots movements, and novel coalitions promoting sustainability. However, most of these new seeds are still being established and have a limited or uncertain potential for replication, and most offer only incremental rather than transformative development. Therefore, although these initiatives provide weak yet real signals for alternative futures, they also suggest that much more needs to be done to support the needed transformation toward sustainable and equitable development.

Keywords: sustainable development; innovative solutions; bioeconomy; new business; horizon scanning

Citation: Medina, G.; Pereira, C.; Ferreira, J.; Berenguer, E.; Barlow, J. Searching for Novel Sustainability Initiatives in Amazonia. *Sustainability* **2022**, *14*, 10299. https://doi.org/10.3390/su141610299

Academic Editors: Nallapaneni Manoj Kumar, Md Ariful Haque and Sarif Patwary

Received: 23 June 2022
Accepted: 15 August 2022
Published: 18 August 2022

Publisher's Note: MDPI stays neutral with regard to jurisdictional claims in published maps and institutional affiliations.

1. Introduction

The question of what constitutes "sustainable development" in Amazonia is of planetary significance, yet it remains uncertain and contested. This is because fostering sustainable development has often focused on balancing the often-competing interests of industrial agriculture and infrastructure expansion with the needs and rights of forest peoples and the conservation of forests and other Amazonian ecosystems. The present challenge could not be greater, as a period of reduced deforestation (2005–2012) has been reversed, with a return to the rapid advance of the agricultural frontier and growing pressures on forest peoples in the Brazilian Amazon [1], with increased rates of deforestation [2], conflicts with local communities, and inequality and poverty [3]. These social and environmental processes are set within a political context where the consolidation of a strong lobby in favor of large-scale agricultural operations [4] has weakened environmental governance [5,6].

Despite the alarming recent trends, many of these social and environmental problems have been occurring in the Amazon for decades. Nonetheless, some of the potential strategies for ameliorating these threats are well-known. The Amazon has been a laboratory of development initiatives for over 50 years [7], and the many established approaches used to address long-standing social and environmental issues include the demarcation of indigenous territories and extractive reserves as a means of protecting forest communities and maintaining traditional, relatively sustainable uses of forest resources [8–10]; market-based initiatives such as the soy moratorium for preventing agribusiness-related deforestation [11–13]; sustainable development projects based on payments for environmental services (PES), community forestry, etc. [14–16]; and government-promoted environmental law-enforcement and development initiatives such as public procurements from family farmers [2]. The successes and failures of many of these approaches are well-known [15], and recent syntheses suggest that many of them will continue to play a prominent role in the coming decades [17].

While established initiatives are increasingly well documented [7], there are two reasons why novel initiatives, distinct from the established set, may be emerging. First, recent societal changes and environmental challenges in the Amazon may have created a fertile space for innovation, driven by rapid changes in social conditions, technologies, and awareness of climate crises in recent years [17,18]. Second, the breadth of approaches to sustainability itself have broadened, with recent evaluations including climate and advocacy coalitions [19,20], forest restoration [21], public–private partnerships [22], and the bioeconomy [23]. Given this context, it is imperative to evaluate whether a new generation of sustainable development initiatives is emerging, as these could potentially be some of the seeds of change promoting promising future scenarios [24].

We addressed this challenge by conducting a cross-sectoral search for recently implemented initiatives broadly (and some contentiously) relevant to sustainable development in the Brazilian Amazon, with the aim of revealing a practitioner's perspective of whether there are new seeds of change. Hence, our focus is on seeds that have been put into practice, rather than just being ideas, and forms a first step in scanning for novel initiatives that could be signals of a more sustainable future for Amazonia. Specifically, this study investigates: (1) What are the emerging sustainability initiatives identified by development practitioners, who is promoting them, and what are their main novel characteristics? and (2) What are the key features of these new initiatives in terms of their level of maturity, potential for scaling up, complementarity with existing initiatives, and transformative capacity? We use these results to discuss how these novel initiatives may complement—or clash with—more-established development initiatives, and we outline the capacities of our identified seeds to either transform Amazonia through radical change or maintain the status quo, seeking a sustainable future within the existing social, political, and economic structures.

2. Theoretical Framework

The seeds concept can help us to understand the different components of a better future that people want and to recognize the processes that may foster the emergence and growth of solutions that fundamentally change human–environment relationships. Bennett et al. (2016) define these seeds as existing initiatives (social, technological, economic, or social–ecological ways of thinking or doing) that represent a diversity of worldviews, values, and regions but are not currently dominant or prominent in the world [24].

Our premise is that scanning for next-generation seeds is insightful because it can provide signals for pathways towards a more sustainable future, even if those signals are currently weak [25]. Identifying seeds of change can be a valuable first step in the decision-making process by creating a comprehensive and transparent basis for subsequent assessments of evidence and effectiveness and contextualised considerations for the practical implementation of different development options [26]. Sustainable development broadly relates to combining concerns for environmental and socioeconomic issues, but there are diverse understandings of what sustainable development looks like in practice. These differences are rooted in contrasting attitudes towards change and means of change, which vary across particular areas of thought [27]. Hopwood et al.'s seminal paper proposed that pluralistic understandings of sustainable development become clearer when 'mapping' initiatives along two axes: a gradient of environmental concern (from virtually none to eco-centred) and a gradient of concern for human well-being and equality.

Concern for well-being and equality among sustainable development initiatives in Amazonia is, we expect, highly variable. Hopwood's first grouping is the status quo, the view that sustainable development can be achieved within present structures. This approach emphasizes top-down management and incremental change through existing decision-making structures and hence would conceive that sustainable development in

Amazonia does not require changes in the distribution of political power, economic resources, or land. Such incremental initiatives, even if not addressing (or attempting) transformations, may still have a positive role by engaging with more downstream points of intervention [28]. To reiterate, transformative development is associated with a change in paradigm that is preceded by significant shifts in the locus of authority over policy and experimentation with new forms of development [29].

Transformative initiatives can benefit from 'leverage points' in complex systems, where relatively small changes can lead to potentially transformative systemic changes that transform the system's rules, values, and paradigms [30]. In this sense, any transformative seeds of change are a source of social innovations that rely on alternative views of nature and social relations [31]. Scholars emphasize the role of civil society in ways in which it can 'unsettle established practices and challenge the state' or how a political culture favourable to sustainability might be nurtured [32]. Speaking to the notion that seeds require nourishment, promising social and technical innovations with the potential to change unsustainable trajectories need to be nurtured and connected to broad institutional resources and responses [33]. Transformations also imply the need to create 'change agents' who can help accelerate change even in difficult circumstances [34]. Transformational change is defined as 'shifts in power relations, discursive practices, and incentive structures that lead away from unsustainable and unjust exploitation' [35].

Following this theoretical framework, we defined an initiative as the project, practice or process mentioned by the interviewed expert and seeds as categories used by the research team to group similar initiatives. New seeds were defined as emerging initiatives that exist and that represent a diversity of worldviews, values, and regions but are not currently dominant or prominent in the world [24]. Finally, we defined 'seeds of transformational change' as initiatives that are associated with a change in the development paradigm [29].

3. Materials and Methods

We used a solution-scanning (or horizon-scanning) approach to make a first evaluation of the new initiatives. Solution scanning involves listing all the known options for addressing a particular problem [36], and it is the first stage of the subject-wide synthesis of evidence [25]. While a complete review of the evidence base for all available initiatives would be preferable, the scale and duration of such reviews are often impractical [26], especially where the solutions are not yet fully developed [37]. Such complete reviews are impractical for something new, as there is, almost by definition, scant or no literature on 'new seeds'.

For scanning the new initiatives in the Amazon, we used a classical snowballing approach as a means for selecting new initiatives. Specifically, interviewees were asked to describe the initiative they were more familiar with and also to mention other cases and experts who could be further contacted by the research team. Every interviewed person was asked to list the new initiatives they had heard about and then describe the one he/she knew best. The research team followed the received suggestions until the moment when new mentioned initiatives (or seeds that were equivalent to them) were already described in our database.

We used two complementary approaches for selecting practitioners and initiatives: (1) a list of key experts, in which we began with people we had personal contact with and asked about contact of other people they knew, and (2) a list of initiatives, in which we also began with people connected to the initiatives we knew and followed the received suggestions. This study is based on 52 interviews with diverse experts including forest peoples, rural smallholders, and their representatives (such as farmers' unions); NGO managers; government agents; and scholars. Interviews were conducted between July and December 2021, resulting in 50 initiatives fully described and 101 different initiatives listed but not described. All but two respondents described one initiative he/she considered to be novel.

A definition of new initiatives was provided upfront to the interviewees, who then listed the initiatives that they considered to fall under these criteria. Interviewees then selected one case to be described in detail. New initiatives were defined as novel on-going efforts towards sustainable development that were different from long-standing initiatives. We invited the interviewees to use their own definition/understanding of sustainable development.

All interviews were conducted with adults (>18 years) from across a range of sectors and geographic sub-regions in Brazilian Amazonia (Table 1). The sampling bias towards adult male respondents who work for NGOs and are based outside the Amazon or in the states of Pará and Amazonas was an outcome of the snowball approach adopted. To protect the identity of the interviewees, we de-identified participants by creating different alpha-numerical codes for each initiative and respondent. Given that the COVID-19 pandemic severely restricted travel and opportunities for in-person contact in Amazonia, interviews were conducted remotely using unrecorded video calls on computers.

Table 1. Profile of the 50 interviewed sustainable development practitioners who described the initiatives reported in this study.

Variables	Profile	Number of Interviewed Experts
Sector	Grassroots (rural smallholders and their representatives)	6
	NGOs	23
	Government	8
	Private	7
	Academia	6
Gender	Man	35
	Woman	15
Focus area	National level (Brazil)	15
	Regional (Amazonia)	11
	Pará State	11
	Amazonas State	11
	Other Brazilian Amazonian states	2

In order to characterize the initiatives, the research team followed an interview protocol structured around four main topics: (1) the initiative's main features and then the authors' evaluations of an initiative's (2) relative power/capacity to address existing threats (i.e., varying degrees of environmental and social concerns as identified by the institution(s) leading the initiative); (3) realistic opportunities for complementing other sustainable development solutions (i.e., those not identified as novel seeds and instead likely to be dominant or prominent for attempts to resolve particular social or environmental problems); and (4) potential for scaling up in terms of area covered and people potentially benefiting from it. Key features of the initiatives included who is promoting the initiative, its aim, and what made it novel from the interviewee's perspective (i.e., unconventional and innovative rather than replicating existing ideas in Amazonia in a new location). The initiative's opportunities for complementing existing solutions was assessed based on its area of influence and whether it spatially overlaps with protected areas, rural settlements, etc. Finally, each initiative's potential for scaling up was assessed with questions regarding its launch date and its growth since then (i.e., assuming that past growth is a reliable indicator of future growth potential). For each topic, a group of questions was asked following a standardized research protocol.

We grouped these initiatives after the interviews into categories based on the problems initiatives are setting out to solve, the solutions they are attempting to achieve, and the main sector involved. Single-sector grouping proved infeasible since some sectors such as NGOS are involved in so many of these activities. Since some initiatives would fit in different seeds, we acknowledge other forms of categorization. Given the specificities of the initiatives, some categories were given subcategories.

4. Results

4.1. The New Initiatives

This section presents the new initiatives identified by interviewees as well as their main promoters and claimed novelties. Figure 1 presents the spatial distribution of the described initiatives based on the location of the headquarters of their main institutional promoter.

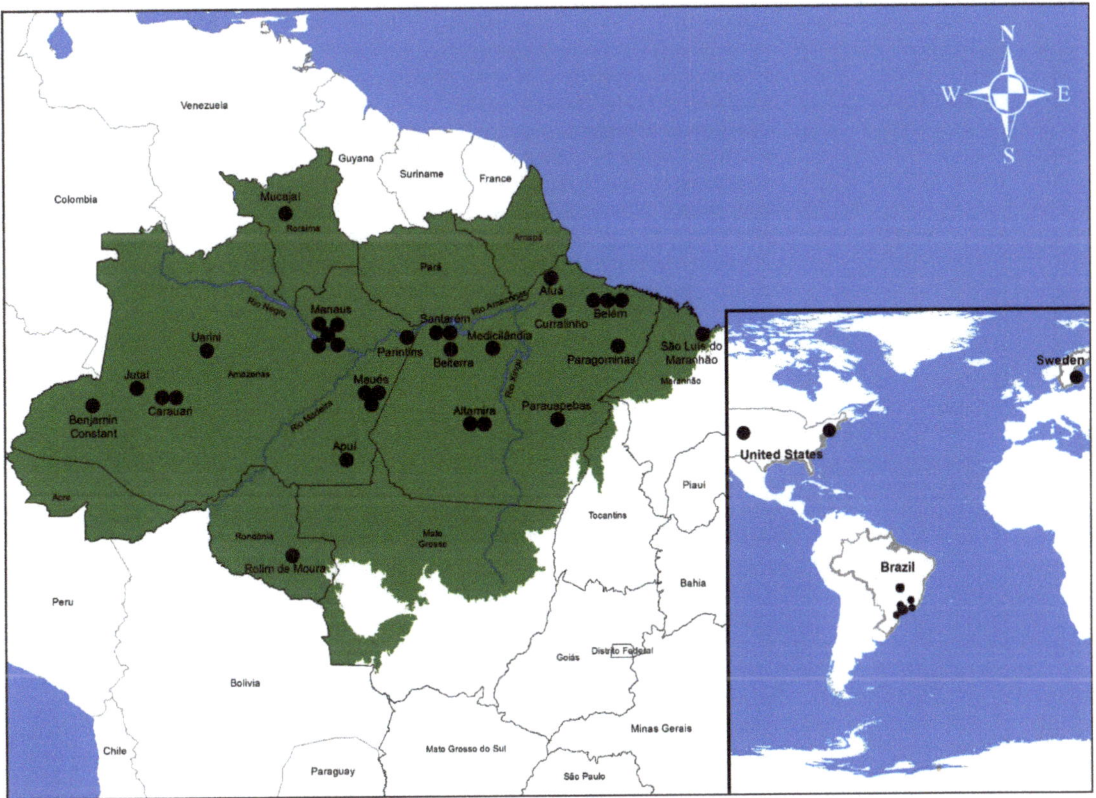

Figure 1. Locations of the described new sustainable development initiatives in the Brazilian Amazon.

4.1.1. What Kinds of New Initiatives Did the Interviewees Identify?

The interviews resulted in 50 described and 101 similar initiatives that were listed but not described. The start dates of these initiatives were from 2007 to 2021, with 44% starting just three years before the interviews, and 88% were within the last six years. The identified new initiatives were summarized into seven groups:

(1) **Eco-business opportunities**—Initiatives related to creating new business opportunities for products from the Amazon. These seeds encompass bioeconomy business incubators; processing high-value products for niche markets; and innovative marketing hubs. These seeds include 13 described and 22 listed initiatives.

(2) **Environmental and social accountability in agribusiness supply chains**—Initiatives related to improving traceability and farming practices in the agribusiness commodities supply chains such as for soybeans and cattle. This group of seeds encompasses transparency and improved field practices and includes 3 described and 22 listed initiatives.

(3) **Private investment in sustainable development**—Initiatives by private companies either as a private investment for profit or responding to environmental or market conditions. These seeds encompass environmental/business conditions and funds and investments and include 7 described and 22 listed initiatives.

(4) **Territorial protection by grassroots movements**—Initiatives by farmers' unions or indigenous associations. These seeds encompass territorial protection and socioeconomic development initiatives and includes 4 described and 18 listed initiatives.

(5) **Subnational governmental policies**—Initiatives by autonomous federal agencies (mainly the ones that are part of the judicial power and not dependent on the President's office or a particular ministry in Brasilia) as well as by state and municipal governments. These seeds encompass initiatives related to socioeconomic development, technological development, and environmental protection. They include six described and seven listed initiatives.

(6) **Coalitions that promote rights and environment**—Sets of different stakeholders collaborating in search of sustainable development alternatives. These seeds encompass either international or national coalitions and include four described and five listed initiatives.

(7) **Civil society activism**—Initiatives by individuals or third-sector organizations. These seeds encompass initiatives related to digital activism; forest restoration and payments for environmental services; alliances for governance; and other efforts. They include 13 described and 24 listed initiatives.

4.1.2. Which Institutions Were Promoting the New Initiatives?

The described new initiatives were promoted by different place-based stakeholders such as governments (n = 10); NGOs (n = 21); rural communities, individuals, and social movements (n = 10); and private companies (n = 9). Some of the initiatives' costs were covered with people's own out-of-pocket money (n = 12), but a considerable number of them relied on external support (n = 38).

Key external sponsors include international cooperative organisations (n = 16) such as USAID and the Norwegian NORAD; private companies (n = 11) such as the banks Bradesco, Itaú, and Santander, and the cosmetics company Natura; Brazilian state and federal governments (n = 5) such as the states of Amazonas and Pará and the Ministry of Agriculture, or other sources (n = 6).

4.1.3. What Did the Interviewees Claim as Novelty?

We used our groupings to examine novelty based on the assessments of the interviewees. Based on the initiatives' main perceived innovations (Figure 2), we summarized seven types of novelty presented in the described initiatives:

(1) **Bioeconomy as a business opportunity for sustainable development**—Interviewed sustainable development practitioners emphasized the current effort to promote bioeconomy mainly through eco-business. The 13 described and 22 listed eco-business opportunities were mainly focused on tapping the green market by exploring the bioeconomy as a business opportunity for sustainable development.

(2) **New use of technology**—New technologies and solutions such as blockchain, crowdfunding, big data, and cell phone apps were used to connect development projects and sponsors and to increase transparency in supply chains in initiatives such as Nature Invest and *Do Pasto ao Prato*. Digital activists included NGOs and unions that organized on different kinds of social media as well as in demonstration campaigns in favour of conservation, fighting deforestation, and denouncing illegalities. In some ways, these efforts can be seen as an extension of previous initiatives that occurred decades ago such as Sting's activism with the Kayapo people and traditional media campaigns but using social media. However, there are two key differences. First, these latest initiatives use social media for the large-scale mobilization of civil society as a means for demanding actions from governments, private-sector companies, etc.

Second, our interviews revealed an emergence of the influencer/activist on social media, especially by indigenous women.

Seed	Initiative	Type of iniative	Year	Main promoter	Main sponsor	Novelties according to the respondent
Eco-business opportunities	Bioeconomy business-incubators	Plataforma Parceiros pela Amazônia	2017	PPA	USAID	Impact entrepreneurship
		Centro de Emprededorismo da Amazônia	2015	CEA	CLUA	Modern methodologies in training young entrepreneurs
		Aceleradora de impacto AMAZ	2019	Idesam	PPA USAID	Acceleration of existing business
		Beiradão de Oportunidades	2016	PSA	Fundação Telefônica Vivo	Youth entrepreneurial education
		Desafio Conexus	2018	CONEXSUS	Fundo Vale	Community business
	Processing of high-value products for niche markets	CacauWay	2010	Coopatrans	Funcacau - Gov. Pará	Cocoa agro-industrialization by cooperative farmers
		IG da Farinha	2015	Mamiruá/FAS	Fundo Amazônia - Norway	Geographic identification targeting nich markets
		Guaraná Agroflorestal	2019	ASPROC	USAID	Commercialization of pirarucu inspected for food safety
		Comércio ribeirinho	2015	ASCAMPA	Own	Commercialization of ecological/agroforestry guarana
		InovaSocioBio Amazonas	2020	SEDECTI Gov. AM	MAPA - Gov. BR	Strengthening of the nut, guarana, pirarucu and honey supply chains
	Innovative marketing hubs	Amazônia Hub	2019	Amazônia Hub	Own	Digital business platform with products from Amazonia
		Amazônia Ativa	2017	BVRio	Norway - NORAD	Digital business platform with products from Amazonia
		Rede Origens Brasil	2016	IMAFLORA	Fundo Amazônia - Norway	Sustainable business connecting communities and companies
Environmental and social accountability in agribusiness supply chains	Transparency	Do Pasto ao Prato	2020	SEI-TRASE	Norway - NORAD	App to inform the consumer about the origin of the meat
		Boi na Linha	2019	Imaflora	Moore Fondation	Forum for designing a meat origin transparency system
	Field practices	Pecuária Verde na Fazenda Marupiara	2006	Particular	Own	Sustainable cattle ranching intensification
Private investment in sustainable development	Environmental / business conditions	Condicionantes do PBA	2017	Norte Energia	Consórcio Belo Monte	Income generation for indigenous people
		Projeto agriculturas de conservação	2015	Vale/ICMBIO	Vale	Support to settlers of Conservation Units buffer areas
		Projeto Ocara / Samsung	2021	Samsung/UEA	Samsung	Tax incentive for investments in local development areas
	Funds and investments	Fundo JBS pela Amazônia	2020	Fundo JBS	JBS	Financing projects with social impact and connected to markets
		Meta Florestal do Fundo Vale	2019	Vale	Vale	Restoration of 100,000 ha of forests, including the Amazon
		Conselho Consultivo da Amazônia	2020	Bradesco/Itaú/Santander	Bradesco/Itaú/Santander	Investments to leverage sustainable development
		Remata	2020	Particular	Own	Large-scale agroforestry system
Territorial protection by grassroots movements	Protection of traditional community territories	Protocolos de consulta	2017	STTR/MPF	Vale	Use of law by communities with areas affected by companies
		Coletivos de ordenamento territorial	2007	STTR	Fase	Local organization for land tenure regularization
		Articulação indígena nacional	2018	APIB	Own	Reaction to threats to rights with the use of the internet
	Development	Fundo Solidário Porto do Açaí (FSPA)	2016	FSPA	Own	Fund for promoting best practices for açaí management
		Catrapoa	2019	MPF	Own	Including local food in the PNAE school meals
	Socioeconomic development	Curso em Agroecologia Sateré Mawé	2018	IFAM	Own	Indigenous education with agroecology principles
		Política de Des. dos povos tradicionais	2007	Governo Federal	Own	Territorial recognition of traditional communities
	Technological development	Parque Tecnológico do Alto Solimões	2020	Governo AM-SEDECTI	Governo Federal	Development of biocosmetics and herbal medicines
		Piscicultura de Igapó	2017	Governo AM	IFAM	Multiphase production applied to family fish farming
Subnational governmental development initiatives	Environmental protection	Selo Verde	2020	SEMAS/PA	Norway-NORAD	Transparency on areas producing cattle with deforestation
	International	GCF Task Force	2008	University of Colorado	USA	Governors' Climate and Forests Task Force
		Science Panel for the Amazon	2019	SDSN	Moore Fundation	Panel of scientists for sustainable development proposals
	National	Uma concertação pela Amazônia	2019	Instituto Arapyaú/Natura	Natura	Private sector initiative to network for development
		Coalizão Clima Florestas e Agricultura	2015	IPAM	Own	Movement towards a new low carbon economy
Coalitions that promote rights and environment	Digital activism	Política por Inteiro	2020	Política por inteiro	Konrad Adenauer Stiftung	Announces deregulatory changes related to the Amazon
		Brigada Digital	2021	Greenpeace Brasil	Greenpeace	Campaigns by digital brigade
		Clima de Eleição	2020	Instituto Clima de Eleição	Konrad Adenauer Stiftung	Campaign on the climate crisis aimed at parliamentarians
	Forest restoration and payments for environmental services	Viveiro Cidadão	2017	Ecoporé	Petrobrás	Women and youth in restoration with agroforestry systems
		Projeto Assentamentos Sustentáveis	2014	IPAM	Fundo Amazônia Norway	Support for sustainable development of rural settlements
	Alliances for governance	Programa territorial do Médio Juruá	2017	SITAWI	USAID	Territorial management connecting companies, local actors, gov.
		Agroecologia nos municípios	2019	ANA Amazônia	Own	Market expansion for agroecological products in city halls
		Mulheres pela Agroecologia	2015	AMABELA	Fundo Dema	Women's empowerment through agroecology
		Aliança Des. Sus. do Sul do Amazonas	2017	WWF	WWF	Alliance between NGOs, grassroots organizations and city halls
Civil society activism	Other	Rádio poste (individual)	2017	Individual	Own	Communication with urban/rural population
		Nature Invest (crowdfunding)	2020	Nature Invest	Own	Environmental crowdfunding for small projects
		Alerta indígena Covid 19 (health)	2020	IPAM	Instituto Clima e Sociedade	App for monitoring cases of COVID 19 in indigenous peoples
		Engaja na Amazônia (youth)	2017	Engajamundo	Instituto Clima e Sociedade	Training young people linked to the climate agenda

Figure 2. An overview of the new sustainable development initiatives identified in the Brazilian Amazon.

(3) **New connections among actors and across boundaries**—The interviews revealed a number of new coalitions that promote rights and environmental protection, including new multi-stakeholder and relatively egalitarian assemblies and decision-making forums. These contrast with former initiatives that were more limited in scope and less horizontal in terms of power balance. Much of this current mood comes from the fact that key stakeholders identified clear limits in individual actions and realized the need and the benefits of cooperation, which includes an opportunity to exert soft power and influence dynamics. International coalitions range from academic coalitions such as the Science Panel for the Amazon and political coalitions such as the GCF Task Force, initially promoted by the republican former governor of California Arnold Schwarzenegger and the Leaf Coalition recently promoted by the democratic national US government. National coalitions include private-sector initiatives to network for development such as *Uma Concertação pela Amazônia* and *Coalizão Brasil Clima Florestas e Agricultura*. The novelty is the mutual effort to cooperate in multi-stakeholder forums.

(4) **Integration of new practices into production**—Technological development initiatives include the use of biocosmetics and herbal medicines at the *Parque Científico e Tecnológico do Alto Solimões* and multiphase fish production by family farmers by the *Piscicultura de Igapó* initiative. Agribusiness sectors are also trying to respond to external demands for sustainable farming and cattle ranch intensification via initiatives that improve transparency and governance in agricultural supply chains.

(5) **New private sector actors investing in sustainable initiatives**—The private sector is promoting large-scale financial investments in support of sustainable development projects in the Amazon. These include private investments by banks, corporations, and individuals being made in forest restoration and sustainable agriculture projects such as the investments the *Conselho Consultivo da Amazônia* recommended to banks such as Bradesco, Itaú, and Santander.

(6) **New use of or creation of legislation**—Communities are promoting the local governance of territories and the use of consultation protocols (previously used mainly by indigenous peoples) as a means of territorial protection. Governmental subnational actions encompass socioeconomic development initiatives such as *Catrapoa* promoted local food in the PNAE school meals and indigenous education with agroecology principles such as the technical course in agro-ecology (*Curso técnico em agroecologia*) in the Sateré-Mawé Indigenous reserve

(7) **Emergence of grassroots actions without significant or permanent external support**—Development initiatives include funds such as the Solidarity fund of the Acai port (*Fundo Solidário Porto do Açaí*) established by local communities for financing better farming practices and market access.

Note 1: Even though some initiatives would fit different solutions, we mentioned each initiative only once in a specific category.

Note 2: Similar mentioned but not described initiatives are (101 in total): **(1) Eco-business opportunities (22 initiatives):** 1.1 Bioeconomy business-incubators: Sitawi; Amazônia UP; Ecocentro; Rainforest Social Business School; Certi Amazônia. 1.2 Processing of high-value products for niche markets: Cacau orgânico—CEPOTX; Guaraná—Sateré-Mawé; Mahá Biocosméticos; Chocolates De Mendes; Mel—Peabiru; Amazônia 4.0; Café Apui; AmazonMel; Projeto Castanha da RESEX do Rio Unini; Inatú Amazônia. 1.3 Innovative marketing hubs: Flor de Jambu; Jirau da Amazônia; Polo BioAmazonas; Observatório da Castanha da Amazônia (OCA); Amazônia 100%; Design & Madeira Sustentável; Manioca. **(2) Environmental and social accountability in agribusiness supply chains (13 initiatives):** 2.1 Transparency: Soja na Linha; Conecta; Plataforma Nice planet—SMGeo; Plataforma da JBS—rastreabilidade do gado; Sirflor (Pará)—regularização de propriedades; Plataforma de adequação ambiental do Imac (MT); Fornecedores Indiretos na Pecuária (Amigos da Terra); PrevisIA; Trase; Lucida. 2.2 Field practices: Pecuária Verde; JBS Net Zero; Pecuariando. **(3) Private investment in sustainable development (12 initiatives):** 3.1 Environmental/business conditions: Programa Prioritário de Bioeconomia; Waimiri Atroari e Parakanã—Eletronorte; Borracha—Michelin. 3.2 Funds and investments: Centro de Orquestração de Inovações (COI)—WTT; The Good Food Institute; KPTL—Fundo de Floresta e Clima; Emerge Amazônia; Impact Hub; Fundo para o Desenvolvimento Sustentável e a Bioeconomia da Amazônia; Althelia biodiversity fund; Mov investimentos; Parcerias corporativas—Café Suruí com contrato com a Três Corações. **(4) Territorial protection by grassroots movements (18 initiatives):** 4.1 Territorial protection: Campanha Não Abra Mão da Sua Terra; Protocolos de consulta sobre Asfaltamento da BR 319; Levante popular; Observatório de Segurança e Soberania Alimentar; Terra sem Males (RO). 4.2 Development: Fundo Dema; ACOSPER/UNICAFES; Projeto CAR Participativo/STTR Santarém; Programa Jurisdicional de REDD+; Finapop—MST; Flores do Campo (Grupo de mulheres de Mojuí dos Campos); Grupo de Jovens do CNS; Casa Familiar Rural de Boa Vista dos Ramos; Associações de mulheres em defesa da agricultura familiar do Rio Canaticu (Marajó); Banco Comunitário do Rio Canaticu (Moeda Social Yaça); Canindé (RO)—mobilização indígena; Comitê Chico Mendes; Surara—Coletivo de mulheres indígenas do Tapajós. **(5) Subnational government initiatives (7 initiatives):** 5.1 Socioeconomic development; Fundo de Apoio a Cacauicultura do Estado do Pará—FUNCACAU; GTI Saúde Indígena. 5.2 Technological development: Instituto Mamirauá; Rações alternativas para piscicultura—IFAM; Manejo florestal—IFT junto à Verde para Sempre; 5.3 Environmental protection: Força Tarefa Amazônia—MPF; Amazônia Protege—MPF. **(6) Coalitions that promote rights and environment (5 initiatives):** 6.1 International: The Lowering Emissions by Accelerating

Forest finance (LEAF) Coalition. 6.2 National: Consórcio de Governadores da Amazônia Legal; Amazônia 2030; Grupo Carta de Belém; Fórum Nacional Permanente em Defesa da Amazônia. **(7) Civil society activism (24 initiatives):** 7.1 Digital activism: Portal Proteja;Rede de Podcasts do Xingu; SOS Amazônia; Engajamento de influenciadores—Alok; Uma gota no oceano; Motosserra de Ouro para Arthur Lira; Floresta sem cortes; Negritar—Coletivo de audiovisual; Coletivo Jovem de Meio Ambiente Pará; 342 Amazônia; Amazônia Alerta; 7.2 Forest restoration and payments for environmental services: Aliança para Restauração da Amazônia; Projeto Bacia do Rio Putumayo-Içá; Acelerador de Agroflorestas e Restauração. 7.3 Alliances for governance: Rede Maniva de Agroecologia; Coletivo do Pirarucu; Legado Integrado da Região Amazônica—LIRA; Projeto Paisagens Sustentáveis da Amazônia. 7.4 Other: Ciência Cidadã para Amazônia; App Castanhadora; Brigada de Incêndio; Instituto Mapinguari (AP); AmIT—Amazon Institute of Technology.

4.2. Features of the New Initiatives

The new initiatives varied in terms of their level of maturity, possibility of complementing existing solutions, potential for scaling up and for transformation. Considering the initiatives' *maturity level*, out of the 50 described, 25 initiatives were reported by the interviewees as still embryonic, 21 as mature, and 4 as ended. Embryonic initiatives still in process of being implemented have not yet had the chance to address existing environmental and/or social challenges and cannot yet be considered viable or long-lasting.

Considering their capacity to *complement* existing development initiatives, 44 initiatives were deemed as building on and complementing existing environmental conservation efforts, while 6 did not include conservation measures and focused only on socioeconomic development, in the sense of concern for improving well-being but not necessarily reducing social inequities. Out of the 44 initiatives we considered to be complementary, 23 were complementary to various environmental conservation efforts, while 21 complemented specific sustainable development initiatives such as the protection of indigenous territories. Out of the 50 described initiatives, 43 claimed to support socioeconomic development by promoting better farming practices (with the perceived 'improvement' reflecting the logic of a particular initiative), sources of income, or organizational skills, while 7 did not include support for socioeconomic development and focused only on conservation. To a large extent, the identified initiatives complemented previous efforts and helped their maintenance/consolidation, such as by supporting the preservation of indigenous lands and conservation units and the development of rural settlements.

Regarding their potential to *replicate and scale up*, 28 initiatives were reported as having a replication potential limited to specific contexts, while 22 were deemed to be replicable across the whole of Amazonia. Lower-cost initiatives based on local efforts showed greater potential for replication in practice. However, initiatives such as accelerators (with greater investments) can also be promoted through public policies.

Regarding the initiative's *transformative capacity* as a whole (i.e., our analysis of whether an initiative sought radical change or instead, reform or incremental change and maintenance of the status quo), the interviewed practitioners considered that 17 presented small contributions (incremental tweaks), 29 would deliver significant contributions but without structural change (reformist adaptations), and 4 promoted structural change (radical transformations). Despite this variation, all were considered by interviewees to have some potential value for sustainable development in the Amazon. Interviewees were not asked to assess the capacity of their initiatives to deliver so-called win-wins (fulfilling socioeconomic and environmental objectives); these are hard to achieve in practice, and it is plausible that a new initiative could, for example, be effective in reducing forest degradation but simultaneously ignore or worsen social inequities.

5. Discussion

The large number of initiatives reported as potential new seeds of change reveal the emergence of diverse, often contradictory, pathways towards sustainable development in the Amazon. Indeed, these initiatives are moving towards different development end-goals with respect to the weight of concern for the environment and improving human well-being and equity. Our study assessed some of these promising efforts to conceive and promote eco-business models, accountability in agribusiness supply chains, private sustainable development, grassroots rights, governmental policies, coalitions, and civil society projects. Taken together, these seeds of change connect very powerful stakeholders such as multinational companies, state governors, and large NGOs that often can count on important investments made by both domestic and international sponsors.

A key positive aspect of these new initiatives is that they often complement or build upon well-described solutions such as extractive reserves [8]. Some of these new solutions were already mentioned in other studies such as climate and advocacy coalitions [20], forest restoration [21], private investments [22], and the bioeconomy [23]. This study has attempted to systematically map out stakeholders' own perspectives on these new potential solutions, adding to previous efforts listing established initiatives in the Amazon [7].

5.1. Seeds' Capacity to Address Existing Environmental and Social Challenges

The new seeds are developing at a time in which the Brazilian Amazon is again experiencing growing rates of deforestation [2], growing number of conflicts with local communities [1], and increased inequality and poverty. Both classical and new sustainable development initiatives are overwhelmed by the current context.

Our assessment of the new initiatives' features reveals some key limitations regarding their ability to address existing challenges. Most initiatives are not yet completely established and have limited potential for scaling up. In this sense, we agree with previous studies' assessment that promising initiatives by themselves, despite their success in transforming local spaces, are often insufficient for advancing sustainable development at broader societal scales because the required political and environmental changes are often beyond their reach [7].

Most of these new initiatives were perceived as delivering incremental change, without upsetting the status quo of the distribution of resources and decision-making powers and falling short of offering a realistic prospect for transformative, radical change. Such incremental changes reflects literature stating that most initiatives do not address (or attempt) transformations but instead go for more downstream points of intervention [38]. In contrast, transformative initiatives tend to be context-specific and depend on 'leverage points' where relatively small changes can lead to potentially transformative systemic changes [30]. Brockhaus makes a strong case that achieving sustainable development in Amazonia and elsewhere requires confronting entrenched patterns of inequality and power relations [35]. Development success therefore requires addressing the power imbalances between different kinds of state and non-state actors. Ignoring these kinds of insights from political ecology [39] creates the conditions whereby certain kinds of seeds of sustainable development may actually reinforce and reproduce inequalities. These outcomes are difficult to predict, and incremental new initiatives could still have a positive role in supporting established initiatives.

5.2. Weak but Real Signals for Promising Future Scenarios

Scanning next-generation seeds is a first step towards decision-making processes that create a comprehensive and transparent basis for subsequent assessments of evidence [26]. Some of the new initiatives we have described can be seen as seeds of change that provide some weak yet rea, signals for promising future scenarios that potentially change human–environment relationships [24]. The history of the Amazon is rich in place-based solutions such as extractive reserves [9,10] that were tried out in a specific region and under specific circumstances and then expanded. The diverse approaches create an 'ecosystem' of new initiatives, increasing the chances that one or more will mature into longer-lasting and far-reaching sustainable development initiatives.

The fact that some seed types are dominant now does not necessarily reflect their capacity to bring about sustainable development but may be for circumstantial reasons. This may be the case for the growing number of efforts for forest restoration given to the UN Decade on Ecosystem Restoration as well as the 'fever' for eco-business solutions promoted by powerful stakeholders such as USAID and private companies. In the 2000s, Amazonia experienced great donor-driven support for forest management, while in the 2010s, the focus was on payments for environmental services [14]. Our assessment suggests that the bioeconomy and forest restoration driven by carbon capture could become the new donor-driven conceptual solutions for the Amazon, while bottom-up civilian science initiatives were also prevalent (Figure 2).

5.3. Implications for Practitioners and Policy Makers

This study reveals that most of the initiatives deemed promising and innovative by the interviewed practitioners are exogenous development concepts such as the bioeconomy that are promoted by external stakeholders such as NGOs and governmental agencies and are often sponsored by external donors. At the same time, the list of identified seeds includes grassroots initiatives mainly focused on territorial protection. Other studies highlight that endogenous, locally developed farming and governance systems should also form the basis for sustainable development in the Amazon [10]. These findings highlight the importance of considering top-down vs. bottom-up management approaches and bringing scholarly and policy attention back to local populations and their resource-use systems.

These lessons are fundamental for practitioners, policy makers, and donor agencies that should reconsider their focus on externally driven top-down concepts to provide increased support for helping communities to improve their existing systems, which are still being developed and can benefit from external support for improvements in technological, economic, environmental, and social aspects [40]. Endogenous approaches to development grounded in local practices and needs can become a viable option for sustainable development in the Amazon. For such approaches, attention should be given to farmer-led technological innovation, local governance, and the recognition of marginalized local knowledge.

5.4. Limitations and Further Research

The methodology used (the snowball sampling approach and the interview method) focused on the practitioners' perspectives in a qualitative evaluation of initiatives related to sustainability development and was neither comprehensive nor representative, meaning numbers should not be evaluated quantitatively. Although we did include quantitative questions on the area covered by the initiatives as well as on the number of beneficiaries, the high diversity of approaches towards sustainable development meant it was not possible to use the data collected to assess the capacity of the different initiatives to address existing threats. Some initiatives focus on raising awareness on specific issues and target different stakeholders in different areas, making it hard to identify their actual impact on the ground. Even initiatives focusing on specific areas and targeting specific communities found it difficult to measure their impacts since their actions are often part of other initiatives and dynamics.

6. Conclusions

The results of this study reveal the existence of new initiatives showing proof of the concept of seeds of changes that, in many cases, are complementary to existing well-known initiatives such as conservation areas, environmental law enforcement, and governments efforts to promote local development. Although these initiatives were new, their capacity to address existing threats is likely to be highly variable; most of the initiatives are difficult to evaluate as they were, by their nascent nature, not yet mature. Yet the range of seeds suggests there is a growing ecosystem of alternatives for sustainable development in the Amazon that could contribute to future solutions. However, the new initiatives are only one part of the long-term continued efforts for sustainable development in the Amazon, and they must occur alongside other actions such as ensuring the recognition of forest peoples' access to land and resources, maintaining all of Amazonia's biodiversity, avoiding dangerous tipping points that could alter the nature of the system itself, and providing more attention to and support for endogenous, locally based development initiatives led by Amazonian rural communities and family farmers.

Author Contributions: G.M.—Methodology, investigation—interviews, data curation, formal analysis, writing—original draft. C.P.—Methodology, investigation—interviews, section on the origin of the initiatives, conclusion. J.F.—Conceptualization, methodology—scanning the new initiatives. E.B.—Conceptualization, methodology—scanning the new initiatives. J.B.—Conceptualization, funding acquisition, methodology, project administration, writing—original draft, writing—review & editing, classification of the initiatives. All authors have read and agreed to the published version of the manuscript.

Funding: This research was funded by UKRI's Global Challenges Research Fund.

Informed Consent Statement: Informed consent was obtained from all subjects involved in the study.

Data Availability Statement: Not applicable.

Acknowledgments: We would like to thank all the development practitioners who collaborated on this study.

Conflicts of Interest: The authors declare no conflict of interest.

References

1. Villén-Pérez, S.; Moutinho, P.; Nóbrega, C.C.; de Marco, P. Brazilian Amazon gold: Indigenous land rights under risk. *Elem. Sci. Anthr.* **2020**, *8*, 31. [CrossRef]
2. Trancoso, R. Changing Amazon deforestation patterns: Urgent need to restore command and control policies and market interventions. *Environ. Res. Lett.* **2021**, *16*, 041004. [CrossRef]
3. Tregidgo, D.; Barlow, J.; Pompeu, P.S.; Parry, L. Tough fishing and severe seasonal food insecurity in Amazonian flooded forests. *People Nat.* **2020**, *2*, 468–482. [CrossRef]
4. Aguiar, D. *Dossiê Crítico da Logística da Soja*; FASE: Rio de Janeiro, Brazil, 2021.
5. Britto, G.; Romero, J.P.; Freitas, E. La gran brecha: Complejidad económica y trayectorias de desarrollo del Brasil y la República de Corea. *Rev. La CEPAL* **2019**, *127*, 217–241. [CrossRef]
6. De Souza, L.E.V.; Fetz, M.; Zagatto, B.P.; Pinho, N.S. Violence and Illegal Deforestation: The Crimes of 'Environmental Militias' in the Amazon Forest. *Capital. Nat. Social.* **2021**, *33*, 1–21. [CrossRef]
7. Brondizio, E.S.; Andersson, K.; de Castro, F.; Futemma, C.; Salk, C.; Tengö, M.; Londres, M.; Tourne, D.C.M.; Gonzales, T.S.; Molina-Garzón, A.; et al. Making place-based sustainability initiatives visible in the Brazilian Amazon. *Curr. Opin. Environ. Sustain.* **2021**, *49*, 66–78. [CrossRef]
8. Jaffé, R.; Nunes, S.; Dos Santos, J.F.; Gastauer, M.; Giannini, T.C.; Nascimento, W., Jr.; Sales, M.; Souza, C.M., Jr.; Souza-Filho, P.W.; Fletcher, R.J., Jr. Forecasting deforestation in the Brazilian Amazon to prioritize conservation efforts. *Environ. Res. Lett.* **2021**, *16*, 084034. [CrossRef]
9. Maciel, R.C.G.; Cavalcanti, F.C.d.; de Souza, E.F.; de Oliveira, O.F.; Filho, P.G.C. The 'Chico Mendes' extractive reserve and land governance in the Amazon: Some lessons from the two last decades. *J. Environ. Manag.* **2018**, *223*, 403–408. [CrossRef]
10. Medina, G.; Pokorny, B.; Campbell, B. Forest governance in the Amazon: Favoring the emergence of local management systems. *World Dev.* **2022**, *149*, 105696. [CrossRef]
11. Neimark, B.; Osterhoudt, S.; Alter, H.; Gradinar, A. A new sustainability model for measuring changes in power and access in global commodity chains: Through a smallholder lens. *Palgrave Commun.* **2019**, *5*, 1. [CrossRef]
12. Newton, P.; Agrawal, A.; Wollenberg, L. Enhancing the sustainability of commodity supply chains in tropical forest and agricultural landscapes. *Glob. Environ. Chang.* **2013**, *23*, 1761–1772. [CrossRef]
13. Rajão, R.; Soares-Filho, B.; Nunes, F.; Börner, J.; Machado, L.; Assis, D.; Oliveira, A.; Pinto, L.; Ribeiro, V.; Rausch, L.; et al. The rotten apples of Brazil's agribusiness. *Science* **2020**, *369*, 246–248. [CrossRef] [PubMed]
14. Pham, T.T.; Moeliono, M.; Yuwono, J.; Dwisatrio, B.; Gallo, P. REDD+ finance in Brazil, Indonesia and Vietnam: Stakeholder perspectives between 2009–2019. *Glob. Environ. Chang.* **2021**, *70*, 102330. [CrossRef]
15. Pokorny, B.; Johnson, J.; Medina, G.; Hoch, L. Market-based conservation of the Amazonian forests: Revisiting win-win expectations. *Geoforum* **2012**, *43*, 387–401. [CrossRef]
16. Medina, G.; Pokorny, B. Financial assessment of community forest management: Lessons from a cross-sectional study of timber management systems in the Brazilian Amazon. *Int. For. Rev.* **2022**, *24*, 197–207. [CrossRef]
17. Nobre, C. *Amazon Assessment Report 2021*; UN Sustainable Development Solutions Network: New York, NY, USA, 2021.

18. Fang, K. Moving away from sustainability. *Nat. Sustain.* **2022**, *5*, 5–6. [CrossRef]
19. Fernández-Llamazares, Á.; Díaz-Reviriego, I.; Luz, A.C.; Cabeza, M.; Pyhälä, A.; Reyes-García, V. Rapid ecosystem change challenges the adaptive capacity of Local Environmental Knowledge. *Glob. Environ. Chang.* **2015**, *31*, 272–284. [CrossRef]
20. Aamodt, S.; Stensdal, I. Seizing policy windows: Policy Influence of climate advocacy coalitions in Brazil, China, and India, 2000–2015. *Glob. Environ. Chang.* **2017**, *46*, 114–125. [CrossRef]
21. Urzedo, D.I.; Neilson, J.; Fisher, R.; Junqueira, R.G.P. A global production network for ecosystem services: The emergent governance of landscape restoration in the Brazilian Amazon. *Glob. Environ. Chang.* **2020**, *61*, 102059. [CrossRef]
22. Furumo, P.R.; Lambin, E.F. Scaling up zero-deforestation initiatives through public-private partnerships: A look inside post-conflict Colombia. *Glob. Environ. Chang.* **2020**, *62*, 102055. [CrossRef]
23. Nobre, I.; Nobre, C. Projet Amazonie 4.0. *Futuribles* **2020**, *434*, 95. [CrossRef]
24. Bennett, E.M.; Solan, M.; Biggs, R.; McPhearson, T.; Norström, A.V.; Olsson, P.; Pereira, L.; Peterson, G.D.; Raudsepp-Hearne, C.; Biermann, F.; et al. Bright spots: Seeds of a good Anthropocene. *Front. Ecol. Environ.* **2016**, *14*, 441–448. [CrossRef]
25. Plieninger, T.; Kohsaka, R.; Bieling, C.; Hashimoto, S.; Kamiyama, C.; Kizos, T.; Penker, M.; Kieninger, P.; Shaw, B.J.; Sioen, G.B.; et al. Fostering biocultural diversity in landscapes through place-based food networks: A 'solution scan' of European and Japanese models. *Sustain. Sci.* **2018**, *13*, 219–233. [CrossRef]
26. Petrovan, S.O.; Aldridge, D.C.; Bartlett, H.; Bladon, A.J.; Booth, H.; Broad, S.; Broom, D.M.; Burgess, N.D.; Cleaveland, S.; Cunningham, A.A.; et al. Post COVID-19: A solution scan of options for preventing future zoonotic epidemics. *Biol. Rev.* **2021**, *96*, 2694–2715. [CrossRef]
27. Hopwood, B.; Mellor, M.; O'Brien, G. Sustainable development: Mapping different approaches. *Sustain. Dev.* **2005**, *13*, 38–52. [CrossRef]
28. Rees, W.E. Achieving Sustainability: Reform or Transformation? *J. Plan. Lit.* **1995**, *9*, 343–361. [CrossRef]
29. Hall, P. Policy paradigms, social learning, and the State: The case of economic policymaking in Britain. *Comp. Polit.* **1993**, *25*, 275–296. [CrossRef]
30. Dorninger, C.; Abson, D.J.; Apetrei, C.I.; Derwort, P.; Ives, C.D.; Klaniecki, K.; Lam, D.P.M.; Langsenlehner, M.; Riechers, M.; Spittler, N.; et al. Leverage points for sustainability transformation: A review on interventions in food and energy systems. *Ecol. Econ.* **2020**, *171*, 106570. [CrossRef]
31. Futemma, C.; de Castro, F.; Brondizio, E.S. Farmers and Social Innovations in Rural Development: Collaborative Arrangements in Eastern Brazilian Amazon. *Land Use Policy* **2020**, *99*, 104999. [CrossRef]
32. Meadowcroft, J. Engaging with the politics of sustainability transitions. *Environ. Innov. Soc. Transit.* **2011**, *1*, 70–75. [CrossRef]
33. Westley, F.; Olsson, P.; Folke, C.; Homer-Dixon, T.; Vredenburg, H.; Loorbach, D.; Thompson, J.; Nilsson, M.; Lambin, E.; Sendzimir, J.; et al. Tipping Toward Sustainability: Emerging Pathways of Transformation. *Ambio* **2011**, *40*, 762–780. [CrossRef] [PubMed]
34. AtKisson, A. *The Sustainability Transformation*; Routledge: Abingdon, UK, 2012.
35. Brockhaus, M.; di Gregorio, M.; Djoudi, H.; Moeliono, M.; Pham, T.T.; Wong, G.Y. The forest frontier in the Global South: Climate change policies and the promise of development and equity. *Ambio* **2021**, *50*, 2238–2255. [CrossRef] [PubMed]
36. Sutherland, W.J.; Burgman, M. Policy advice: Use experts wisely. *Nature* **2015**, *526*, 317–318. [CrossRef]
37. Orr, A.; Ahmad, B.; Alam, U.; Appadurai, A.; Bharucha, Z.P.; Biemans, H.; Bolch, T.; Chaulagain, N.P.; Dhaubanjar, S.; Dimri, A.P.; et al. Knowledge Priorities on Climate Change and Water in the Upper Indus Basin: A Horizon Scanning Exercise to Identify the Top 100 Research Questions in Social and Natural Sciences. *Earth's Future* **2022**, *10*, e2021EF002619. [CrossRef]
38. Abson, D.J.; Fischer, J.; Leventon, J.; Newig, J.; Schomerus, T.; Vilsmaier, U.; von Wehrden, H.; Abernethy, P.; Ives, C.D.; Jager, N.W.; et al. Leverage points for sustainability transformation. *Ambio* **2017**, *46*, 30–39. [CrossRef]
39. Peluso, N.L.; Vandergeest, P. Writing Political Forests. *Antipode* **2020**, *52*, 1083–1103. [CrossRef]
40. Lee, C.L.; Strong, R.; Dooley, K.E. Analyzing precision agriculture adoption across the globe: A systematic review of scholarship from 1999–2020. *Sustainaibility* **2021**, *13*, 10295. [CrossRef]

 *sustainability*

Article

Techno-Economic Analysis of Integrated Solar Photovoltaic Winnower-Cum Dryer for Drying Date Palm Fruit

Surendra Poonia [1], Anil Kumar Singh [1], Dilip Jain [1], Nallapaneni Manoj Kumar [2,3,*] and Digvijay Singh [4]

1   Division of Agricultural Engineering and Renewable Energy, ICAR—Central Arid Zone Research Institute, Jodhpur 342003, Rajasthan, India
2   School of Energy and Environment, City University of Hong Kong, Kowloon, Hong Kong
3   Center for Circular Supplies, HICCER–Hariterde International Council of Circular Economy Research, Palakkad 678631, Kerala, India
4   Center of Excellence for Energy and Eco-Sustainability Research, Uttaranchal University, Dehradun 248007, Uttarakhand, India
*   Correspondence: mnallapan2-c@my.cityu.edu.hk or nallapanenichow@gmail.com

Abstract: Date palm (*Phoenix dactylifera* L.) fruits are widely grown in rural areas of arid Rajasthan of India. The grown date palm fruits are generally dried in forced convection mode. However, given the socio-economic status of farmers, dryer facility affordability has become crucial. Additionally, there is a critical need for a simple winnower, especially with its operation. To address the highlighted issues with the dryer and winnower and given a location already receiving abundant solar radiation, a solar photovoltaic (PV) winnower cum-dryer was designed and developed. The developed winnower cum-dryer was tested in actual conditions to realize the performance. First, the drying experiment for dehydrating date palm fruits and, second, the winnower experiment for separating grains from straw were carried out. The date palm fruits used for experimentation have a moisture content of 65% on a wet basis. During the drying trial, the dryer reduced this moisture content by 39% in 6 days. In contrast, in the open sun drying, it took 8 days. The drying chamber's temperature gradient was reduced to 2–3 °C from 6–8 °C in the system provided with a preheater, resulting in uniform drying. The observed effective moisture diffusivity and the dryer's efficiency are 4.34×10^{-9} m^2·s^{-1} and 16.1%, respectively. A high IRR of 57.4% and a shorter payback period of 2.10 years were found in the economic analysis, indicating that the dryer is cost-effective. The winnower operation results suggest that about 200–300 kg grains could be separated daily when used as a winnower without natural wind. Overall, the developed winnower cum-dryer produced better-quality dried date palms in a shorter time than open drying by efficiently using solar energy and separating the grains from straw to enhance the utility throughout the year.

Keywords: *Phoenix dactylifera* L.; date palm drying; solar dryer; PV winnower; economic evaluation; hybrid solar dryer

Citation: Poonia, S.; Singh, A.K.; Jain, D.; Kumar, N.M.; Singh, D. Techno-Economic Analysis of Integrated Solar Photovoltaic Winnower-Cum Dryer for Drying Date Palm Fruit. *Sustainability* **2022**, *14*, 13686. https://doi.org/10.3390/su142013686

Academic Editor: Muhammad Ikram

Received: 30 September 2022
Accepted: 20 October 2022
Published: 21 October 2022

Publisher's Note: MDPI stays neutral with regard to jurisdictional claims in published maps and institutional affiliations.

1. Introduction

Date palm (*Phoenix dactylifera* L.) is one of the first fruit trees in the world to be grown and the only crop capable of withstanding high temperatures and low humidity levels throughout the bearing stage. This fruit crop is significantly grown in most arid regions of the planet. Regarding nutritional value, due to its high sugar content, the date palm fruit is considered a high-energy food (300 calories/100 g), is low in fat, and is an excellent source of iodine, iron, calcium, and potassium [1–3]. In Indian conditions, among the date palm cultivars, Khadrawy is one of the critical cultivars grown commercially, mainly in the country's northwestern region. Once this fruit is transformed into chuhara (dry) form, it is excellent for ingestion. The primary issue with the dates grown in this area is that their maturation period falls in June, just before the monsoon season begins. Date fruits that have been exposed to rain develop fruit rot and deterioration, rendering them unfit for consumption as fresh dates. According to statistics, heavy rains cause up to 60% of the dates to rot. To avoid spoiling, dates are therefore picked early and sold before the rainy season. Given the aforementioned information, it is imperative to close the gap using some effective, realistic, and useful methods. Therefore, turning these immature dates into goods with the extra value could increase marketability [4,5]. Drying is the option mostly used by farmers to remove the moisture present in date palms, which is conventional. It is recommended to avoid traditional drying, as it has several detrimental consequences on plant materials, including shrinkage, discoloration, and vitamin oxidation. Given these conditions, it is imperative to come up with a drying technology that avoids all the issues. However, in general, when it comes to the agriculture in rural areas of developing nations, especially in India, drying and winnowing are the two essential tasks that farmers follow when aiming at a marketable product—drying reduces the moisture content, and winnowing removes the straw from grains. However, these two do not come in one system, and a farmer needs to have two separate systems—one for drying and the other for winnowing.

Drying enhances the storability of the date palm product by reducing the loss, and doing this at the farm level would be more productive [6,7]. However, most farmers rely on farming only and cannot afford costly devices; instead, they rely on open drying under the sun. This is commonly seen in India, and the central problems are product defiling and non-uniform drying. Mechanical or electric drying in the forced convection mode is suggested as an alternative to open sun drying to avoid defiling and non-uniform drying. However, the mechanical or electric drying was quite prominent only for industrial purposes, and it is a bit expensive. The small and marginal farmers cannot afford this hi-tech facility, especially people living in the Thar desert of India, where date palm farming is common. Regarding the solar energy potential, India is in a stronger position. The desert area is fortunate to have access to alternative energy sources including biogas, wind, and solar electricity. In India's dry region, solar energy ranges from 5.8 to 6.3 kWh m^{-2} day^{-1}, on average [8].

Open sun-drying can be replaced with solar dryers because of their low operating costs [6,7,9,10]. The solar dryer is very profitable, cheap, and eco-friendly, and it is cost-effective for rural people. It can be used in sun-drying to ensure the quality product [11]. Numerous studies demonstrate that the solar-based drying technology is preferable to open drying for the generated dried foods [12–15]. The thermal and PV equipment can support current energy sources. In comparison to solar thermal collectors and side-by-side PV modules, PV/T collectors produce more energy per unit of surface area [16]. The dryer is utilized in the forced convection mode, and the PV powers the fan. No traditional source is necessary. The integrated photovoltaic thermal hybrid system has a respectable return on investment [17]. According to Tonui and Tripanagnostopoulos, various inexpensive modifications could enhance the efficiency of air-cooled PV/T solar collectors [18]. Slices of tomato were dried, reducing their moisture content from 95% to 9% (wet basis) after 10 h of solar drying in an indirect-type hybrid solar dryer, and the overall drying efficiency of the system was estimated as 41% [19]. For drying grapes, Barnwal and Tiwari employed a photovoltaic/thermal greenhouse dryer [20]. In order to dry Indian jujube, a hybrid PV/T-forced solar convection dryer was constructed [6,7]. They reported that the forced convection mode's average thermal efficiency was higher (16.7%) than the mode's average thermal efficiency (15.6%). The greenhouse solar dryer is very effective in elevating the temperature by 10–14 °C over the surrounding temperature, and the dryer also has a quick payback period of 1.5–2.1 years [21]. Cerci and Hurdogan investigated a hybrid drying system for the low-temperature drying of peanuts [22]. Tuncer et al. revealed that the payback period of the PVT hybrid solar dryer varied between 2.98 and 3.51 years [23]. The optimal conditions for the drying of basil leaves in a hybrid solar dryer were a 63.8 °C air temperature and a bed thickness of 2 cm, and the exergy efficiency varied in the range of 31.78–86.55% [24].

CAZRI, Jodhpur, India, has developed a unique solar dryer that is used for both drying and winnowing simultaneously [25]. In this mixed mode dryer, the fan was used to circulate air inside the drying bin, and provisions were made to let the dehydration process continue at night through natural circulation. Although it was found to provide satisfactory results in drying different vegetables and fruits, retaining the color and aroma of the product, there had been a thermal gradient of 5–6 °C or more across the height inside the drying bin, leading to non-uniform drying, and, therefore, different trays needed to be reshuffled [25,26]. Moreover, the villagers find it difficult to clean the threshed material if there is a lull in natural winds. Sometimes, electrical or manual-driven fans are employed for winnowing. The intermittence of electricity makes the farmer handicapped in using the electrical device at the time of requirement. In contrast, the manual system is not only less efficient but adds to human drudgery. However, it was observed to have limited use only after the harvest, particularly in arid regions. Therefore, the utility of the PV dryer was extended by using it as a PV winnower [27]. As per the literature review, significantly less research exists on date palm solar drying; many studies reported on green transition and the use of renewables as a means for sustainable development, both at the centralized and decentralized levels [28–31].

Therefore, in this paper, a study was undertaken to develop a preheater to overcome the problem of the thermal gradient, assess the drying attributes of date palm and the thin-layer drying process for the solar drying of date palm fruit, and compare it with open sun-dried date palm fruits. There existed a temperature gradient of 5–6 °C inside the drying chamber, and drying trays had to be reshuffled. An economic analysis of a PV winnower cum hybrid solar dryer has been carried out.

2. Materials and Methods

2.1. Description of the Photovoltaic Winnower Cum Dryer

The components of the PV winnower cum dryer unit include a PV module-mirror booster assembly, a suitable winnower, a heating tunnel that is used to pre-heat air, and a solar drying cabinet that is specifically developed to utilize solar energy effectively. The interfacing arrangements made in the design will allow the product to be dried more quickly by using the winnower's fan. As shown in Figure 1, the winnower is simply a trapezoidal chamber with a PV-driven blower, a hopper, and a guide for feeding the material that is to be dried and for boosting the airspeed.

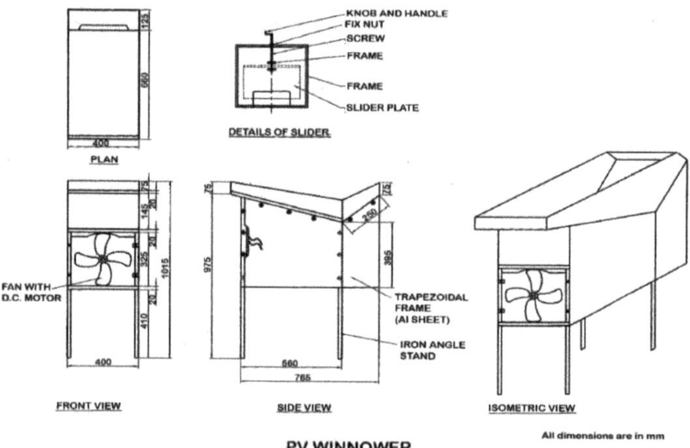

Figure 1. A design of the photovoltaic (PV) winnower.

The system comprises a three-sided aluminum sheet covering an iron-angle frame. A direct-current (DC) motor (35 W)-run fan is fixed at one of the vertical openings on the frame and is operated with an output of a PV module. A specifically made tray with a slope and a slit covers the top, having arrangements to reduce or increase the opening size. This tray serves as a hopper while also facilitating airflow toward the product that is descending from the slit. A dryer of the size 2006 mm × 1400 mm was made using 22 gauge galvanised sheet metal. It consists of twelve trays, each being 870 mm × 620 mm in size, made of angle frames and wire meshes made with stainless steel. The chamber is provided with a door facility. As shown in Figure 2, the dryer comprises two clear glass windows (980 mm × 680 mm) that are 4 mm thick—one was provided on the front side, and another glass window (450 mm × 700 mm) is fixed on the east side of the cabinet.

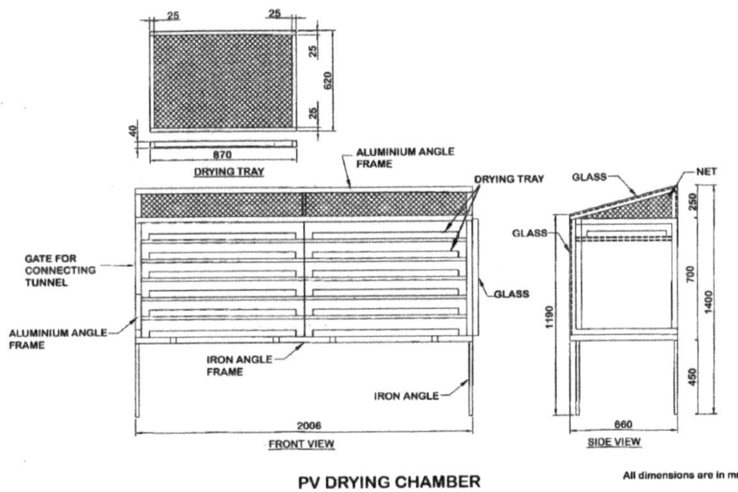

Figure 2. Schematic sketch with neat dimensions showing the design of the photovoltaic (PV) hybrid solar dryer.

The dryer is about 45 cm above the floor. Four holes are provided for ventilation at the base with detachable caps. The dryer functions in both natural and forced circulation modes in east and north directions. The dryer is provided with GI wire mesh to ensure ventilation. Proper provisions have been made to use the fan as a dryer and winnower. Initially, the 35 Wp PV panel confirmed the dryer's operation for more than six hours daily. Later, a booster was provided to capture more solar radiation (22.6% and 35.4% from 9:00 to 11:00 h). The booster had 1.5 and 2 times more length than the length of the PV panel. An additional 2–3% cost of the booster can ensure 20% more power [25]. The device is installed as shown in Figure 3. The device worked on the principle of the greenhouse effect, where the solar radiation was converted into long-wave radiation and was trapped inside, giving rise to the inside temperature. The PV panel and fan were provided to use the system in the forced convection mode to ensure the faster drying of date palm fruits.

2.2. Experimental Procedure

The on-field trials were conducted from 13 to 18 June 2020 under clear skies at the CAZRI in Jodhpur, India (26°18′ N and 73°04′ E). In total, 20 kg of date palm fruit was purchased for the drying experiment from CAZRI's horticultural block in Jodhpur. Twenty kilograms of date palm were separated and evenly dispersed on the left- and right-side trays, and the experiments were conducted between the hours of 8:00 and 18:00. In these studies, a thermopile pyranometer was used to record the total solar radiation intensity (Gs) on a horizontal surface hourly. Using a DTM-100 thermometer with point-contact thermocouples with an accuracy of 0.1 °C, the temperatures inside the dryer chamber are measured. The temperature of the surrounding air was measured using a mercury thermometer with a 0.1 °C precision. Every 60 min, the moisture content of the drying product was determined using an electronic digital balance (Testing Instrument Pvt. Ltd

Faridabad, India) with a 0.001 g precision. The sample trays were used to collect 100 g of the sample.

Figure 3. Picture depicting the installation of the PV/T hybrid solar dryer with a winnower at the CAZRI solar yard in India.

2.3. Moisture Content and Moisture Ratio

The moisture, which is typically referred to as the initial moisture (M_i) content of the collected fruits, can be estimated using Equation (1).

$$Mi = \left(\frac{Wi - Wf}{Wi}\right) \times 100 \tag{1}$$

Equation (2) was used to estimate the drying rate [6,7]

$$DR = \frac{\Delta M}{\Delta t} \tag{2}$$

Equation (3) was used to estimate the moisture ratio.

$$MR = \frac{M - Me}{Mo - Me} \tag{3}$$

2.4. Effective Moisture Diffusivity

Using Equation (4), the effective moisture diffusivity (D_{eff}) is estimated, and it is defined by Fick's second law [6]:

$$MR = \frac{M - M_e}{M_o - M_e} = \frac{8}{\pi^2} \sum_{n=1}^{\infty} \frac{1}{n^2} \exp\left(-\frac{n^2\pi^2 D_{eff} t}{4r^2}\right) \tag{4}$$

Equation (4) is further simplified and can be expressed in a logarithmic form (see Equation (5)), and its slope is given in Equation (6).

$$\ln(MR) = \ln\left(\frac{8}{\pi^2}\right) - \left(\frac{\pi^2 D_{eff} t}{4r^2}\right) \tag{5}$$

$$Slope = \left(\frac{\pi^2 D_{eff}}{4r^2} \right) \tag{6}$$

2.5. Thermal Efficiency (η)

The thermal efficiency is the ratio of the moisture removal to the heat gained. Equation (7) is used to estimate the thermal efficiency [6,10,32]:

$$\eta = \frac{ML}{A \int_0^\theta H_T d\theta} \times 100 \tag{7}$$

2.6. Economic Analysis of the PV Hybrid Solar Dryer

The dryer's economic viability was examined using five economic factors: net present value (NPV), payback period (PBP), benefit–cost ratio (BCR), annuity (A), and internal rate of return (IRR).

2.6.1. Net Present Value (NPV)

Comparing the worth of future benefits to the cost of the initial investment while accounting for the appropriate interest rate is the goal of calculating the net present value. The capital cost of a company, or the net present value (NPV), was determined using a 12% interest rate to calculate the present value of anticipated future cash flow. The aforementioned rate was applied to all cash inflows and outflows. The NPV of solar devices was worked out using Equation (8) [33]. Here, the initial cost (IC) = INR 33,089/-, a = (0.12), and n = 10 years. The gross benefits from the sale of products (E) = INR 19,800/-, and the maintenance cost of the dryer (M) = INR 1166/-

$$NPV = \frac{(E - M)}{a} \left[1 - \left(\frac{1}{1+a} \right)^n \right] - IC \tag{8}$$

2.6.2. The Benefit–Cost Ratio (BCR)

The benefit–cost ratio was calculated as the original cost divided by the net present value, as shown in Equation (9).

$$BCR = \frac{IC + NPV}{IC} \text{ and } BCR = 1 + \frac{NPV}{IC} \tag{9}$$

2.6.3. Annuity (A)

The project's annuity (A) shows the average net annual returns, determined using Equation (10).

$$A = \frac{NPV}{\sum_{t=1ton} \left(\frac{1}{1+a} \right)^n} \tag{10}$$

2.6.4. Payback Period (PBP)

The amount of time needed to recover the initial investment made in a project or investment was calculated using the payback period formula; see Equation (11).

$$PBP = \frac{\log \frac{(E-M)}{a} - \left(\log \frac{(E-M)}{a} - IC \right)}{\log(1+a)} \tag{11}$$

different seasons for dehydrating various drying materials. The data of this experiment were found by the mathematical model for air heating and for predicting the temperature in the bin to a reasonable extent. In this experiment, the color and aroma of the product were maintained. Figure 6 displays the variation in the moisture ratio for the drying time. Figure 6 clearly shows that both solar drying and open sun drying were found to have a decreasing moisture ratio. Additionally, when compared to open-air sun drying, the required moisture ratio of (0.3) was accomplished quickly with solar drying.

Figure 7 shows the date palm after solar and open sun drying, respectively. From Figure 7, it is clear that the quality of fruit in solar drying is much better than that in open sun drying.

Figure 7. Date palm photographs after drying.

3.2. Effective Moisture Diffusivity (D_{eff})

Equation (1) was used estimate the effective moisture diffusion coefficient (Deff) of date palm fruits against drying time. The Deff was then extrapolated. The straight lines were generated by linear regression, with good correlation coefficients r^2 ranging from 0.990 to 0.9957, and they sufficiently reflect the drying behavior over the moisture ratio range (0MR1.0), a range that comprises the majority of the drying process. Figure 8 ln (MR) graph against time was used to calculate the effective moisture diffusivity. The experiment measured the moisture diffusivity of date palm fruits at 30–65 °C for a loading rate of 20 kg in solar drying, which is 4.34×10^{-9} m^2·s^{-1}, and when the same fruits are dried under the open sun, the observed moisture diffusivity is 3.22109 m^2·s^{-1}. Because of this, the majority of food ingredients have Deff values that fall within the broad range of 10–12 to 10–8 m^2·s^{-1} [6,7,39,40]. This is comparable to the outcomes for the thin-layer drying of date palm at 50–80 °C (from 7.53×10^{-9} to 1.11×10^{-8} m^2·s^{-1}) [40], for the convective drying of pumpkin at 30–70 °C (4.08×10^{-8} to 2.35×10^{-7} m^2·s^{-1}) [41], and for the thin-layer solar drying of Indian jujube (*Zizyphus mauritiana*) fruits at 50–70 °C (3.34×10^{-7}) [6,7]. The use of varied drying temperatures, the physical or chemical pretreatment, the moisture content and sample variety, the composition, and the geometrical properties of drying materials could all contribute to the difference in D_{eff} for different biological materials.

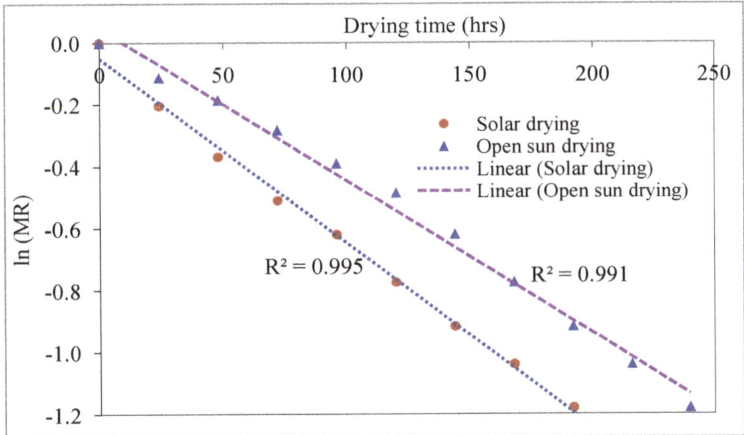

Figure 8. Date palm drying under sunlight and open sun: ln (MR) vs. drying time.

3.3. Dryer Overall Efficiency

According to Equation (4), which yielded a result of 16.1%, the dryer's effectiveness is dependent on the amount of drying time, the temperature, and the sun radiation. Due to the fact that unbound moisture is initially removed and is influenced by surface area, it was originally high and decreased in response to a decrease in moisture content. The outcome was consistent with earlier studies that found that a forced convective, flat-plate solar heat collector dryer for drying cauliflower had an average thermal efficiency of 16.5% [42]. Similarly, the average thermal efficiency is around 12.1% for drying ber (Zizyphusmauritiana) fruit using a low-cost solar dryer [10], and it is 16.7% for drying Indian jujube using a PV/T hybrid solar dryer [6,7].

3.4. Performance Evaluation of the Winnower

As a winnower, the airspeed at the exit of the winnower was found to vary from 1.6 m/s to 3.5 m/s with a single PV panel, while it was 2.3 to 3.7 m/s with PV panel-reflector assembly. In contrast, with two panels, it works quite satisfactorily for extended hours. The system works better with PV panel reflector assemblies compared to the PV panel. Still, with two PV panels, it works better for extended hours and provides a scope to store the surplus energy in the battery for use at night. The device was tested for the winnowing of different threshed materials such as pearl millet, mung bean, moth bean, cluster bean, and mustard. The system could be operated even with a single panel for a winnowing of pearl millet. An additional mirror reflector of an extended length could be used more comfortably to clean mustard and cluster beans. However, if the system needs to be used for cleaning and illumination in the night, two PV modules with a storage battery are preferable. On average, 900 kg of cluster bean, threshed manually, could be winnowed by operating it for 5.5 h each for four days; 350 kg of pearl millet could be winnowed by operating it for 3.5 h a day for two days; and 450 kg of mustard could be winnowed by operating it for 8.5 h in two days, indicating that, on average, 34–52 kg of cleaned grains/seeds could be separated in an hour from threshed material weighing three to four times more. The device helps to save manpower in completing an important post-harvest agricultural task, especially during those days when natural winds are not available.

3.5. Analysis of Economic Viability

The calculations, as per the parameters shown in Section 2.5, were carried out to assess the dryer's viability from an economic standpoint. The dryer unit had a 33,089 INR initial investment (Pi). The annual maintenance cost (M), which is INR 1166, was calculated as

5% of the initial cost (INR 23,300), and it includes minor replacements, if required. A 10% deduction from the initial investment was made for the salvage value.

The yearly benefit is mainly accrued from drying date palm, sangri (Pods of Prosopis cineraria), and Lasora (Cordiamyxa), the labor-saving in winnowing operations, and the benefits gained through lightning. The benefit accrued from drying date palm and sangri was INR 3500/each, and the labor-saving in winnowing amounted to INR 9000/-, and that from lighting amounted to INR 300/-. Thus, the benefits amounted to INR 19,800/year, whereas the present value of the life cycle cost was INR 33,089, including the initial cost and the battery replacement charge, the fan replacement, and the drying chamber replacement cost. By discounting the values to the present, the entire cash inflow and outflow for the production of the solar dryer's net present value was determined. The dryer has a ten-year lifespan and an initial cost of INR 33,089, or 12% of the total cost. Sales of the goods will generate a profit of INR 19,800. The dryer's net present value (NPV) was calculated using Equation (15), which indicates that the investment in the solar dryer had an NPV of INR 113,023. Based on the NPV, it is determined that the fabrication of the dryer is more affordable than that of the hybrid solar/biomass dryer [43], the hybrid PV/T greenhouse dryer [20], and the PVT hybrid solar dryer [6,7].

Equation (9) has been used to determine the benefit–cost ratio by dividing the present value of the benefit stream by the current price of the cost stream. It comes out, for the dryer, as 4.41. Swain et al. examined the effectiveness of copra drying equipment powered by biomass [44]. The findings showed that it required 22 h for biomass-fired equipment to reduce the initial moisture content from 57.4% (Wb) to 6.8%. (Wb) [44]. For two dryers tested for producing high-quality copra, the cost–benefit ratio was determined to be 1.4 and 1.19, respectively. The cost–benefit ratio is estimated to be 1.86 for the PVT hybrid solar dryer [6]. The dryer's annuity was calculated using Equation (7), which shows that its average net yearly return is INR 14,415. The payback period, which is 2.10 years, is shorter than the dryer's anticipated lifespan of around 10 years. A solar tunnel dryer's payback period is four years for basic mode dryers and three to four years for optimum mode dryers [14]. An analysis of the cost of a hybrid photovoltaic greenhouse dryer was conducted in [37]. Grapes have been dried using a hybrid PV/T integrated greenhouse dryer in the forced mode of operation [20]. With an initial expenditure of INR 27,400, the system's payback period is approximately 1.25 years. The payback period of the PVT hybrid solar dryer is 2.26 years, with an initial investment of INR 14,000 [6]. The internal rate of return can be determined using Equation (9). Table 1 provides the values of the NPV at various discount rates.

Table 1. Net present values for different rates of discount/interest (*i*).

	Interest Rate *i* (%)		
	12	40	60
Net present value (INR)	113,023	13,485	−2032

From Table 1, it may be deduced that the NPV is INR 113,023 at a 12% interest rate. The net present value is INR 13,485 at a 40% interest rate. However, with a 60% interest rate, the NPV is negative (i.e., NPV = INR −2032/-). The internal rate of return (IRR) is as high as 57.4%. Compared to the cost of capital, the IRR is higher. The option with the highest IRR may be viewed as the superior choice when all other factors are held constant and may be used as the decision-making criterion. This conclusion is based, in part, on the fact that a greater IRR denotes a lower risk [33]. The important economic parameters are given in Table 2.

Table 2. Economic parameters of the solar PV dryer system.

Parameters	Values
Benefit–cost ratio	4.41
Net present value	113,023
Annuity	14,415
Internal rate of return (percent)	57.38
Payback perid (years)	2.10

4. Conclusions

This paper presented a novel design of a solar photovoltaic powered dryer cum winnower used for drying date palm and separating grains from straw. The date palm grown in rural arid regions of Rajasthan was used for studying the drying experiment. The characteristics of the date palm fruit dried in a photovoltaic hybrid solar dryer were analyzed using a mathematical model for the pre-air heating tunnel to reduce the thermal gradient in the drying chamber. The proposed system provided better a moisture diffusivity and efficiency. Additionally, the economic evaluation of the hybrid solar dryer unit also indicated a favorable IRR and a shorter PBP, suggesting it to be cost-effective. The winnower also showed a good potential for the grain separation from straw. Using such a novel system considerably reduced the drying time, and, at the same time, it was fully operational using renewable energy consumption. Using such PV hybrid forced solar convection dryers at remote locations/rural areas can ensure value addition in dried produce and reduce pollution.

This dryer could be used for dehydrating various fruits and vegetables in arid regions and for determining their cost-economics and CO_2 mitigation potential. The heat generated by the PV panel can be diverted to the drying chamber using a fan. This can be carried out by covering the back position of the PV panel, where the temperature rises to 70 °C. Thus, PV panels could be used in the hybrid mode.

Author Contributions: Conceptualization, S.P. and N.M.K.; Data curation, S.P., A.K.S., D.J., N.M.K. and D.S.; Formal analysis, S.P., A.K.S., D.J. and N.M.K.; Funding acquisition, S.P. and N.M.K.; Investigation, S.P. and A.K.S.; Methodology, S.P., A.K.S., D.J. and N.M.K.; Project administration, S.P. and N.M.K.; Resources, D.J., N.M.K., and D.S.; Software, S.P. and D.J.; Supervision, N.M.K.; Validation, S.P. and N.M.K.; Visualization, S.P., A.K.S. and D.J.; Writing—original draft, S.P., A.K.S. and D.J.; Writing—review & editing, N.M.K. All authors have read and agreed to the published version of the manuscript.

Funding: This research received no external funding.

Institutional Review Board Statement: Not applicable.

Informed Consent Statement: Not applicable.

Data Availability Statement: Not applicable.

Conflicts of Interest: The authors declare no conflict of interest.

Nomenclature

M_i	Initial moisture content of the sample (w.b.) %
W_i	Initial weight of the sample (g)
W_f	Final weight of the sample (g)
ΔM	Loss of mass of fruit (kg water·kg^{-1} dry matter)
Δt	Interval of time (min)
DR	Drying rate
M	Moisture content of the sample at a given time
M_0	Initial moisture content of the sample
Me	Equilibrium moisture content of the sample (kg water. kg^{-1} solids)
D_{eff}	Effective diffusivity coefficient (m^2·s^{-1})
r	Half thickness of the sample (m)
n	Positive integer
t	Drying time (s)
η	Efficiency of the solar dryer (%)
A	Absorber area (m^2)
HT	Solar radiation on the horizontal plane (J.m^{-2}·h^{-1})
L	Latent heat of vaporization (J·kg^{-1})
M	Mass of moisture evaporated from the product (kg)
θ	Period of test (h)
a	Compound interest rate per annum
IC	Initial cost of the dryer (INR)
E	Gross benefits from the sale of products (INR)
M	Maintenance cost of the dryer (INR)
n	Number of years
NPV	Net present value (INR)
BCR	Benefit–cost ratio
A	Annuity (INR)
PBP	Payback period (years)
IRR	Internal rate of return (%)

References

1. Ahmed, I.A.; Ahmed, A.W.K.; Robinson, R.K. Chemical composition of date varieties as influenced by the stage of ripening. *Food Chem.* **1995**, *54*, 305–309. [CrossRef]
2. Al-Turki, S.; Shahba, M.A.; Stushnoff, C. Diversity of antioxidant properties and phenolic content of date palm (*Phoenix dactylifera* L.) fruits as affected by cultivar and location. *J. Food Agric. Environ.* **2010**, *8*, 253–260.
3. Uchoi, J.; Nikhumbhe, P.H.; Kumar, A.; Patidar, A.; Harish, G.D. Impact of inclined solar drier for dehydration quality in khadrawy dates during doka maturity stage at north western arid India. *Int. J. Curr. Microbiol. Appl. Sci.* **2020**, *9*, 119–125. [CrossRef]
4. Kulkarni, S.G.; Vijayanand, P.; Aksha, M.; Reena, P.; Ramana, K.V.R. Effect of dehydration on the quality and storage stability of immature dates (*Pheonixdactylifera*). *LWT-Food Sci. Technol.* **2008**, *41*, 278–283. [CrossRef]
5. Sagarika, N.; Kapdi, S.S.; Sutar, R.F.; Patil, G.B.; Akbari, S.H. Study on drying kinetics of date palm fruits in greenhouse dryer. *J. Pharmacogn. Phytochem.* **2019**, *8*, 2074–2079.
6. Poonia, S.; Singh, A.K.; Jain, D. Mathematical modelling and economic evaluation of hybrid photovoltaic-thermal forced convection solar drying of Indian jujube (*Zizyphusmauritiana*). *J. Agric. Eng.* **2018**, *55*, 74–88.
7. Poonia, S.; Singh, A.K.; Jain, D. Design development and performance evaluation of photovoltaic/thermal (PV/T) hybrid solar dryer for drying of ber (*Zizyphusmauritiana*) fruit. *Cogent Eng.* **2018**, *5*, 1–18. [CrossRef]
8. Poonia, S.; Singh, A.K.; Jain, D. Design development and performance evaluation of concentrating solar thermal desalination device for hot arid region of India. *Desalination Water Treat.* **2020**, *205*, 1–11. [CrossRef]
9. Purohit, P.; Kumar, A.; Kandpal, T.C. Solar drying vs. Open Sun drying: A framework for financial evaluation. *Sol. Energy* **2006**, *80*, 1568–1579. [CrossRef]
10. Poonia, S.; Singh, A.K.; Santra, P.; Jain, D. Performance evaluation and cost economics of a low-cost solar dryer for ber (*Zizyphusmauritiana*) fruit. *Agric. Eng. Today* **2017**, *41*, 25–30.
11. Sharma, A.; Chen, C.R.; Vu Lan, N. Solar-energy drying systems: A review. *Renew. Sustain. Energy Rev.* **2009**, *13*, 1185–1210. [CrossRef]
12. Mahapatra, A.K.; Imre, L. Role of solar agricultural drying in developing countries. *Int. J. Ambient. Energy* **1990**, *2*, 205–210. [CrossRef]

13. Sodha, M.S.; Chandra, R. Solar drying systems and their testing procedures: A Review. *Energy Convers. Manag.* **1994**, *35*, 219–267. [CrossRef]
14. Ekechukwu, O.V.; Norton, B. Review of solar energy drying systems II: An overview of solar drying technology. *Energy Convers. Manag.* **1999**, *40*, 615–655. [CrossRef]
15. Hossain, M.A.; Woods, J.L.; Bala, B.K. Optimisation of solar tunnel drier for drying of chilli without color loss. *Renew. Energy* **2005**, *30*, 729–742. [CrossRef]
16. Zondag, H.A. Flat-plate PV-Thermal collectors and systems: A review. *Renew. Sustain. Energy Rev.* **2008**, *12*, 891–959. [CrossRef]
17. Huang, B.J.; Lin, T.H.; Hung, W.C.; Sun, F.S. Performance evaluation of solar photovoltaic/thermal systems. *Sol. Energy* **2001**, *70*, 443–448. [CrossRef]
18. Tonui, J.K.; Tripanagnostopoulos, Y. Air-cooled PV/T solar collectors with low-cost performance improvements. *Sol. Energy* **2007**, *81*, 498–511. [CrossRef]
19. Sharma, M.; Atheaya, D.; Kumar, A. Exergy, drying kinetics, and performance assessment of Solanumly copersicum (tomatoes) drying in an indirect type domestic hybrid solar dryer (ITDHSD) system. *J. Food Process. Preserv.* **2022**, e16988. [CrossRef]
20. Barnwal, P.; Tiwari, G.N. Grape drying by using hybrid photovoltaic–thermal (PV/T) greenhouse dryer: An experimental study. *Sol. Energy* **2008**, *82*, 1131–1144. [CrossRef]
21. Philip, N.; Duraipandi, S.; Sreekumar, A. Techno-economic analysis of greenhouse solar dryer for drying agricultural produce. *Renew. Energy* **2022**, *199*, 613–627. [CrossRef]
22. Çerçi, K.N.; Hürdoğan, E. Performance assessment of a heat pump assisted rotary desiccant dryer for low temperature peanut drying. *Biosyst. Eng.* **2022**, *223*, 1–17. [CrossRef]
23. Tuncer, A.D.; Khanlari, A.; Afshari, F.; Sözen, A.; Çiftçi, E.; Kusun, B.; Şahinkesen, I. Experimental and numerical analysis of a grooved hybrid photovoltaic-thermal solar drying system. *Appl. Therm. Eng.* **2023**, *218*, 119288. [CrossRef]
24. Parhizi, Z.; Karami, H.; Golpour, I.; Kaveh, M.; Szymanek, M.; Blanco-Marigorta, A.M.; Marcos, J.D.; Khalife, E.; Skowron, S.; Adnan Othman, N.; et al. Modeling and optimization of energy and exergy parameters of a hybrid-solar dryer for basil leaf drying using RSM. *Sustainability* **2022**, *14*, 8839. [CrossRef]
25. Pande, P.C. Design and development of PV winnower-cum-dryer. In Proceedings of the International Congress on Renewable Energy (ICORE 2006), Hyderabad, India, 8–11 February 2006; Sastry, E.V.R., Reddy, D.N., Eds.; Allied Publishers Pvt. Ltd.: Mumbai, India, 2006; pp. 265–269.
26. Pande, P.C.; Singh, A.K.; Dave, B.K.; Purohit, M.M. A preheated Solar PV dryer for economic growth of farmers and entrepreneurs. In Proceedings of the International Congress on Renewable Energy (ICORE 2010), Chandigarh, India, 3 December 2010; Solar Energy Society of India: New Delhi, India, 2010; pp. 68–73.
27. Pande, P.C. A Solar PV winnower. In Proceedings of the 2003, 26th National Renewable Energy Convention and International Conference in New Millennium, Coimbatore, India, 17–19 January 2003; pp. 22–26.
28. Ikram, M.; Zhang, Q.; Sroufe, R.; Shah SZ, A. Towards a sustainable environment: The nexus between ISO 14001, renewable energy consumption, access to electricity, agriculture and CO_2 emissions in SAARC countries. *Sustain. Prod. Consum.* **2020**, *22*, 218–230. [CrossRef]
29. Chand, A.A.; Prasad, K.A.; Mar, E.; Dakai, S.; Mamun, K.A.; Islam, F.R.; Kumar, N.M. Design and Analysis of Photovoltaic Powered Battery-Operated Computer Vision-Based Multi-Purpose Smart Farming Robot. *Agronomy* **2021**, *11*, 530. [CrossRef]
30. Solangi, Y.A.; Tan, Q.; Mirjat, N.H.; Valasai, G.D.; Khan MW, A.; Ikram, M. An integrated Delphi-AHP and fuzzy TOPSIS approach toward ranking and selection of renewable energy resources in Pakistan. *Processes* **2019**, *7*, 118. [CrossRef]
31. D'Adamo, I.; Gastaldi, M.; Morone, P.; Rosa, P.; Sassanelli, C.; Settembre-Blundo, D.; Shen, Y. Bioeconomy of sustainability: Drivers, opportunities and policy implications. *Sustainability* **2021**, *14*, 200. [CrossRef]
32. Leon, A.M.; Kumar, S.; Bhattacharya, S.C. A comprehensive procedure for performance evaluation of solar food dryers. *Renew. Sustain. Energy Rev.* **2002**, *6*, 367–393. [CrossRef]
33. Singh, A.K.; Poonia, S.; Jain, D.; Mishra, D.; Singh, R.K. Economic evaluation of a business model of selected solar thermal devices in Thar Desert of Rajasthan, India. *Agric. Eng. Int. CIGR J.* **2020**, *22*, 129–137.
34. Mennouche, D.; Boubekri, A.; Chouicha, S.; Bouchekima, B.; Bouguettaia, H. Solar drying process to obtain high standard "deglet-nour" date fruit. *J. Food Process Eng.* **2017**, *40*, e12546. [CrossRef]
35. Boubekri, A.; Benmoussa, H.; Mennouche, D. Solar drying kinetics of date palm fruits assuming a step-wise air temperature change. *J. Eng. Sci. Technol.* **2009**, *4*, 292–304.
36. Boubekri, A.; Benmoussa, H.; Courtois, F.; Bonazzi, C. Softening of over dried 'DegletNour' dates to obtain high-standard fruits: Impact of rehydration and drying processes on quality criteria. *Dry. Technol.* **2010**, *28*, 222–231. [CrossRef]
37. Domyaz, I. Drying kinetics of black grapes treated with different solutions. *J. Food Eng.* **2006**, *76*, 212–217. [CrossRef]
38. Doymaz, İ. The kinetics of forced convective air drying of pumpkin slices. *J. Food Eng.* **2007**, *79*, 243–248. [CrossRef]
39. Babalis, S.J.; Belessiotis, V.G. Influence of drying conditions on the drying constants and moisture diffusivity during the thin-layer drying of figs. *J. Food Eng.* **2004**, *65*, 449–458. [CrossRef]
40. Falade, K.O.; Abbo, E.S. Air-drying and rehydration characteristics of date palm (*Phoenix dactylifera* L.) fruits. *J. Food Eng.* **2007**, *79*, 724–730. [CrossRef]
41. Hashim, N.; Daniel, O.; Rahamann, E. A preliminary study: Kinetic model of drying process of pumpkins (*Cucurbita Moschata*) in a convective hot air dryer. *Agric. Agric. Sci. Procedia* **2014**, *2*, 345–352. [CrossRef]

42. Kadam, D.M.; Samuel DV, K. Convective flat-plate solar heat collector for cauliflower drying. *Biosyst. Eng.* **2006**, *93*, 189–198. [CrossRef]
43. Dhanushkodi, S.; Vincent, H.W.; Kumarasamy, S. Life cycle cost of solar biomass hybrid dryer systems for cashew drying of nuts in India. *Environ. Clim. Technol.* **2015**, *15*, 22–33. [CrossRef]
44. Swain, S.; Din, M.; Chandrika, R.; Sahoo, G.P.; Roy, S.D. Performance evaluation of biomass fired dryer for copra drying: A comparison with traditional drying in subtropical climate. *J. Food Process. Technol.* **2014**, *5*, 294. [CrossRef]

 *sustainability*

Perspective

Apparel Consumer Behavior and Circular Economy: Towards a Decision-Tree Framework for Mindful Clothing Consumption

Sarif Patwary [1,†], Md Ariful Haque [2,*,†], Jehad A. Kharraz [2,†], Noman Khalid Khanzada [2,†], Muhammad Usman Farid [2,†] and Nallapaneni Manoj Kumar [2,3,4,*,†]

1   Kontoor Brands, Inc., 1250 Revolution Mills Dr Ste 300, Greensboro, NC 27405, USA
2   School of Energy and Environment, City University of Hong Kong, Kowloon, Hong Kong
3   Center for Product Life Extension, HICCER—Hariterde International Council of Circular Economy Research, Palakkad 678631, Kerala, India
4   Center for Resource Recovery, HICCER—Hariterde International Council of Circular Economy Research, Palakkad 678631, Kerala, India
*   Correspondence: mahaque3-c@my.cityu.edu.hk (M.A.H.); mnallapan2-c@my.cityu.edu.hk (N.M.K.)
†   These authors contributed equally to this work.

Abstract: The apparel consumer, one of the vital stakeholders in the apparel supply chain, has a significant role to play in moving the clothing industry in a sustainable direction. From purchasing and care practice to donation and disposal, every step of their decisions impacts the environment. Various internal and external variables influence those decisions, including culture, customs, values, beliefs, norms, assumptions, economy, gender, education and others. Therefore, we believe having a scientific understanding is very important, because consumers need to be aware of what makes eco-conscious apparel behavior; only then will the circular transition be eased. However, the key concern is whether the apparel consumers are aware of this knowledge or not. Therefore, we formulated a prospective study from a life cycle thinking point of view with a key focus on synthesizing apparel consumer behavior concerning clothing acquisition, maintenance and disposal through the circular economy lens. Hence, a circular economy lens framework is proposed, followed by three research questions' (RQ) formulation: RQ1. What is the current norm of clothing acquisition, maintenance and disposal behavior?; RQ2. Is apparel consumer clothing acquisition, maintenance and disposal behavior circular-driven?; RQ3. What is the sustainable way of clothing acquisition, maintenance and disposal? These questions are followed by circular economy lens framework development for apparel consumers. Second, following the research questions, state-of-the-art literature-driven decisions were gathered to form constructive consumer-centric decisions over the apparel lifecycle. Third, building on this synthesis, a critical discussion is offered, following the decision-tree approach to inform relevant behavioral guidelines for consumers and other stakeholders in the apparel supply chain. Overall, our findings on apparel consumer behavior through the circular economy lens could serve as new guidelines for consumers to exercise mindful clothing consumption behavior.

Keywords: apparel; clothing industry; apparel consumer behavior; circular economy lens; mindful clothing consumption; sustainable fashion; green fashion; circular textiles; sustainable garments; green consumer; life cycle thinking of textiles; responsible consumption

Citation: Patwary, S.; Haque, M.A.; Kharraz, J.A.; Khanzada, N.K.; Farid, M.U.; Kumar, N.M. Apparel Consumer Behavior and Circular Economy: Towards a Decision-Tree Framework for Mindful Clothing Consumption. *Sustainability* **2023**, *15*, 656. https://doi.org/10.3390/su15010656

Academic Editor: Antonis A. Zorpas

Received: 21 November 2022
Revised: 21 December 2022
Accepted: 27 December 2022
Published: 30 December 2022

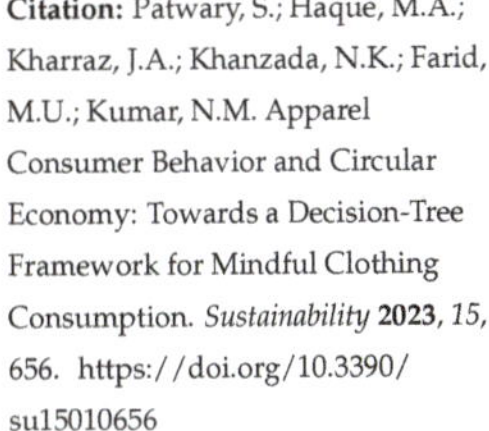

1. Introduction

Consumer behavior related to apparel products is mainly associated with acquisition (purchasing), maintenance (keeping, using, and care) and disposition of clothes (everything after the primary owner's use) [1]. Sustainable apparel behavior involves putting social and environmental considerations into clothing acquisition, maintenance and disposal decisions. Thus, the environmentally sustainable behavior of apparel consumers can be understood from three perspectives: eco-conscious acquisition, eco-conscious care and eco-conscious disposal behavior. These behaviors involve a range of psychological equations guided by one's values, beliefs, aspirations, assumptions, financial conditions, education, family history and culture [2]. Eco-conscious acquisition is guided by consumer awareness, attitude, concern and commitment. Eco-conscious care is determined by the number of uses, laundering frequency, washing methods, drying and ironing methods [3]. Eco-conscious disposal behavior can be understood from the intention to recycle (upcycling and downcycling), donate and reuse, throw away (landfill or incineration) and keep in the closet [2,3]. The most significant environmental benefits are achieved by consuming less (i.e., responsible consumption) and extended use of garments, broadly falling under the core principles of the circular economy. Consumer has direct control over consuming less, extended use and sustainable care practice. Outside consumers' direct influence, the other sustainable options are rechanneling, recycling, energy recovery from incineration and landfilling. Out of all these, landfilling (throwing away) is the worst of all types of clothing's end-of-life fate [4]. However, consumers can influence the fate of the unwanted apparel by being responsible in disposition (for example, donating or dropping in collection bin). Therefore, responsible consumer behavior is vital to realize a circular vision across the supply chain.

However, today, clothing consumers are fast fashion-oriented; this started around the early 2000s [3,5]. Cheap and low-quality materials characterize fast fashion. Due to cheap, poor-quality materials, apparel loses its appeal quickly. At the same time, due to the rise in purchasing power, consumers can afford to buy new clothes many times a week. By swiftly offering new collections and crafting planned obsolescence, brands allure consumers to refill their wardrobes by throwing away used clothes that still have their useful life left. Along with them, brands' attractive marketing strategies, traditional and social media, opinion leaders, bloggers, celebrities and peers play an essential role in influencing consumers to consume fast fashion [6,7]. As a result, clothing consumption has doubled in the last decade, whereas consumers keep clothing half as long as they did 15 years ago [8]. Their usual apparel purchasing decisions are mainly driven by fit, color, style, durability and easy care [9]. Mainstream consumers do not care about the dark side of clothing (i.e., how it impacts our environment and society). One of the main reasons for this behavior is that the clothing supply chain is complex, and consumers have a poor understanding of climate change issues and the underlying science associated with them. Lately, circular economy as a concept for managing resources and mitigating climate change issues was suggested [10]. Since then, it has become a mainstream strategy in the framework of waste management, particularly concerning different products and waste streams [11]. Nevertheless, some concerns been expressed; these are mainly related to monitoring and evaluating waste prevention activities, which are very critical [12]. The apparel sector, also facing such concerns, needs critical waste prevention activities.

Why Circular Thinking Is Important for Apparel Consumers, and Key Contributions

Today, most consumers lack an understanding of how clothing is made, and the impact of their consumption [13–15]. They have limited knowledge of sustainable care practices [16]. They also have a poor understanding of how their disposal behavior affects the environment negatively [17]. They do not know where to dispose of clothing, or how [18]. Therefore, it is important to examine what factors influence apparel consumers' eco-conscious behavior and what behavior guidelines can be made. While previous studies have identified the factors influencing eco-conscious apparel acquisition [14,15], care [19,20], and disposal behavior [17,21], no previous study has synthesized this information for consumers to holistically understand their sustainable apparel behavior. Moreover, previous studies fell short in providing consumers with straightforward guidelines, without a vision of what makes eco-conscious acquisition, care of and disposal of clothing. Though circular economy thinking from the perspective of apparel consumers is articulated in the literature [22–26], only a few studies have comprehensively explored this concept. For instance, a "sustainability bias" has emerged, through assessing consumer attitudes towards the fashion sector from the perspective of the bioeconomy and the circular economy [23,24].

The circular premium concept promotes circular strategy in the fashion industry by bringing all the stakeholders under one roof with a clear vision for transition [25]. Very recently, Papamichae et al., 2022 presented a study on building a new mindset in tomorrow's fashion development, wherein they suggested that this should be through circular strategy models, and introduced a clear vision and new strategy involving customers, businesses, and policymakers [26]. From these very recent studies, it is obvious that consumer involvement is crucial, raising questions such as 'Is the consumers' mindset based on scientific understanding of the circular economy?', and 'To what extent do consumers consider circular fashion?' The answers to these questions are still debatable and have no solid evidence. As a whole, it is clearly seen that consumers lack proper scientific understanding and relevant guideline. We believe having a scientific understanding is very important, because consumers need to be aware of what makes eco-conscious apparel behavior; only then will the circular transition be eased. Therefore, we formulated a prospective study from a life cycle point of view, with a key focus on synthesizing apparel consumer behavior concerning clothing acquisition, maintenance and disposal through a circular economy lens.

To realize the proposed aim, three research questions (RQs) are formulated:

RQ1. What is the current norm of clothing acquisition, maintenance and disposal behavior?
RQ2. Is apparel consumer clothing acquisition, maintenance and disposal behavior circular-driven?
RQ3. What is the sustainable way of clothing acquisition, maintenance and disposal?

The key contribution of this study is the understanding of the apparel consumer behavior under the circular economy lens framework and decision tree approach, followed by a synthesis of state-of-the-art literature-driven decisions of apparel consumer behavior concerning clothing acquisition, maintenance and disposal to inform relevant behavioral guidelines for consumers and other stakeholders in the apparel supply chain.

2. Framework and Methodology

2.1. Circular Economy Lens Framework for Apparel Consumers

The circular economy lens framework shown in Figure 1 for apparel consumers is formulated to better understand consumer behavior related to apparel products mainly associated with three things: 1. Acquisition (purchasing), 2. Maintenance (keeping, using, and care), and 3. Disposal (everything after primary owner's use). Figure 1a shows the general consumer psychology framework where there is a dual possibility, i.e., consumers may or may not have a solid understanding of circular economy principles. The framework in Figure 1b drives consumers' psychology, mainly with scientific understanding of the circular economy with appropriate awareness. Hence, the key difference between the proposed and apparel psychology framework in ref [1] is broader understanding of circular economy principles, such as 'regenerate nature', 'minimizing or eliminating waste', 'eliminating pollution', 'circular products', and 'circular materials'-related decisions among consumers, who play a significant role in moving the apparel industry towards a sustainable path.

2.2. Methodology

Based on the framework presented in Figure 1, a keyword search methodology was adopted for carrying out the critical study. First, a brainstorming discussion was performed at the authors' level in multiple sessions to select the keywords while keeping the research questions (RQ1, RQ2, and RQ3) in mind, followed by verification and improvements with expert's opinions. The contacted experts broadly fall under the category of consumers with some awareness, textile production and recycling industry personnel, non-profit advocacy teams and other senior figures in the subject area from academia. The selected keywords include 'sustainable apparel(s)', 'sustainable clothing industry', 'apparel consumer behavior', 'circular economy in textiles', 'mindful clothing consumption', 'sustainable fashion', 'green fashion', 'circular textiles', 'sustainable garments', 'green clothing consumer', 'life cycle thinking of textiles', 'responsible clothing consumption'. These are used in line with the circular economy principles keywords, as shown in Figure 1b.

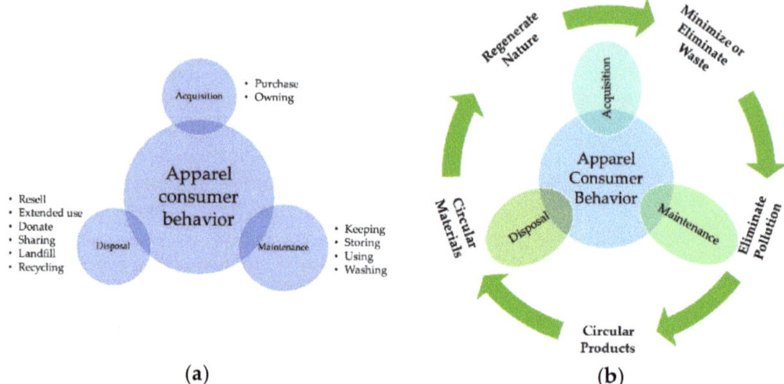

Figure 1. Framework for understanding consumer clothing behavior, (**a**). General consumer psychology framework; (**b**). Circular economy lens framework for apparel consumers.

Second, these keywords were used in various databases (including Google Scholar, Web of Science, ScienceDirect, Scopus and other related databases of gray literature). Third, the obtained articles from the keyword search are further filtered and processed to pick the most relevant literature for this study. Additionally, duplication in the articles between the indexed databases was removed. Fourth, the filtered articles were grouped per the consumer psychology framework categories, i.e., clothing acquisition, clothing maintenance, and clothing disposal. Fifth, these articles were studied critically under three groups for addressing the RQ1, RQ2, and RQ3. The detailed steps of the methodological process of the review are shown in Figure 2.

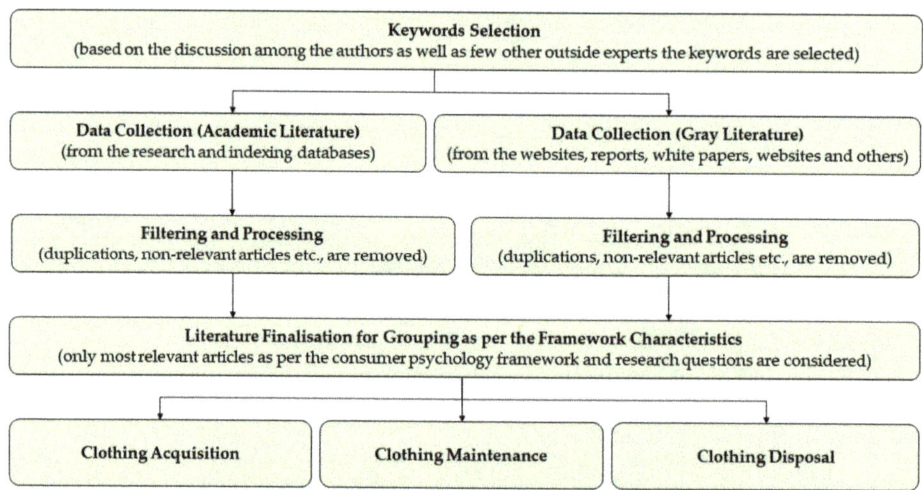

Figure 2. Literature review methodology showing the data collecting process.

3. Results and Discussion

From the literature review, the brief answers to the research questions are that the current form of clothing acquisition, maintenance and disposal behavior is not sustainable, even among highly knowledgeable consumers, suggesting a lack of life cycle perspectives and circular thinking. It is also clear that there is no standard sustainable way that is followed among consumers in terms of clothing acquisition, maintenance and disposal. In the below sections, our findings are explained in detail, along with the discussion leading to mindful clothing consumption.

3.1. Clothing Acquisition

Consumer knowledge and awareness of sustainable apparel influence their purchasing decisions. However, it is not true that highly knowledgeable consumers will always buy sustainable clothing. There exist many other factors that impact the decision. For example, a knowledgeable consumer might have a financial limitation in buying sustainable apparel (by paying a higher price) [19]. Moreover, it is difficult for consumers to research and identify sustainable clothing during their purchase. Often, they rely on their perception of the brands they are purchasing from, and justify their purchase on the basis of the reputation of those brands [19]. As Harris et al. [15] mentioned, consumers fall short of demonstrating sustainable behavior because (1) clothing sustainability is complex, and consumers lack knowledge, (2) consumers are diverse in their concerns, and (3) sustainability is less important in consumer purchase decision criteria.

Concerning obstacles, Hiller Connell [14] identified two internal and four external barriers. Internal barriers include knowledge about eco-conscious apparel acquisition (ECAA) and attitudes about environmentally preferable apparel (EPA), while external barriers include limited availability of EPA, economic resources, less enjoyable secondhand stores and society's expectations. Consumers have limited knowledge of what materials are more environmentally friendly, how apparel is manufactured and its associated impact on the environment. On the other hand, spending extra money, putting extra work into acquiring EPA, and less social acceptance play their part in consumers not purchasing EPA. Likewise, consumers perceive financial, performance-related, psychological and social risks while making responsible decisions [27]. While forming purchase decisions, consumers carefully consider possible monetary loss, functional deficit, compromise of self-image and social unacceptance.

Among the enablers, ethical commitment to apparel purchase and ethical values were mentioned by Niinimäki [28]. However, product attributes are key to attracting consumers to buy EPA [14,28]. If EPA cannot compete with fast fashion in terms of attributes and price, the process of consumer acquisition of EPA would be slow. Creating competitive, sustainable fashion that has the same appeal as fast fashion is easier than making consumers aware of environmental issues and driving them to act. Six overarching values drive consumers' sustainable apparel acquisition: self-expression, self-esteem, responsibility, protecting the planet, sense of accomplishment and social justice [29]. On the other hand, consumer knowledge of green industry initiatives and green brands, beliefs relating to corporate responsibility, subjective norms, motivation to research, search and buy green apparel and attitudes toward purchasing green clothing were found to influence purchase intention and purchase behavior of green textiles and apparel [30]. Consumer demography (i.e., geography, age, etc.) influences sustainable clothing purchases [31]. For instance, younger consumers show more favorable attitudes toward environmentally responsible clothing consumption [32]. Consumers' belief that they can positively impact the environment through their buying of sustainable apparel (a term called 'perceived consumer effectiveness') positively impacts the purchase intention of sustainable apparel [33]. Additionally, consumers' belief that a particular product matches their personal style and value (a term called 'perceived personal relevance') positively impacts the purchase intention of sustainable apparel [33]. Similarly, the subjective perception of ease or difficulty of engaging in any behavior (termed 'perceived behavior control') is associated with sustainable apparel purchases [34].

However, as stated earlier, environmental knowledge does not often translate into behavior [35]. For example, a study found that understanding clothing production's environmental impact did not positively influence environmentally friendly consumption behavior [36]. Similarly, peers' sustainable choices may not influence purchase decision-making [37,38]. Therefore, it is understandable that knowledge, values, beliefs, attitudes, commitment, subjective norms, demographics, external factors and different types of internal and external barriers play a significant role in consumer intention and decision-making toward acquiring EPA. These variables can drive consumers to purchase sustainable apparel in favorable circumstances.

As shown in Figure 3, consumers perceive various risks, hold different values, face many barriers and are driven by many enablers. Subjective norms (i.e., peer pressure), product attributes and brand reputation also play an important role in consumers' purchase decisions. For instance, while perceiving the financial risk of acquiring EPA (i.e., in relatively costlier products), consumers demonstrate poor intention to purchase. Similarly, the bad reputation of EPA brands leads to poor intention to purchase. Conversely, consumer knowledge of the environmental impact of clothing production, positive attitudes towards EPA and higher subjective norm score lead to higher purchase intention. From the above discussion, it is evident that there needs to be an interplay of many favorable factors in order for consumers to make EPA purchasing decisions. However, it can reasonably be said that consumer knowledge of the environmental impact of products, their attitudes towards purchasing EPA, willingness to research and ability to afford are the main drivers of sustainable clothing acquisition [31]. These drivers might also be the precursor of other enablers (such as perceived consumer effectiveness) and personal values [31].

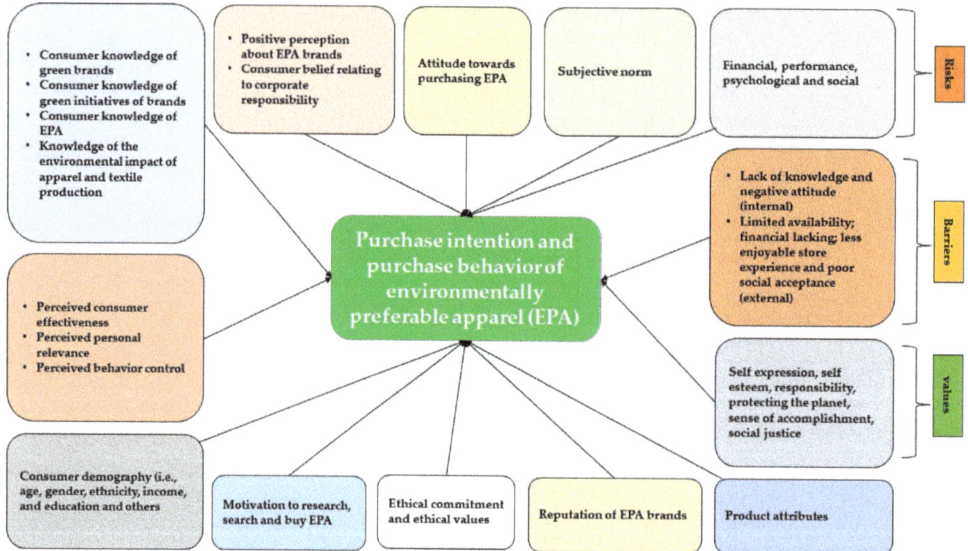

Figure 3. Factors affecting sustainable clothing acquisition. Authors own compilation based on referred literature [19,27–31,33–36,38].

3.2. Clothing Maintenance

Clothing maintenance is a significant environmental hotspot for many clothing types (such as cotton) [16,39]. Laundering culture and frequency are the main defining factors of clothing maintenance. Clothing maintenance is mainly influenced by everyday habits, customs, social norms and culture [21,40]. Most consumers do not know the impact of their clothing maintenance activities. A very negligible portion of consumers might see the impact; however, they do not necessarily act due to the attitude–behavior gap and lack of infrastructure. The attitude–behavior gap exists where consumers have the knowledge, but cannot act due to various internal and external limitations. On the other hand, lack of infrastructure also hinders sustainable behavior. For example, in most developing countries, such as India and Bangladesh, hand washing and line drying are prevalent. Therefore, the reduced impact from clothing maintenance of those consumers is due to social norms and infrastructure, rather than awareness. On the other hand, machine washing and drying are the social norm in the United States and other developed countries. As a result, the environmental impact of consumer clothing maintenance is simply a result of cultural norms.

A slight modification of consumer behavior in the use phase might bring significant environmental benefits. For instance, the elimination of tumble drying and ironing along with washing in low-temperature settings might lead to a 50 percent reduction in the global climate change impact of clothing products [16]. A lot of other factors determine the environmental effects associated with clothing care, for example, types of clothing cared for, lifetime number of washes, washing machine type (i.e., efficiency, front-loading/top-loading), washing machine setting (i.e., cold or hot), geographical location, cultures, etc. [16,39,40]. In the case of an automatic washing machine, the environmental impact is determined by the machine type (i.e., horizontal vs. vertical loading), age of the machine, temperature setting, load size and the number of washes. In contrast, the impact of manual washing is determined primarily by the water and chemicals used. The wash cycles vary by country and size of the household, and so do the energy and water consumption. Japan was found to carry out the most significant number of wash cycles per household, followed by North America and Australia [40]. The average water consumption per washing cycle also varies by the type of washing machine used. The vertical axis machine requires twice as much water as the horizontal axis machine per cycle. Water consumption per wash cycle is the highest for North America, followed by South Korea and Japan [40]. Electricity consumption per wash cycle was found to be greatest for Turkey, followed by East Europe, West Europe and North America [40]. The annual electricity consumption per household for North America from clothing wash was reported as about 124.3 kWh [40]. This variation in wash cycles and energy and water consumption suggests that different kinds of interventions are needed for different geographical locations to make clothing care habits sustainable. Changing consumer clothing care practice is not easy, but rather ingrained in multiple layers of knowledge, cultures, habits, customs and geography.

Furthermore, benchmarking the environmental impact of clothing care requires knowing the number of times consumers wash different types of apparel. Most literature assumed either 25 or 50 wash cycles [16,39,41]. However, in order to update the data, Daystar et al. [3] surveyed 6000 respondents from China, Germany, Italy, Japan, the United Kingdom and the United States to characterize the use of T-shirts, knit collared shirts and woven pants. They determined the global average of total washes per lifetime as 17.3, 22.2, and 23.5 washes for T-shirts, knit collared shirts and woven pants, respectively. Therefore, it seems that the assumption of 25 cycles is logical. The average first-life use period was determined as 37, 40, and 42 months for t-shirts, knit collared shirts and woven pants, respectively. This result suggests a greater overall maintenance impact for the T-shirt as it has a shorter lifetime. Still, diverse types of apparel and more geographical locations need to be included in future studies to upgrade the global average. In addition, a detailed analysis of the environmental impact of consumer clothing maintenance activity is needed. Such analysis should consider a variety of existing washing machine types, their settings and variation in consumers' maintenance habits.

3.3. Clothing Disposal

Waste generation from throwaway clothes is a big problem that has its root in fast fashion. Fast fashion is produced in shorter lead times, typically made with low-quality and inexpensive materials and built-in planned obsolescence [3,5]. The low price of garments and increased individual purchasing power entices consumers to buy a lot of fast fashion, often impulsively [42]. However, they lose interest in the products quickly because of fast fashion's low quality and obsoleteness. As a result, most of these items are thrown away long before their actual usability ends; this is termed "throwaway culture" [18]. The average American throws away 82 pounds of clothes yearly [43]. In 2015, the United States generated about 16 million tons of textile waste, of which 65.7% went into landfills, 19% to the incinerator, and 15.3% was recycled [44]. An average UK consumer throws away about 66 pounds of clothing and textiles (a total reported as 2.35 million tons), of which 74% go to the landfill, 13% to incinerators and 13% to material recovery [16]. The average European Union consumer generates 57 pounds of textile waste [45]. Globally, 91 million tons of clothing are thrown away yearly; this is equivalent to one garbage truck of clothing every second [46]. The environmental cost of this massive amount of clothing waste is enormous in terms of groundwater pollution (from leachate), greenhouse gas (GHG) emissions and land occupation.

Three scenarios might arise during consumer decision-making of garment disposition: (1) keep it (i.e., reuse, downcycling, etc.), (2) permanently dispose of it (throwaway, giveaway, etc.), and (3) temporarily dispose of it (loan, rent, etc.) [1]. Based on Jacoby's [1] classification, the factors impacting the decision to dispose of garments can be grouped into three categories: psychological attributes of the decision-maker (personality, attitudes, learning, etc.), the intrinsic value of the product (condition, fit, durability, etc.) and factors extrinsic to the product (finances, fashion change, legal, etc.). Table 1 presents the key factors affecting clothing disposal behavior [10], as published in literature from 1980–2013.

Table 1. Factors affecting clothing disposal behavior.

Agenda	Focus
Destinations	Mainly focuses on where clothes go after disposal. Primary channels identified as charity, giving away to friends and family, donations, etc.
Motivations	Focuses on the reasons behind choosing specific disposal methods. The main motivations identified are the convenience of recycling, donating as a form of helping others and social and environmental concerns.
Disposal reasons	Focuses on why consumers dispose of their garments. Disposal reasons can be categorized into wear and tear, fit or size, fashion, taste or boredom, and other reasons.
Demographics	Focuses on the effect of gender differences on clothing disposal behavior.

However, there is a research gap in understanding the clothing disposal behavior of consumers. A recent study by Bernardes et al. [47] reviewed 51 studies concerning clothing disposal behavior. They reported that studies mainly examined how clothing is disposed of, not why they disposed of them. Based on their review, they proposed immediate investigations in four research directions: (1) investigating the decision-making process of consumers' clothing disposal, (2) examining sustainable disposal behavior, (3) exploring external factors, and (4) improving the current methodology of understating the issue. Nevertheless, changing consumer behavior related to discarding clothes and waste generation is challenging. The problem is ingrained into cultural, social, and national practice, and it requires both individual and institutional efforts to bring the desired change. Clothing disposal is a vital consumer behavior because disposal creates the demand for consuming virgin materials, along with incurring a significant environmental cost. Thus, responsible behavior in the disposal phase involves actions associated with diverting waste from landfills or incinerators through repairing, donating, reusing, repurposing and recycling.

3.3.1. Clothing Donation

Studies reported both self-oriented reasons and others-oriented reasons behind clothing donations [48]. Self-oriented reasons are freeing up closet space, and remaining guilt-free [44,49] and others-oriented reasons are social and environmental concerns and helping others [50,51]. The primary motivation for donating clothes is to free up closet space [49]. Cloth donation is not primarily influenced by social consciousness, and consumers do not regard donating clothes as valuable as donating money or food [49]. Consumers keep expensive and high-quality items as long as they can. They try to donate those items they do not want to keep anymore. They throw away those items even after one-time use, long before their useful life ends [18]. The subjective evaluation of the quality of the garment and the sentimental value attached to it play a significant role in deciding what to donate and what not to donate. If the sentimental value is higher, consumers tend not to donate the item regardless of physical condition, for example, an item that reminds a memory or incident. Consumers also hesitate to donate intimate items, for example, underwear [49]. Consumers feel guilty about how much clothing they own and their limited use of them [49]. Putting in environmental terms, Morlet et al. [46] reported that the clothing industry loses nearly USD 500 billion each year due to the un-utilization of clothes.

During donation, consumers choose close family members and friends as their first choices [49]. Charity donation is another common method of sustainable clothing disposal [18]. The convenience of the donation channel is an important factor in determining where the clothes would be donated [49]. Through the overall act of donation, consumers gain both hedonistic (i.e., good feeling) and utilitarian value (i.e., freeing up closet space) from donating clothes [49].

On the other side of the spectrum, a consumer who shops for donated clothes from secondhand, thrift store, and vintage shops presumably does so for both self-oriented reasons (to look different and unique) and others-oriented reasons (economy, sustainability and recycling) [52,53]. There exist attitudinal and contextual barriers to acquiring secondhand apparel [54]. Attitudinal barriers include consumers' evaluation of secondhand shops as unhygienic, unattractive, less socially desirable, etc. In contrast, contextual barriers include unappealing store ambience, unattractive product offerings and the price mix [54]. Among the motivations, waste-efficiency and economy were found to be important [55]. Whatever the case, secondhand clothes need to compete with mainstream fast fashion products to fulfill the basic attributes of clothes, such as price, style, fit and attractiveness.

Donation is important for extending product life. Extending product life potentially saves virgin materials and reduces waste. Previous studies estimated the extent of saving new items from secondhand use of clothing. For instance, Patwary [56] estimated that using 100 pieces of secondhand clothing could substitute between 63–73 pieces of new clothing for US consumers. Based on this substitution rate, the study reported that reusing 100 pieces of 100% cotton t-shirt may reduce an estimated 1.48 kg CO_2, eq. of greenhouse gas (GHG) emissions. Similarly, Nørup et al. [57] estimated that every 100 pieces of secondhand clothing used substitutes around 45 pieces of new clothing. In the case of the United States, clothing donation might potentially divert 10 million tons from landfills and 3 million from the incinerator, assuming those clothes were still in their useful life [44]. Considering the global scale of textile waste, the number would still be significant if 10% of all textile waste were still in usable condition. Therefore, clothing donation should be encouraged.

3.3.2. Clothing Reuse and Recycle

The prolonged use of a garment can potentially reduce the overall environmental impact of the supply chain. Prolonged use can be direct reuse after mending or repairing, or reuse by others through sharing/donation, as long as the products retain some value throughout it. Prolonged use of garments would reduce the associated manufacturing need and hence minimize environmental impact from the production phase. Allwood et al. ([16], p40) reported, "Extending the life of clothing so that demand for new products is reduced by 20% leads to a reduction of about 20% in all measures in the producing country". Other studies also reported the highest energy and CO_2 equivalent savings from the direct reuse of clothing [56,58]. Reducing clothing requiring 1 kg of virgin cotton fibers through secondhand clothing or reusing might save 65 kWh. In the case of polyester, it might save up to 90 kWh [59]. This suggests that establishing a reuse mechanism for synthetic fibers (i.e., polyester, nylon) is more important than natural fibers. Fisher et al. [58] estimated the environmental benefit of reusing cotton t-shirts and woolen jumpers in the UK. They reported that direct reusing (e.g., from a charity shop and eBay) saves approximately 6.6 lbs. CO_2 eq for a cotton t-shirt and 8.8 lbs. for a woolen jumper. Sandin and Peters [60] reviewed the published literature focusing on the environmental impact of textile reusing and recycling. Their review lent strong support to the idea that reuse and recycling are better choices than incineration and landfill, with reusing being a better option than recycling. However, there are cases where reusing and recycling might not be environmentally beneficial. For example, if the use of recycled garments does not reduce the purchasing of new clothes (i.e., low replacement rate), if the recycling is powered by fossil energy and if the avoided production as a result of the reuse is environmentally clean.

4. Towards a Decision-Tree Framework for Mindful Clothing Consumption

Mindful clothing consumption is essential as it reduces the burden of waste from overconsumption. It was observed from the results that clothing acquisition does not solely depend on life cycle or circularity thinking, as there were many other factors that influenced consumers. On the other hand, sustainable clothing maintenance seems to be relatively more doable among consumers, if provided with the right information and proper infrastructure. The following steps might be recommended to provide simple instructions for sustainable clothing maintenance, as they are observed to play a critical role in extending the useful life of the clothing [40,61].

- Wash less if you can
- Utilize the full load of the machine
- Use a cold setting (30 degrees or less)
- Use liquid detergent (because it is less abrasive)
- Use softener (because it reduces friction)
- Reduce spin speed (it provides less agitation)
- Dump lint fibers into the bin, not in the sink

Besides, extending clothing's lifetime is vital as it reduces the need to buy new clothes. During the care phase, consumers should wash less, utilize a full load of the machine, use a cold temperature setting (30 or less), select the right detergents (i.e., liquid detergent), use softener (to reduce friction and fiber breakage), reduce spin speed, empty residual lint into a bin (not in the sink) and finally, air-dry clothes. If clothing loses its appeal to the primary consumer but still has its useful life left, it should be channeled to others for reuse (donate, swap, garage sell, etc.). Giving away to needy family and friends is a better option than putting garments into donation and recycling bins. The less the clothes travel to be reused, the better. For example, collecting garments from U.S. households and then sorting them in and selling them in the USA is better than collecting garments from the USA and sending them to be sorted and sold in some African countries. If a garment cannot be directly reused, it should be recycled (either upcycled or downcycled).

It is obvious that consumers need to be knowledgeable and responsible. In every stage of their consumption, from acquisition to disposal, they need to go through a decision tree that can be linked to seven forms of sustainable clothing, (1) on-demand and custom-made, (2) green and clean, (3) high quality and timeless design, (4) fair and ethical, (5) repair, re-design and upcycle, (6) rent, lease, and swap and (7) second hand and vintage [62]. Besides, consumer decisions should be guided by the 7Rs of fashion: reduce, rent, repair, repurpose, recycle, reuse and resell [63]. First, consumers need to be mindful of consumption (i.e., asking if consumption is necessary). Second, they should rent from clothing banks or other channels if possible. Third, they should perform a minor repair if that extends the functional life of the item. Fourth, if repairing compromises functionality, they should repurpose the item (such as creating other products through upcycling or downcycling). Fifth, they should channel it to recycling if they cannot repair or repurpose it (because of a lack of skills, financial resources or other things). Sixth, if they do not have any emotional or functional attachment to the product, they should resell it or finally donate it for others to reuse. Based on the above discussion, the following decision tree shown in Figure 4 can be produced for consumers and other relevant stakeholders as related to sustainable clothing consumption.

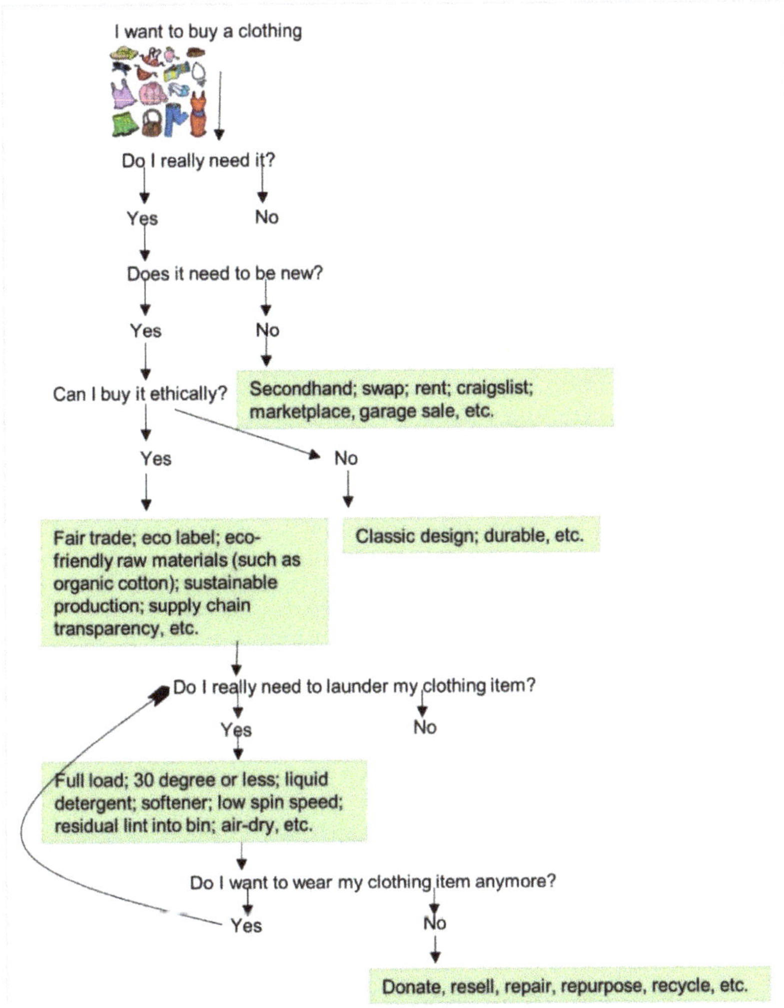

Figure 4. Decision tree for sustainable clothing consumption. Authors own creation following Parischa's study in reference [64]. The green color represents sustainable consumption practices.

Needless to say, it is challenging to change consumer habits. Unless consumers are environmentally concerned and have available resources (in terms of money, time, willingness, policy, incentive and infrastructure) to act, it is difficult for them to adopt sustainable practices. Particularly, changing clothing care behavior would be challenging unless they realize the real impact of their behavior (for example, if the electricity bill goes up significantly, or they need to pay for the water they use to wash clothes). If consumers understand the difference in their clothing care practice in terms of money, labor and time, they might change their behavior. So, consumers need to be educated about the benefit of sustainable care practices both in monetary and environmental terms [21]. Additionally, improved technology can offer options to reduce the environmental impact of laundering; however, it would always be in the consumers' hands to choose those technologies and their useful options. Therefore, consumers are key in reducing the environmental impact of the care phase [65].

Above all, consumer knowledge of the impact of clothing is considered 'the best hope for sustainability' in the clothing industry" ([66], p. A454). Therefore, they need to be educated on how to acquire, care for and dispose of clothing sustainably. Alternatively, a brand-focused mindset of consumers might help to deal with a lack of knowledge [21]. For instance, if brands are held liable to produce apparel sustainably, consumers can easily follow any brand without dealing with complex knowledge of sustainability. This approach seems easier than changing the habits and norms of consumers because brands and retailers operate within specific policy frameworks. Egels-Zanden and Hanson [67] found that improved company transparency positively impacts consumer willingness to buy products from that company. Therefore, if consumers purchase from a sustainable and transparent brand, they can rest assured to some extent. A report published in 20220 by McKinsey & Company and Global Fashion Agenda also suggests the same on how fashion industry should act, and how consumers should purchase to urgently cut the greenhouse gas emissions [68]. However, this comes with a clear understanding of opportunities and policy implications for the industries [69]. By doing so, value chain flexibility can be achieved, which ultimately favors the metrics related to sustainable development goals. Nevertheless, a sector-wise ethics and sustainability guideline needs to be set up by the legislating body, because an individual initiative might put a brand at a disadvantageous position [21]. Among all three perspectives of sustainable clothing behavior, it seems that it would be easier to bring a change in sustainable clothing care behavior than purchasing and disposal [21]. However, barriers to sustainable clothing are ingrained at an individual level, social and cultural level and industry level [15]. Therefore, it is not possible to bring change overnight. It would take interventions on all three levels, and obviously, it would be a slow process.

5. Conclusions

This study reviewed published studies to synthesize the existing body of knowledge related to sustainable clothing consumption behavior. We synthesized studies into three main phases of clothing consumption: acquisition, maintenance and disposal. While reviewing relevant information, the study examined typical clothing consumption behavior and sustainable approaches. It is understood that the current form of clothing acquisition, maintenance and disposal behavior was not sustainable even among the highly knowledgeable consumers, suggesting the current patterns are not circular and we need many efforts to make apparel consumers better understand the good practices. It is also clear that there is no standard sustainable way that is followed among consumers in terms of clothing acquisition, maintenance and disposal. Building on the synthesis, we provided a decision tree for guiding consumers to exercise sustainable clothing consumption behavior. Therefore, it is not possible to bring change overnight. It would take interventions at the individual, social, cultural and industry levels, and obviously, it would be a slow process. Additionally, it is observed that consumer behavior is dynamic, so we believe future studies should update the decision tree as new knowledge emerges. Additionally, studies should test the decision tree to educate and measure consumer knowledge of sustainable clothing consumption.

Author Contributions: Conceptualization, S.P.; methodology, S.P., N.M.K. and M.A.H.; formal analysis, S.P., M.A.H., J.A.K., N.K.K., M.U.F. and N.M.K.; investigation, S.P.; resources, S.P.; data curation, S.P., M.A.H., J.A.K., N.K.K., M.U.F. and N.M.K.; writing—original draft preparation, S.P., M.A.H., J.A.K., N.K.K., M.U.F. and N.M.K.; writing—review and editing, S.P., M.A.H., J.A.K., N.K.K., M.U.F. and N.M.K.; visualization, S.P.; funding acquisition, S.P., M.A.H. and N.M.K. All authors have read and agreed to the published version of the manuscript.

Funding: This research received no external funding.

Institutional Review Board Statement: Not applicable.

Informed Consent Statement: Not applicable.

Data Availability Statement: Not applicable.

Conflicts of Interest: The authors declare no conflict of interest.

References

1. Jacoby, J. Consumer psychology: An octennium. *Annu. Rev. Psychol.* **1976**, *27*, 331–358. [CrossRef]
2. Kant Hvass, K. Post-Retail Responsibility of Garments—A Fashion Industry Perspective. *J. Fash. Mark. Manag.* **2014**, *18*, 413–430. [CrossRef]
3. Daystar, J.; Chapman, L.L.; Moore, M.M.; Pires, S.T.; Golden, J. Quantifying apparel consumer use behavior in six countries: Addressing a data need in life cycle assessment modeling. *J. Text. Appar. Technol. Manag.* **2019**, *11*, 1–25.
4. Directive EC. Directive 2008/98/EC of the European Parliament and of the Council of 19 November 2008 on waste and repealing certain Directives. *Off. J. Eur. Union* **2008**, 312.
5. Cachon, G.P.; Swinney, R. The Value of Fast Fashion: Quick Response, Enhanced Design, and Strategic Consumer Behavior. *Manag. Sci.* **2011**, *57*, 778–795. [CrossRef]
6. House of Commons Environmental Audit Committee. Fixing Fashion: Clothing Consumption and Sustainability. UK Parliament. Available online: https://publications.parliament.uk/pa/cm201719/cmselect/cmenvaud/1952/full-report.html (accessed on 29 July 2020).
7. LeHew, M.L.A.; Patwary, S. Investigating Consumption Practices of Sustainable Fashion Bloggers: Leading the Way or Leading Astray? In Proceedings of the 3rd International Conference of the Sustainable Consumption Research and Action Initiative (SCORAI), Copenhagen, Denmark, 27–30 June 2018.
8. Remy, N.; Speelman, E.; Swartz, S. *Style That's Sustainable: A New Fast-Fashion Formula*; McKinsey & Co.: Atlanta, GA, USA, 2016.
9. Lynch, M. The power of conscience consumption. *J. Cult. Retail. Image* **2009**, *2*, 1–9.
10. Zorpas, A.A.; Doula, M.K.; Jeguirim, M. Waste Strategies Development in the Framework of Circular Economy. *Sustainability* **2021**, *13*, 13467. [CrossRef]
11. Zorpas, A.A. Strategy development in the framework of waste management. *Sci. Total Environ.* **2020**, *716*, 137088. [CrossRef]
12. Zorpas, A.A.; Lasaridi, K. Measuring waste prevention. *Waste Manag.* **2013**, *33*, 1047–1056. [CrossRef]
13. Kim, H.S.; Damhorst, M.L. Environmental concern and apparel consumption. *Cloth. Text. Res. J.* **1998**, *16*, 126–133. [CrossRef]
14. Connell, K.Y.H. Internal and External Barriers to Eco-Conscious Apparel Acquisition. *Int. J. Consum. Stud.* **2010**, *34*, 279–286. [CrossRef]
15. Harris, F.; Roby, H.; Dibb, S. Sustainable Clothing: Challenges, Barriers and Interventions for Encouraging More Sustainable Consumer Behaviour: Sustainable Clothing. *Int. J. Consum. Stud.* **2016**, *40*, 309–318. [CrossRef]
16. Allwood, J.M.; Laursen, S.E.; de Rodriguez, C.M.; Bocken, N.M. Well dressed? The present and future sustainability of clothing and textiles in the United Kingdom. *J. Home Econ. Inst. Aust.* **2015**, *22*, 42.
17. Morgan, L.R.; Birtwistle, G. An Investigation of Young Fashion Consumers' Disposal Habits. *Int. J. Consum. Stud.* **2009**, *33*, 190–198. [CrossRef]
18. Birtwistle, G.; Moore, C.M. Fashion Clothing—Where Does It All End Up? *Int. J. Retail Distrib. Manag.* **2007**, *35*, 210–216. [CrossRef]
19. Goworek, H.; Fisher, T.; Cooper, T.; Woodward, A.; Hiller, A. Sustainable clothing consumption: Exploring the attitude-behavior gap. In Proceedings of the International Centre for Corporate Social Responsibility Conference, Nottingham, UK, 26 April 2012.
20. Pakula, C.; Stamminger, R. Electricity and water consumption for laundry washing by washing machine worldwide. *Energy Effic.* **2010**, *3*, 365–382. [CrossRef]
21. Ha-Brookshire, J.E.; Hodges, N.N. Socially responsible consumer behavior? Exploring used clothing donation behavior. *Cloth. Text. Res. J.* **2009**, *27*, 179–196. [CrossRef]
22. D'Adamo, I.; Lupi, G.; Morone, P.; Settembre-Blundo, D. Towards the circular economy in the fashion industry: The second-hand market as a best practice of sustainable responsibility for businesses and consumers. *Environ. Sci. Pollut. Res.* **2022**, *29*, 46620–46633. [CrossRef]
23. Colasante, A.; D'Adamo, I. The circular economy and bioeconomy in the fashion sector: Emergence of a "sustainability bias". *J. Clean. Prod.* **2021**, *329*, 129774. [CrossRef]
24. D'Adamo, I.; Colasante, A. Survey data to assess consumers' attitudes towards circular economy and bioeconomy practices: A focus on the fashion industry. *Data Brief* **2022**, *43*, 108385. [CrossRef]
25. D'Adamo, I.; Lupi, G. Sustainability and resilience after COVID-19: A circular premium in the fashion industry. *Sustainability* **2021**, *13*, 1861. [CrossRef]
26. Papamichael, I.; Chatziparaskeva, G.; Pedreno, J.N.; Voukkali, I.; Candel, M.B.A.; Zorpas, A.A. Building a new mind set in tomorrow fashion development through circular strategy models in the framework of waste management. *Curr. Opin. Green Sustain. Chem.* **2022**, *36*, 100638. [CrossRef]
27. Kang, J.; Kim, S.H. What Are Consumers Afraid of? Understanding Perceived Risk toward the Consumption of Environmentally Sustainable Apparel. *Fam. Consum. Sci. Res. J.* **2013**, *41*, 267–283. [CrossRef]
28. Niinimäki, K. Eco-Clothing, Consumer Identity and Ideology. *Sustain. Dev.* **2010**, *18*, 150–162. [CrossRef]

29. Lundblad, L.; Davies, I.A. The Values and Motivations behind Sustainable Fashion Consumption: Motivations behind Sustainable Fashion Consumption. *J. Consum. Behav.* **2016**, *15*, 149–162. [CrossRef]

30. Sampson, K.L. Consumer Analysis of Purchasing Behavior for Green Apparel. Ph.D. Thesis, North Carolina State University, Raleigh, NC, USA, 2009. Available online: https://repository.lib.ncsu.edu/bitstream/handle/1840.16/431/etd.pdf?sequence=1&isAllowed=y (accessed on 31 July 2020).

31. Han, T.I.; Chung, J.E. Korean consumers' motivations and perceived risks toward the purchase of organic cotton apparel. *Cloth. Text. Res. J.* **2014**, *32*, 235–250. [CrossRef]

32. Butler, S.M.; Francis, S. The Effects of Environmental Attitudes on Apparel Purchasing Behavior. *Cloth. Text. Res. J.* **1997**, *15*, 76–85. [CrossRef]

33. Kang, J.; Liu, C.; Kim, S.H. Environmentally Sustainable Textile and Apparel Consumption: The Role of Consumer Knowledge, Perceived Consumer Effectiveness and Perceived Personal Relevance: Environmentally Sustainable Textile and Apparel Consumption. *Int. J. Consum. Stud.* **2013**, *37*, 442–452. [CrossRef]

34. Zheng, Y.; Chi, T. Factors Influencing Purchase Intention towards Environmentally Friendly Apparel: An Empirical Study of US Consumers. *Int. J. Fash. Des. Technol. Educ.* **2015**, *8*, 68–77. [CrossRef]

35. Hiller Connell, K.Y.; Kozar, J.M. Sustainability Knowledge and Behaviors of Apparel and Textile Undergraduates. *Int. J. Sustain. High. Educ.* **2012**, *13*, 394–407. [CrossRef]

36. Brosdahl, D.J.C.; Carpenter, J.M. Consumer Knowledge of the Environmental Impacts of Textile and Apparel Production, Concern for the Environment, and Environmentally Friendly Consumption Behavior. *J. Text. Appar. Technol. Manag.* **2010**, *6*, 1–9.

37. Belleau, B.D.; Summers, T.A.; Xu, Y.; Pinel, R. Theory of reasoned action: Purchase intention of young consumers. *Cloth. Text. Res. J.* **2007**, *25*, 244–257. [CrossRef]

38. Cowan, K.; Kinley, T. Green Spirit: Consumer Empathies for Green Apparel: Consumer Empathies for Green Apparel. *Int. J. Consum. Stud.* **2014**, *38*, 493–499. [CrossRef]

39. Yun, C.; Patwary, S.; LeHew, M.L.A.; Kim, J. Sustainable care of textile products and its environmental impact: Tumble-drying and ironing processes. *Fibers Polym.* **2017**, *18*, 590–596. [CrossRef]

40. Gooijer, H.; Stamminger, R. Water and energy consumption in domestic laundering worldwide–a review. *Tenside Surfactants Detergents* **2016**, *53*, 402–409. [CrossRef]

41. Roos, S.; Sandin, G.; Zamani, B.; Peters, G. *Environmental Assessment of Swedish Fashion Consumption—Five Garments, Sustainable Futures*; Mistra Future Fashion: Stockholm, Sweden, 2015.

42. Bhardwaj, V.; Fairhurst, A. Fast fashion: Response to changes in the fashion industry. *Int. Rev. Retail. Distrib. Consum. Res.* **2010**, *20*, 165–173. [CrossRef]

43. Harmony. The Facts about Textile Waste (Infographic). Available online: https://harmony1.com/textile-waste-infographic/ (accessed on 31 July 2020).

44. U.S. Environmental Protection Agency. Advancing Sustainable Materials Management. Available online: https://www.epa.gov/sites/production/files/2018-07/documents/2015_smm_msw_factsheet_07242018_fnl_508_002.pdf (accessed on 31 July 2020).

45. Laitala, K.; Boks, C. Sustainable clothing design: Use matters. *J. Des. Res.* **2012**, *10*, 121–139. [CrossRef]

46. Morlet, A.; Opsomer, R.; Herrmann, S.; Balmond, L.; Gillet, C.; Fuchs, L. *A New Textiles Economy: Redesigning Fashion's Future*; Ellen MacArthur Foundation: Cowes, UK, 2017.

47. Bernardes, J.P.; Ferreira, F.; Marques, A.D.; Nogueira, M. Consumers' clothing disposal behavior: Where should we go? In Proceedings of the 2nd International Textile Design Conference (D_TEX 2019), Lisbon, Portugal, 19–21 June 2019; CRC Press: Boca Raton, FL, USA, 2019.

48. Park, M.; Cho, H.; Johnson, K.K.; Yurchisin, J. Use of behavioral reasoning theory to examine the role of social responsibility in attitudes toward apparel donation. *Int. J. Consum. Stud.* **2017**, *41*, 333–339. [CrossRef]

49. Fenitra, R.M.; Handriana, T.; Usman, I.; Hartani, N.; Premananto, G.C.; Hartini, S. Sustainable clothing disposal behavior, factor influencing consumer intention toward clothing donation. *Fibres Text.* **2021**, *28*, 7–15.

50. Bianchi, C.; Birtwistle, G. Consumer clothing disposal behaviour: A comparative study. *Int. J. Consum. Stud.* **2012**, *36*, 335–341. [CrossRef]

51. Joung, H.M.; Park-Poaps, H. Factors motivating and influencing clothing disposal behaviours. *Int. J. Consum. Stud.* **2013**, *37*, 105–111. [CrossRef]

52. Reiley, K. The Vintage Clothing Market, Consumer, and Wearer in Minneapolis/St. Paul, MN. Master's Thesis, University of Minnesota, Minneapolis, MN, USA, 2003. Available online: https://books.google.com.hk/books/about/The_Vintage_Clothing_Market_Consumer_and.html?id=f-xpvgAACAAJ&redir_esc=y (accessed on 31 July 2020).

53. Reiley, K.; DeLong, M. A consumer vision for sustainable fashion practice. *Fash. Pract.* **2011**, *3*, 63–83. [CrossRef]

54. Hiller Connell, K.Y. Exploration of second-hand apparel acquisition behaviors and barriers. In Proceedings of the 2009 ITAA, Bellevue, WA, USA, 28–31 October 2009; International Textile and Apparel Association: Bellevue, WA, USA, 2009.

55. Laitala, K.; Klepp, I.G. Motivations for and against second-hand clothing acquisition. *Cloth. Cult.* **2018**, *5*, 247–262. [CrossRef] [PubMed]

56. Patwary, S.U. An Investigation of the Substitution Rate and Environmental Impact Associated with Secondhand Clothing Consumption in the United States. Ph.D. Thesis, Kansas State University, Manhattan, KS, USA, 2020.

57. Nørup, N.; Pihl, K.; Damgaard, A.; Scheutz, C. Replacement rates for second-hand clothing and household textiles–A survey study from Malawi, Mozambique and Angola. *J. Clean. Prod.* **2019**, *235*, 1026–1036. [CrossRef]
58. Fisher, K.; James, K.; Maddox, P. *Benefits of Reuse Case Study: Clothing*; Waste and Resource Action Programme: Banbury, UK, 2011.
59. Woolridge, A.C.; Ward, G.D.; Phillips, P.S.; Collins, M.; Gandy, S. Life cycle assessment for reuse/recycling of donated waste textiles compared to use of virgin material: An UK energy saving perspective. *Resour. Conserv. Recycl.* **2006**, *46*, 94–103. [CrossRef]
60. Sandin, G.; Peters, G.M. Environmental Impact of Textile Reuse and Recycling—A Review. *J. Clean. Prod.* **2018**, *184*, 353–365. [CrossRef]
61. Hann, S.; Darrah, C.; Sherrington, C.; Blacklaws, K.; Horton, I.; Thompson, A. *Reducing Household Contributions to Marine Plastic Pollution: Report for Friends of the Earth*; Report for Friends of the Earth; Eunomia Research & Consulting Ltd.: Bristol, UK, 2018.
62. Brismar, A. Seven Forms of Sustainable Fashion. Available online: https://www.greenstrategy.se/sustainable-fashion/seven-forms-of-sustainable-fashion/ (accessed on 31 July 2020).
63. Fashion Takes Action. The 7Rs of Fashion. Available online: https://fashiontakesaction.com/7rs/ (accessed on 31 July 2020).
64. Parischa, A. Slow Fashion, Collaborative Consumption, and Mindfulness. Available online: https://sophia.stkate.edu/cgi/viewcontent.cgi?article=1010&context=apparel_fac (accessed on 29 September 2022).
65. Piontek, F.M.; Müller, M. Literature reviews: Life cycle assessment in the context of product-service systems and the textile industry. *Procedia CIRP* **2018**, *69*, 758–763. [CrossRef]
66. Claudio, L. Waste Couture: Environmental Impact of the Clothing Industry. *Environ. Health Perspect.* **2007**, *115*, 7. [CrossRef]
67. Egels-Zandén, N.; Hansson, N. Supply chain transparency as a consumer or corporate tool: The case of Nudie Jeans Co. *J. Consum. Policy* **2016**, *39*, 377–395. [CrossRef]
68. Berg, A.; Magnus, K.H.; Kappelmark, S.; Granskog, A.; Lee, L.; Sawers, C.; Polgampola, P.; Lehmann, M. *Fashion on Climate: How the Fashion Industry Can Urgently Act to Reduce Its Greenhouse Gas Emissions*; McKinsey & Company and Global Fashion Agenda: Atlanta, GA, USA, 2020.
69. D'Adamo, I.; Gastaldi, M.; Morone, P.; Rosa, P.; Sassanelli, C.; Settembre-Blundo, D.; Shen, Y. Bioeconomy of sustainability: Drivers, opportunities and policy implications. *Sustainability* **2021**, *14*, 200. [CrossRef]

sustainability

Article

Assessing Bioeconomy Development Opportunities in the Latvian Policy Planning Framework

Krista Laktuka [1],* , Dagnija Blumberga [1] and Stelios Rozakis [2]

[1] Institute of Energy Systems and Environment, Faculty of Electrical and Environmental Engineering, Riga Technical University, LV-1048 Riga, Latvia
[2] Bioeconomy and Biosystems Economics Laboratory, School of Chemical and Environmental Engineering, Technical University of Crete, 73100 Chania, Greece
* Correspondence: krista.laktuka@rtu.lv

Abstract: The broad spectrum of bioresource use makes it challenging to interconnect strategic objectives and policy planning documents without compromising a coherent development vision. Bioeconomy development directions have been defined at the EU and Latvian levels. Nevertheless, to facilitate their implementation, the goals must be consistent with those specified in relevant national policy planning documents and vice versa. To determine whether internationally defined bioeconomy objectives are implemented in Latvian policy planning documents and what priority is given to them, a mixed methods approach was used—a systematic literature review combined with a keyphrase assignment approach. The results are summarized in an illustrative screening matrix and aggregated using the TOPSIS method to identify in which policy planning documents bioeconomy objectives are prioritized and to what extent. The results have shown a high prioritization of bioeconomy objectives in Latvian policy planning documents, especially in hierarchically higher documents.

Keywords: bioeconomy strategy; policy coherence; policy framework; strategic development

Citation: Laktuka, K.; Blumberga, D.; Rozakis, S. Assessing Bioeconomy Development Opportunities in the Latvian Policy Planning Framework. *Sustainability* **2023**, *15*, 1634. https://doi.org/10.3390/su15021634

Academic Editors: Nallapaneni Manoj Kumar, Md Ariful Haque and Sarif Patwary

Received: 30 November 2022
Revised: 5 January 2023
Accepted: 9 January 2023
Published: 13 January 2023

1. Introduction

The year 2022 marks the 10th anniversary of the first European Bioeconomy Strategy "Innovating for sustainable growth—A bioeconomy for Europe" (further EBS) [1,2]. The EBS sets out a series of objectives aimed at expanding the use of bioresources and underlining the need to move from the "old" to the "new" bioeconomy—knowledge-based and innovative [2,3]. The aim is to improve the current practices in land use and resources sustainably, to reduce emissions during resource extraction and processing, to lessen waste and use by-products to create higher value-added products, to move towards a circular economy, to minimize the use of non-renewable, unsustainable resources and adopt other environmentally friendly practices [2,3]. The bioeconomy has not lost its relevance over the last 10 years. On the contrary, global climate change, the current geopolitical situation, and rising energy prices further emphasize the need for the sustainable use of bioresources and the replacement of fossil fuels [4,5].

The economic impact of the COVID-19 pandemic and the tense geopolitical situation with the active warfare in Ukraine has created a situation where energy and food prices are rising rapidly due to disrupted and uncertain supply. The objectives of the EBS [6] are now central and could be a solution in terms of replacing fossil fuels with renewable resources under the condition of efficient and knowledge-based use of bioresources [5,7]. As the application of bioresources is wide-ranging, the EBS should be seen as a part of a larger equation, which could be solved by attaining coherence between policies affecting the bioeconomy and bioresources management at the international and the EU Member State level. Latvia is no exception. Therefore, bioeconomy development potential in Latvia should be assessed in order to identify the coherence between policy planning documents concerning different domains (external coherence) and also within each policy domain to verify the consistency of expressed policy goals, instruments, and other policy-related signals (internal coherence) [8].

Coherent policy-making across sectors could contribute to environmental sustainability and the development of successful national and regional cooperation mechanisms for forming functional regulation mechanisms and achieving common goals [4,8]. National action plans for the governance of bioresources aligned with international targets could ease food and energy supply risks, promote more rational and efficient use of bioresources, and prevent rapid and unpredictable inflation in Latvia and across the EU [5]. It is essential to identify whether Latvia's policy planning documents are coherent regarding the internal and external policy domains of bioeconomy development opportunities [4,8,9]. Whether the implementation of the overall strategic vision of bioeconomy development is following a top-down approach and maintains coherence at all levels is therefore important to identify [10,11].

2. Materials and Methods

2.1. Systematic Literature Review

A methodology was developed to identify a framework for developing the bioeconomy in Latvia's policy planning documents (Figure 1). The authors first identified internationally important documents that outline the main bioeconomy development trends and priorities. Three documents were selected: the 2009 OECD report "The Bioeconomy to 2030. Designing a Policy Agenda" [12]; EBS (2012) [6] and the EBS Action plan "A sustainable bioeconomy for Europe. Strengthening the connection between economy, society and the environment: updated bioeconomy strategy" (2018) [13]; UN "Transforming our world: the 2030 Agenda for Sustainable Development" (2015) [14,15]. The objectives and action lines for bioeconomy development in these documents were identified through a systematic literature review (SLR). From the identified objectives, keywords were selected for further work with Latvian policy planning documents [16] to determine whether their objectives for bioeconomy development coincide with those set at the international level.

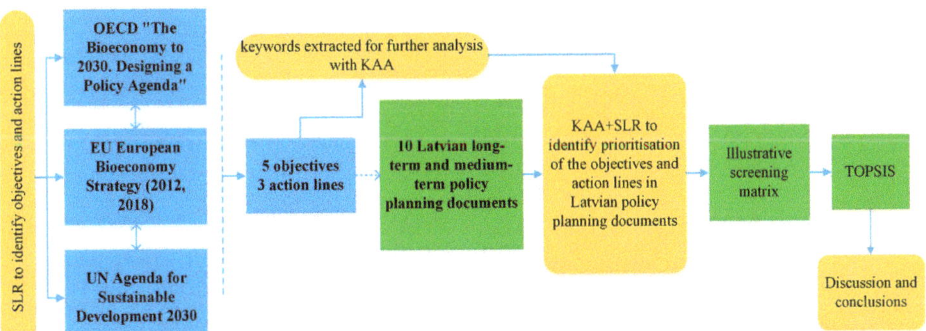

Figure 1. Methodology for assessment of policy planning documents at the international and national level.

The next step was to select Latvia's policy planning documents for further analysis. This step was also necessary to understand whether the documents coincide with the international purpose of the bioeconomy [11,17]. A total of 10 policy strategies and development plans directly related to the bioeconomy were selected (Appendix A) using a hybrid search strategy—SLR in combination with snowballing [18]. The search began with the analysis of the Latvian Bioeconomy Strategy 2030 (further LBS) and expanded the set to policy planning documents in relation to it and to the development of the bioeconomy. Further, a keyphrase/keyword assignment approach (KAA) [16,19,20] in combination with SLR was used to identify the specification of the internationally agreed objectives in the selected

Latvian policy planning documents. This would also allow for assessing the level of priority given to them. The results were presented in an illustrative screening matrix using a Likert-type scale [21], where more mentions of an objective indicated its higher priority. The priority of an objective was determined by the number of times it was mentioned in the policy planning document, as shown in Table 1.

Table 1. Rating scale for the illustrative screening matrix.

Rating	Level of Priority	Interaction
1	not a priority	no mention
2	low priority	1–2 mentions
3	medium priority	3–4 mentions
4	high priority	5–8 mentions
5	essential	9 and more mentions

The SLR has its roots in evidence-based policy and practice. It can be used to address environmental issues and evaluate policies and policy instruments [22–24]. One of the strengths of the SLR is that it allows one to answer a specific question or test a hypothesis [22,24]. Hence, the SLR method was chosen to identify international objectives and later to improve the quality of the illustrative screening matrix, showing the priority given to each of the international objectives identified above in the policy planning documents. SLR can be very time-consuming; therefore, KAA [19,20] was applied to reduce the time needed to review all 10 of Latvia's policy planning documents. The KAA was chosen because it is less time-consuming and helps to revise a document and the issue more closely while maintaining consistency [19,20]. Latvian policy planning documents range in length from 32 pages [25] to 228 pages [26], therefore, looking for pre-assigned keyphrases and keywords (Table 2) identified at the international level helped to maintain the scope and to constrain the study to a concise timeframe.

Table 2. Assigned keyphrases and keywords.

No.	Identified Objectives and Action Lines	Keyphrases and Keywords *
O_1	Ensure food and nutrition security	food security, ensure food, food availability
O_2	Manage natural resources sustainably	natural resources; sustainability; resources; natural
O_3	Reduce dependence on non-renewable, unsustainable resources	dependence; non-renewable; fossil
O_4	Limit and adapt to climate change	climate change; adaptation
O_5	Strengthen European competitiveness and create jobs	employment; jobs; promoting employment; competitiveness
A_1	Strengthen and scale up the biobased sectors, unlock investments and markets	biobased; attracting investment; innovation; investment
A_2	Deploy local bioeconomies rapidly across the whole of Europe	bioeconomy; bioresources; regions
A_3	Understand the ecological boundaries of the bioeconomy	ecological; boundaries; biological

* Keyphrases and keywords searched in documents in Latvian by using the word root.

2.2. Multi-Criteria Decision Analysis

An illustrative screening matrix was developed from the results of SLR and KAA. It shows the priority level of the international bioeconomy development goals in each of Latvia's policy planning documents. To determine which of the ten policy planning documents has the highest level of coherence with the internationally defined bioeconomy development objectives, the authors carried out a multi-criteria decision analysis—the Technique for Order Preference by Similarity to Ideal Solution (TOPSIS).

The advantage of the TOPSIS method is its simplicity and the relatively small amount of data required to apply it [27,28]. The TOPSIS method is used for decision-making in various areas, including the evaluation of strategies by determining the proximity of predefined alternatives to the ideal positive and negative solutions [27–29]. One component of the TOPSIS calculation is the application of weighted criteria values. The calculation is then repeated with equal weights to determine the impact of the weights on the obtained results. [28]. The calculation was performed according to the steps and formulas listed below [30–32].

$$D = \begin{array}{c} A_1 \\ \vdots \\ A_m \end{array} \begin{array}{c} C_1 \; \ldots \; C_n \\ \begin{pmatrix} x_{11} \cdots x_{1n} \\ \vdots \ddots \vdots \\ x_{m1} \cdots x_{mn} \end{pmatrix} \end{array} \tag{1}$$

where:

$A_1 \ldots A_m$—comparable alternatives;

$C_1 \ldots C_n$—criteria according to which the comparison is performed;

x_{ij}—performance/value of alternative A_i (where i is alternative 1 to m) according to criterion C_j (where j from 1 to n).

$$D_{\text{norm}} = \begin{array}{c} A_1 \\ \vdots \\ A_m \end{array} \begin{array}{c} C_1 \; \ldots \; C_n \\ \begin{pmatrix} r_{11} \cdots r_{1n} \\ \vdots \ddots \vdots \\ r_{m1} \cdots r_{mn} \end{pmatrix} \end{array} \tag{2}$$

The next step is to calculate the normalized rating using the formula:

$$r_{ij} = \frac{x_{ai}}{\sqrt{\sum_{a=1}^{n} x_{ai}^2}} \tag{3}$$

When the normalized evaluation of all alternatives according to the criteria specified in Table 2 is obtained, it is necessary to determine the individual weight w_i of each criterion. Weights are determined by meeting a condition—the sum of criterion weights is equal to 1.

Expert evaluation is used to determine the individual weight of each criterion. As criteria weights, expert evaluation was obtained by Dolge et al. [33], analyzing the national bioeconomy strategies of nine EU countries using the TOPSIS method. The identified objectives and action lines in Table 2 coincide with the evaluation criteria set out in the study by Dolge et al. [33] (EBS objectives and EBS Action plan action areas); therefore, in order to ensure continuity and comparability of the studies, it was decided to use the expert evaluation in this research as well (Table 3). The expert evaluation was obtained through an online survey of industry stakeholders involved in any of the primary bioresources production or processing sectors or in scientific research in the field of bioeconomy, climate, and environmental sustainability [33]. The experts were asked to rate each of the criteria to detect the most important ones for the rapid development of the bioeconomy [33]. The weights of the criteria are the average of the 27 experts' responses for each criterion, giving a total of 1 or 100% for all criteria [33] (Table 3). The criterion weights obtained were inserted into the TOPSIS matrix and used for further calculations. In the next step, the criteria weight values w_i obtained from the expert evaluation are multiplied by the normalized values r_{ia} to obtain the normalized weighted value v_{ai}, as shown in Equation (4):

$$v_{ai} = w_i * r_{ia} \tag{4}$$

Table 3. Expert evaluation used for criterion weights [33].

No.	Criteria	Criterion Weights, w_i
1	O_1—Ensure food and nutrition security	0.11
2	O_2—Manage natural resources sustainably	0.18
3	O_3—Reduce dependence on non-renewable, unsustainable resources	0.19
4	O_4—Limit and adapt to climate change	0.12
5	O_5—Strengthen European competitiveness and create jobs	0.10
6	A_1—Strengthen and scale up the biobased sectors, unlock investments and markets	0.13
7	A_2—Deploy local bioeconomies rapidly across the whole of Europe	0.08
8	A_3—Understand the ecological boundaries of the bioeconomy	0.09
	TOTAL:	1.00 (100%)

When the normalized weighted decision matrix is constructed, the ideal positive solution d_a^+ and the ideal negative solution d_a^- are calculated. Initially, the distance to the ideal solution (MAX) and the distance to the anti-ideal solution (MIN) are determined. Distances are determined by formulas:

$$=\text{MAX}(v_{a1}:v_{a3}) \tag{5}$$

$$=\text{MIN}(v_{a1}:v_{a3}) \tag{6}$$

After determining the distance to the ideal and anti-ideal solution, the next step is to determine the ideal positive and ideal negative solution according to the formulas:

$$d_a^+ = \sqrt{\sum_{j=1}^{n}\left(v_i^+ - v_{ai}\right)^2} \tag{7}$$

$$d_a^- = \sqrt{\sum_{j=1}^{n}\left(v_i^- - v_{ai}\right)^2} \tag{8}$$

The relative proximity of the alternative to the ideal solution is calculated as shown in formula No. 9:

$$C_a = \frac{d_a^-}{d_a^+ + d_a^-} \tag{9}$$

The result is equal to values that show the proximity of the alternative to the ideal positive solution and the distance from the ideal negative solution.

To determine the impact of the weights of the criteria set by the expert evaluation (Table 3) on the evaluation of criteria, a re-evaluation of the criteria is performed, assigning equal values to all alternatives by using the same equations described above.

3. Results

3.1. International Policy Framework of Bioeconomy

Three leading policy planning documents were selected for identifying internationally established directions for bioeconomy development. The 2009 OECD report "The Bioeconomy to 2030. Designing a Policy Agenda" [12] lays the foundations for a strategic view of the bioeconomy and the benefits that could arise from the wider use of bioresources and biotechnologies [15]. The report states that the bioeconomy has all the potential needed to ensure long-term economic and environmental sustainability, however, to achieve this, a broad public and national government support is crucial [12]. The OECD report identifies nine vital challenges in the bioeconomy till 2030 [12] and they are summarized in Table 4.

Table 4. Objectives and actions identified at international level to develop bioeconomy [12] (pp. 287–293), [6,14] (pp. 9–11), [13] (pp. 10–22).

OECD "The Bioeconomy to 2030. Designing a Policy Agenda" [12]	UN Agenda for Sustainable Development 2030 [14]	European Bioeconomy Strategy (2012, 2018) [6,13]
reverse the neglect of primary production and industrial applications;	Goal 2: eradicate hunger, achieve food security and improved nutrition, and promote sustainable agriculture;	O1—ensure food and nutrition security;
prepare for a costly but beneficial revolution in healthcare;	Goal 7: ensure universal access to affordable, reliable, sustainable, and modern energy services;	O2—manage natural resources sustainably;
manage the globalization of the bioeconomy;	Goal 8: promote sustained, inclusive, and sustainable economic growth, full and productive employment, and decent work for all;	O3—reduce dependence on non-renewable, unsustainable resources;
turn the economically disruptive power of biotechnology to advantage;		
prepare for multiple futures;	Goal 12: ensure sustainable consumption and production patterns;	O4—limit and adapt to climate change;
maximize the benefits of integration;	Goal 13: take urgent action to combat climate change and its impact;	O5—strengthen European competitiveness and create jobs
reduce barriers to biotechnology innovation;	Goal 14: preserve and sustainably use the oceans, seas, and marine resources to ensure sustainable development;	(A1)—strengthen and scale up the biobased sectors, unlock investments and markets;
create a dynamic dialogue between governments, citizens, and firms;	Goal 15: protect, restore, and promote sustainable use of terrestrial ecosystems, sustainably manage forests, combat desertification, and halt and reverse land degradation and halt biodiversity loss.	(A2)—deploy local bioeconomies rapidly across the whole of Europe;
prepare the foundation for the long-term development of the bioeconomy.		(A3) understand the ecological boundaries of the bioeconomy.

The role of the bioeconomy as a globally significant driver for future development is reinforced and complemented by the plan—"Transforming our world: the 2030 Agenda for Sustainable Development", adopted by the UN in 2015 [14]. The 17 Sustainable Development Goals (further SDGs) include a series of actions aimed not only at eradicating poverty and hunger but also at combating climate change by encouraging responsible and efficient use of resources and other environmentally friendly measures [14]. Seven of the SDGs are more closely linked to bioeconomy development and are summarized in Table 4 [14,25].

In 2012, shortly after the OECD report was published, the EU adopted its first EBS [6] to address ecological, environmental, energy, food supply, and bioresource challenges [15]. As a result, a set of five key objectives for promoting and strengthening the bioeconomy were brought forward (Table 4) [6]. The EBS was designed to complement existing EU policies such as the Common Agricultural Policy and Common Fisheries Policy and invited EU Member States to develop their own national strategies in order to place bioeconomy on their policy agenda [6].

A few years later in 2018, the existing EBS was revised, and the direction of the strategy was adjusted by adding new areas of action [2,6,13]. An updated EBS "A sustainable bioeconomy for Europe. Strengthening the connection between economy, society and the environment: updated bioeconomy strategy", and Action Plan were adopted [13]. The Action Plan identifies three action areas (Table 4), under which a total of 14 sub-activities are identified [2,13]. The EBS Action Plan (2018) is created taking into account and is closely interlinked with the SDGs [2].

The objectives and action lines summarized in Table 4 are listed in the order in which they appear in each document, without attempting to group them thematically or by importance. Table 4 shows the main goals of the documents and, as they do not show conflicting ideas, for further analysis only the objectives and action lines from the EBS and EBS Action Plan will be used in order to reduce the number of keywords to be searched for in the policy planning documents. An additional argument for the keywords being drawn only from the EBS and EBS Action Plan is that the documents adopted by the OECD [12] and the UN [14] have been taken into account in the development of the EBS [6] and later EBS Action Plan [13].

In addition, four keywords were added as the major sources of bioresources: agriculture, forest sector, fisheries, and aquaculture. Hence, it would be possible to determine whether one of these three sectors is being developed more or, on the contrary, neglected.

On average, between 15 and 20 keywords or keyphrases in Latvian were used per policy document, with the potential to indicate the inclusion of the objectives listed in Table 4 or the three bioeconomy-related sectors in policy documents.

3.2. National Policy Framework—The Latvian Bioeconomy Strategy

The LBS was adopted in 2017 [25] in regard to Latvia's highest hierarchical long-term planning document—the Latvian Sustainable Development Strategy 2030 [34]. Latvia's Sustainable Development Strategy 2030 sets a goal "to become the EU leader in the preservation, increase, and sustainable use of natural capital" [25,34], but, to achieve this, the bioeconomy needs to be given a more important role at the national level, and possible directions for development need to be identified. Bioeconomy in Latvia encompasses many economic sectors that can be divided into several groups: primary production of bioresources (agriculture, forest sector, fisheries); processing sectors of bioresources, where operation completely or mainly depend on bioresources; processing sectors of bioresources, where bioresources compete with other raw materials or replace them; service sectors using bioresources [25].

LBS states that Latvia has ample opportunities to successfully develop the bioeconomy and use natural resources sustainably and as efficiently as possible [25]. Through the development of the bioeconomy, land resources could be used in a strategic and sustainable manner and new well-paid jobs could be created [25]. An important future development would be the reduction in waste in manufacturing and processing industries and the substitution of fossil resources for bioresources [25]. Objectives and action directions defined by the LBS are presented in Table 5 [25].

Table 5. Objectives of the Latvian Bioeconomy Strategy 2030 [25] (pp. 5–22).

Latvian Bioeconomy Strategy 2030	
Objectives	**Action Directions**
(1) promotion and preservation of employment in bioeconomy sectors to up to 128 thousand employees	Attractive Entrepreneurial Environment
(2) increasing the added value of bioeconomy products to at least 3.8 billion euros in 2030	Result-oriented Efficient and Sustainable Resource Management
(3) increasing the value of bioeconomy export production to at least 9 billion euros in 2030	Knowledge and Innovations
	Promotion of Manufacturing the Produce in Bioeconomy
	Socially Responsible and Sustainable Development

3.3. Latvian Policy Planning Documents Related to the Bioeconomy

For the analysis of Latvian policy planning documents, 10 long-term and medium-term planning documents (Appendix A) were identified by using SLR in combination with snowballing. Starting with LBS [25], then expanding the selection to the Latvian Sustainable Development Strategy 2030 [34] and documents related to waste management [26], achieving climate neutrality [35], moving towards a circular economy [36], and other thematically related policy planning documents. To determine how the bioeconomy development possibilities are covered and to what extent was prioritized by these documents.

In terms of year of adoption, the earliest document in the set is the Sustainable Development Strategy of Latvia 2030 [34], adopted in 2010, followed by the LBS [25], adopted in 2017 (Appendix A). Other policy planning documents were adopted in 2019 or earlier. Most policy documents in Appendix A set out not only the objectives to be achieved, but also specific action lines and performance indicators. Documents with actions defined in a generic manner, without specific actions, are the Strategy of Latvia for the Achievement of Climate Neutrality by 2050 [35] and the LBS [25]. The Strategy of Latvia for the Achievement of Climate Neutrality by 2050 and LBS are the only policy planning documents in the selection of documents that are "informative reports" and do not have action plans. LBS does not set qualitative or quantitative indicators to measure the achievement of the objectives [25]. An important element in achieving objectives set out in policy planning documents is an interim evaluation to monitor the progress of the implementation. Sustainable Development Strategy of Latvia 2030 [34]; Strategy of Latvia for the Achievement of Climate Neutrality by 2050 [35]; Latvian National Development Plan 2021–2027 [37] and National Energy and Climate Plan for 2021–2030 [38] have incorporated a periodic or mid-term evaluation. Documents that do not include a mid-term assessment are the LBS [25]; Action Plan for the Transition to a Circular Economy 2020–2027 [36], and Environmental Policy Guidelines 2021–2027 [39].

Half of the revised policy planning documents do not have indicative funding for implementation. In addition, the LBS has no indication of the approximate amount or possible sources of funding for the promotion and development of the bioeconomy [25]. Regarding the source of funding, some of the policy planning documents (Appendix A) include a statement that the action lines will be implemented within the existing national budget, putting a particular emphasis on the possibility to attract funding from the EU Structural Funds, as well as other sources of funding, including private finance.

3.4. Implementation of International Objectives in Latvia's Policy Planning Documents

3.4.1. Illustrative Screening Matrix

The results obtained with KAA and SLR on the prioritization of bioeconomy development goals in Latvian policy planning documents were normalized according to Table 1. Acquired ratings were displayed in the illustrative screening matrix (Table 6). The matrix does not analyze the nature of interactions but looks at the priority of objectives (Table 4) in Latvian policy planning documents by counting mentioned keyphrases and keywords in the context of bioeconomy objectives. The assumption is that the more often an objective or action line is mentioned in a policy document, the higher the priority is given to it and the more likely it is to be implemented.

The illustrative screening matrix (Table 6) not only allows one to assess the priorities set in Latvian policy planning documents in relation to internationally defined objectives and action lines but also allows one to estimate the internal and external coherence between different policy domains (Figure 2) [8]. Additionally, vertical interactions can be observed— whether international-level documents are implemented on a national level, and on lower-level planning documents related to the bioeconomy sector [8]. Horizontal interactions show whether there is synergy between the objectives set out on international and local level policy planning documents across external and internal dimensions [8].

Table 6. Illustrative screening matrix.

Long-Term and Medium-Term Planning Documents	Bioeconomy-Related Objectives and Action Lines Stated in European Bioeconomy Strategy (2012, 2018) (Table 4)										Rating Per Document	Agriculture	Forest Sector	Fisheries, Aquaculture	Sum for Sectors
	O1	O2	O3	O4	O5	Sum O	(A1)	(A2)	(A3)	Sum A					
Sustainable Development Strategy of Latvia until 2030 [34]	4	5	2	5	5	21	4	3	5	12		5	2	3	10
Strategy of Latvia for the Achievement of Climate Neutrality by 2050 [35]	4	5	5	5	5	24	5	4	4	13		5	4	2	11
Latvian National Development Plan 2021–2027 [37]	3	5	2	5	5	20	5	5	4	14		2	2	2	6
Latvian National Energy and Climate Plan for 2021–2030 [38]	2	3	3	5	4	17	4	3	1	8		4	4	2	10
Latvian Bioeconomy Strategy 2030 [25]	5	5	5	4	5	24	5	5	3	13		5	5	5	15
Latvia's Adaptation to Climate Change Plan for the Period Until 2030 [40]	2	3	2	5	3	15	2	1	3	6		5	4	4	13
National Waste Management Plan for 2021–2028 [26]	2	5	3	4	1	15	4	1	1	6		2	2	1	5
National Industrial Policy Guidelines for 2021–2027 [41]	3	3	1	5	5	17	5	5	4	14		4	3	1	8
Action Plan for the Transition to a Circular Economy 2020–2027 [36]	3	5	1	1	3	13	5	3	3	11		3	1	1	5
Environmental Policy Guidelines 2021–2027 [39]	1	5	1	5	1	13	2	4	5	11		5	3	2	10
Rating per objectives and action lines	29	44	25	44	37		41	34	33			40	30	23	

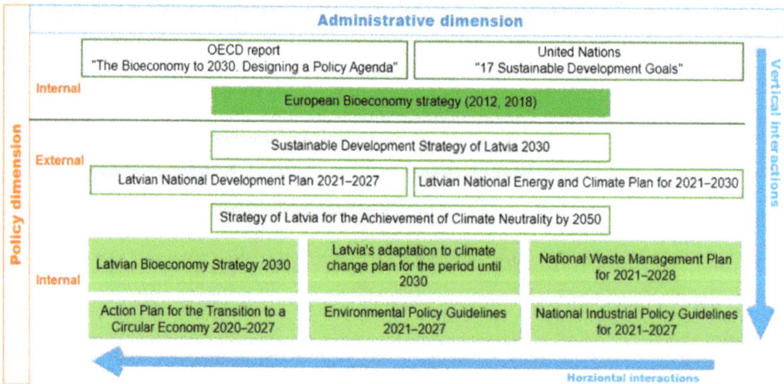

Figure 2. Coherence amongst bioeconomy-related international and national policy planning documents (adapted from [8]).

The illustrative screening matrix indicates that Latvia's long-term and medium-term policy planning documents, in general, prioritize the same objectives and action lines that have been set at the international level by the EU, UN, and OECD. Highest-level policy planning documents such as the Sustainable Development Strategy of Latvia until 2030 [34],

Strategy of Latvia for the Achievement of Climate Neutrality by 2050 [35], Latvian National Development Plan [37], Latvian National Energy and Climate Plan for 2021–2030 [38], and the LBS [25] give high priority to the international bioeconomy objectives. However, it was already expected for the LBS to score the highest out of the set of documents considered because the LBS itself mentions that it has been designed taking into account the objectives set by the EBS [25]. A lower level of prioritization can be observed in policy documents that define strategic development in more specific areas such as waste management [26], circular economy [36], and adaptation to climate change [40], because of having more specific deliverables but on average showing high results in the overall policy framework for bioeconomy development.

The results obtained by adding up the objectives (*O1–O5*), action lines (*A1,A2*), and bioeconomy sectors (agriculture, forest sector, fisheries, and aquaculture) were assessed separately in the illustrative screening matrix (see Table 6). This allowed us to assess the inclusion of the internationally agreed objectives in Latvia's policy planning documents, as well as to identify whether the EBS Action Plan adopted in 2018 is taken into account. The priority given to bioeconomy sectors in each of the documents was also assessed, thus showing which of them is being prioritized.

The evaluation of the policy planning documents (Table 6) by adding up the objectives (*O1–O5*) showed that the LBS [25] and the Strategy of Latvia for the Achievement of Climate Neutrality by 2050 [35] have the highest ranking with 24 points. The following documents are next in order of points—the Sustainable Development Strategy of Latvia until 2030 [34] with 21 points, the Latvian National Development Plan 2021–2027 [37] with 20 points, and close behind with 17 points the Latvian National Energy and Climate Plan for 2021–2030 [38], and the National Industrial Policy Guidelines for 2021–2027 [41]. Latvia's Adaptation to Climate Change Plan for the Period Until 2030 [40] and the National Waste Management Plan for 2021–2028 [26] obtained 15 points each. The lowest scores are shown by the Action Plan for the Transition to a Circular Economy 2020–2027 [36] (13 points) and the Environmental Policy Guidelines 2021–2027 [39] (13 points).

The analysis of the inclusion of action lines from the EBS Action Plan (*A1–A3*) in policy documents showed that none of the documents scored the highest possible score of 15, but both the Latvian National Development Plan 2021–2027 [37] and the National Industrial Policy Guidelines for 2021–2027 [41] scored close with 14 points each. Two policy documents scored highly, with 13 points the Strategy of Latvia for the Achievement of Climate Neutrality by 2050 [35] and the LBS [25]. The Sustainable Development Strategy of Latvia until 2030 [34] obtained 12 points, and both the Action Plan for the Transition to a Circular Economy 2020–2027 [36] and the Environmental Policy Guidelines 2021–2027 [39] scored 11 points. Policy documents with the lowest scores were the Latvian National Energy and Climate Plan for 2021–2030 [38] (8 points), Latvia's Adaptation to Climate Change Plan for the Period Until 2030 [40] (6 points), and the National Waste Management Plan for 2021–2028 [26] (6 points).

The score per sector (agriculture, forest sector, fisheries, and aquaculture) in policy planning documents shows a bit of a different breakdown. The LBS [25] obtained the maximum score of 15 points. The next highest score is reached by Latvia's Adaptation to Climate Change Plan for the Period Until 2030 [40] (13 points), and the Strategy of Latvia for the Achievement of Climate Neutrality by 2050 [35] (11 points). The Sustainable Development Strategy of Latvia [34], the Latvian National Energy and Climate Plan for 2021–2030 [38], and the Environmental Policy Guidelines 2021–2027 [39] have the same score of 10 points. Other policy planning documents have scored less—the National Industrial Policy Guidelines for 2021–2027 [41] (8 points), the Latvian National Development Plan 2021–2027 [37] (6 points), the National Waste Management Plan for 2021–2028 [26], and the Action Plan for the Transition to a Circular Economy 2020–2027 [36] (5 points).

Despite the fact that the EBS was taken into account in the development of the LBS [25], it has not received the highest possible scores, although it shows the greatest consistency with the internationally defined objectives (*O1–O5*) and action lines (*A1–A3*). It should be noted that the LBS scored highest in the bioeconomy sectoral assessment, giving equal priority to all three sectors. Comparatively higher scores in the objective assessment were achieved by higher level policy planning documents as well as the National Industrial Policy Guidelines for 2021–2027 [41], which can be considered a positive trend as it shows that internationally defined bioeconomy development objectives are being taken into account. The presence of the National Industrial Policy Guidelines among the highest scoring documents should be seen as a logical outcome, as a knowledge-based innovative bioeconomy is one of the five knowledge areas (RIS3) identified for Latvia and discussed in more detail in the document [41].

The assessment of the implementation of the actions ($A1$–$A3$) of the EBS Action Plan in policy planning documents has shown similar results, with the National Industrial Policy Guidelines [41] scoring second highest. The Latvian National Energy and Climate Plan for 2021–2030 [38] scored relatively low compared to other hierarchically higher documents, possibly due to its thematic focus on energy and energy efficiency issues, with less attention to ecological boundaries. The bioeconomy sectors (agriculture, forest sector, fisheries, and aquaculture) have received varying attention in the policy planning documents reviewed. As already mentioned, the LBS has given equal priority to all sectors. No clear correlation can be discerned between the prioritization of bioresource extraction sectors in higher and lower-level policy planning documents.

Looking at the priority areas assigned to the objectives related to the development of the bioeconomy in the policy planning documents (Table 6—$O1$–$O5$, ($A1$)–($A3$) vertically) the results indicate that in Latvian policy documents, priority is given to $O2$—"manage natural resources sustainably" (44 points) and $O4$—"limit and adapt to climate change" (44 points). Slightly lower scores are received by ($A1$)—"strengthen and scale up the biobased sectors, unlock investments and markets" (41 points) and $O5$—"strengthen European competitiveness and create jobs" (37 points); ($A2$)—"deploy local bioeconomies rapidly across the whole of Europe" (34 points) and ($A3$)—"understand the ecological boundaries of the bioeconomy" (33 points). The lowest priority was given to objectives $O1$—"ensure food and nutrition security" (29 points) and $O3$—"reduce dependence on non-renewable, unsustainable resources" (25 points).

The priority given to the agriculture, forest sector, fisheries, and aquaculture sectors in Latvia's policy planning documents altogether was assessed to determine whether any bioresource sector is prioritized over others. The assessment shows that the highest priority in the context of the bioeconomy is given to developing the agricultural sector (40 points); the forest sector scores lower with 30 points and the least priority is given to developing fisheries and aquaculture with 23 points.

3.4.2. TOPSIS Results

The TOPSIS criteria were weighted according to expert evaluation [33] (Table 3). The experts determined which of the criteria (Table 2) could play a crucial role in the development of the bioeconomy in Latvia. Thus, the TOPSIS analysis results would reveal which of Latvia's policy planning documents puts the most emphasis on a particular objective.

Therefore, the prioritized bioeconomy development objectives in the policy planning document combined with expert evaluation (Table 3), identifying which of these objectives are most important, were the ideal positive solution. In the evaluation of the Latvian policy planning documents using the TOPSIS method, with criteria weights (Table 3), the LBS [25] (0.98), and the Strategy of Latvia for the Achievement of Climate Neutrality by 2050 [35] (0.98) have the highest score and are the closest to the ideal positive solution for bioeconomy development in Latvia (Figure 3). The Sustainable Development Strategy of Latvia 2030 [34] with 0.58 points and the Latvian National Development Plan 2021–2027 [37] with 0.57 points scored significantly lower; the next closest to the ideal solution was the Latvian National Energy and Climate Plan for 2021–2030 [38] with 0.46 points. The next highest scorers are policy planning documents aimed at developing a specific policy area or sector—the National Waste Management Plan for 2021–2028 [26] (0.38 points); the National Industrial Policy Guidelines for 2021–2027 [41] (0.37 points); and the Environmental Policy Guidelines 2021–2027 [39] (0.23 points). Latvia's Adaptation to Climate Change Plan for The Period Until 2030 [40] and the Action Plan for the Transition to a Circular Economy 2020–2027 [36] received only 0.22 points.

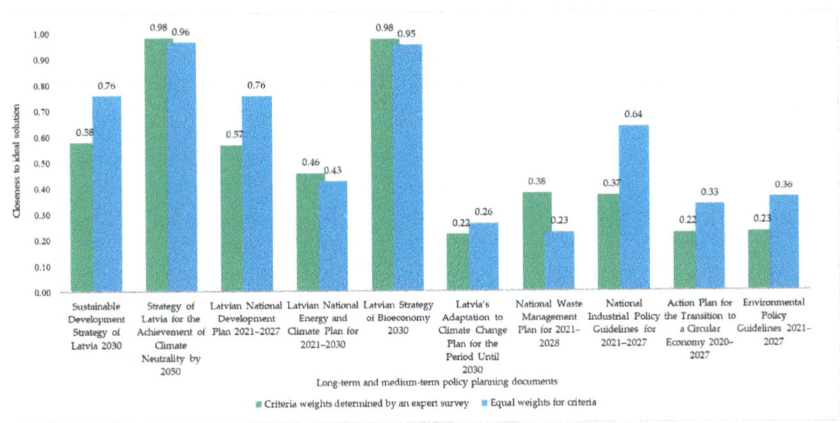

Figure 3. TOPSIS results on prioritized objectives and action lines in the Latvian policy planning documents.

TOPSIS results with applied equal criteria weights show similar results as when applying the criteria weights determined by experts. The policy planning documents closest to the ideal positive solution are the Strategy of Latvia for the Achievement of Climate Neutrality by 2050 [35] (0.96) and the LBS [25] (0.95) (Figure 3). The Sustainable Development Strategy of Latvia 2030 [34] with 0.76 points and the Latvian National Development Plan 2021–2027 [37] with 0.76 points scored significantly higher than in the evaluation with criteria weights set by experts, however, these two documents maintain the third and fourth highest ranking. The National Industrial Policy Guidelines for 2021–2027 [41] showed a better result with equal criteria weights by scoring 26 points higher than in the evaluation with criteria weights determined by expert evaluation (0.64 points). The Latvian National Energy and Climate Plan for 2021–2030 [38] with 0.43 points has almost a similar score as in the previous assessment with weights assigned by experts. Farther from the positive ideal solution are the Environmental Policy Guidelines 2021–2027 [39] (0.36 points), the Action Plan for the Transition to a Circular Economy 2020–2027 [36] with 0.33 points, Latvia's Adaptation to Climate Change Plan for the Period Until 2030 [40] with 0.26 points, and National Waste Management Plan for 2021–2028 [26] with 0.23 points.

The TOPSIS analysis on agriculture, forest sector, and fisheries and aquaculture in Latvian policy planning documents (Figure 4), with equal criteria weights, has shown the following results. One document is the ideal positive solution with 1.00 point—the LBS [25]. Latvia's Adaptation to Climate Change Plan for the Period Until 2030 [40] is the second closest with 0.92 points. Other policy planning documents scored lower in the TOPSIS assessment. The third document that is the closest to the ideal positive solution is the Strategy of Latvia for the Achievement of Climate Neutrality by 2050 [35] (0.52). The fourth is the Latvian National Energy and Climate Plan for 2021–2030 [38] (0.46), and the fifth is the Sustainable Development Strategy of Latvia 2030 [34] (0.45). The Environmental Policy Guidelines 2021–2027 ranked close to this score [39] (0.38). The National Industrial Policy Guidelines for 2021–2027 [41] obtained 0.18 points, the Latvian National Development Plan 2021–2027 [37] 0.08 points, and the National Waste Management Plan for 2021–2028 [26] with 0.03, and the Action Plan for the Transition to a Circular Economy 2020–2027 [36] with 0.01 are the furthest away from the ideal positive solution.

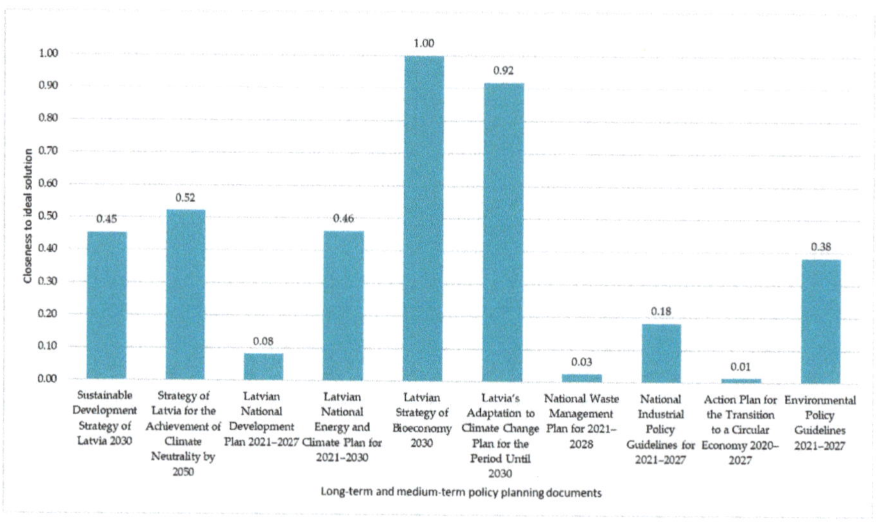

Figure 4. TOPSIS results on agriculture, forest sector, fisheries, and aquaculture.

4. Discussion

The results of the analysis of Latvia's long-term and medium-term policy planning documents by constructing an illustrative screening matrix and a subsequent analysis with TOPSIS indicate a positive trend in the implementation and prioritization of the internationally agreed objectives in Latvia's policy planning. Each of the 10 documents selected for the study could be linked to the international objectives. Notably, the policy planning documents that are higher up the policy planning hierarchy, such as the Sustainable Development Strategy of Latvia until 2030 [34], the Strategy of Latvia for the Achievement of Climate Neutrality by 2050 [35], and the Latvian National Development Plan 2021–2027 [37] performed considerably better than specifically targeted lower-level sectoral plans, as for example the Environmental Policy Guidelines 2021–2027 [39].

Looking at the priority given to each objective and action line in the Latvian policy planning documents, a less thematic elaboration on the objective of $O1$—"ensure food and nutrition security" [6] and $O3$—"reduce dependence on non-renewable, unsustainable resources" can be observed [6]. Food and nutrition safety and food quality are not seen as an issue in Latvia's policy planning, because of a well-developed agricultural sector that is fully capable of meeting current food demand and high EU quality standards [42]. Consequently, it is not considered to be a topical issue that calls for strategic planning at the national level. References to objective $O1$ are found in all the policy planning documents analyzed (Table 6), apart from the Environmental Policy Guidelines 2021–2027 [31].

Despite the negative environmental impact of fossil fuels identified in policy planning documents [26,34,35], there are no concrete actions outlined to phase out fossil fuels. The low priority given to the need to reduce dependence on non-renewable and unsustainable resources ($O3$) could be an indication of the resistance of policymakers to fossil fuel divestment, given the existing infrastructure of fossil energy sources and the year's long low prices of natural gas and oil. This scenario changed rapidly this year.

The development of agriculture and forest sectors was mentioned relatively frequently, whereas the development of fisheries, except for the LBS, received very little attention. Latvia is a water-rich country with a long maritime border, which makes it unclear why fisheries and aquaculture development is given such a low priority in planning documents. The authors suggest that this may be because Latvia's fisheries and aquaculture sectors have historically been based on fishing in the sea [43,44], but the collapse of the Soviet Union and in later years the introduction of EU fishing quotas due to the depletion of significant fish species has led to a stagnation in the development of the fisheries sector [45,46]. However, innovative technologies and a shift towards growing fish and other marine organisms in aquacultures could change the situation [47,48]. The efficient management and use of inland waters and fish, shellfish, and algae these waters contain could be used to produce innovative products [47,48].

Assessment of the Latvian policy planning documents from a technical perspective showed that most of them set specific actions to be taken, and indicators and interim evaluations to track progress (Appendix A). Nevertheless, a critical element for all the policy planning documents is the unclear financing mechanism. The documents mostly indicate that financial resources should be allocated within the existing national budget on an annual basis or applied for from EU Structural Funds or private funding to implement the measures. This raises concerns about the extent to which the objectives and action lines for bioeconomy development could be implemented.

5. Conclusions

The methodology developed in this study allows relatively quick and easy identification of any pre-defined objectives and actions set out in policy documents. It also allows for assessing the level of priority given to such objectives and actions. However, rather than stand-alone research, this methodology can be recommended as a first step in a more in-depth examination of policy planning documents to determine the level of bioeconomy development priorities in them. It can be applied as a valuable help to facilitate the evaluation of a larger set of documents. The main drawback of this methodology is its inability to provide an assessment of direct contradictions that may exist between the elaboration of the objectives and/or the document itself. For a more detailed in-depth study, the documents with the highest or lowest scores determined using this methodology should be selected, depending on the expected outcome.

Author Contributions: Conceptualization, D.B. and S.R.; methodology, D.B. and K.L.; validation, K.L. and S.R.; formal analysis, K.L.; investigation, K.L.; writing—original draft preparation, K.L.; writing—review and editing, D.B. and S.R.; visualization, K.L.; supervision, D.B. and S.R.; project administration, D.B.; funding acquisition, K.L. All authors have read and agreed to the published version of the manuscript.

Funding: The study was prepared with support from the European Social Fund within Project No 8.2.2.0/20/I/008 "Strengthening of PhD students and academic personnel of Riga Technical University and BA School of Business and Finance in the strategic fields of specialization" of the Specific Objective 8.2.2 "To Strengthen Academic Staff of Higher Education Institutions in Strategic Specialization Areas" of the Operational Program "Growth and Employment".

Data Availability Statement: Not applicable.

Conflicts of Interest: The authors declare no conflict of interest.

Appendix A

Table A1. Latvian policy planning documents linked to bioeconomy.

National Policy Planning Documents Related to Bioeconomy	Information about the Document					
	Year	Action Lines to Achieve Objectives	Performance Indicators	Interim Evaluation	Funding Needed	Source of Financial Resources
Sustainable Development Strategy of Latvia 2030 [34]	2010	Specific	Qualitative and quantitative	Yes (every 2 years)	No information	Under available national budget, EU funds, private

Table A1. *Cont.*

National Policy Planning Documents Related to Bioeconomy	Information about the Document					
	Year	Action Lines to Achieve Objectives	Performance Indicators	Interim Evaluation	Funding Needed	Source of Financial Resources
Strategy of Latvia for the Achievement of Climate Neutrality by 2050 [35]	2019	Generic	Qualitative and quantitative	Yes (every 10 years)	Yes	Under available national budget, EU funds, private
Latvian National Development Plan 2021–2027 [37]	2020	Specific	Qualitative and quantitative	Yes (every 2 years)	Yes	Under available national budget, EU funds, private
Latvian National Energy and Climate Plan for 2021–2030 [38]	2020	Specific	Qualitative and quantitative	Yes	Yes	Under available national budget, EU funds, private
Latvian Bioeconomy Strategy 2030 [25]	2017	Generic	Generic	No	No information	Not specified
Latvia's Adaptation to Climate Change Plan for the Period Until 2030 [40]	2019	Specific	Qualitative and quantitative	Yes (mid-term)	Not specified	Under available national budget, EU funds, private
National Waste Management Plan for 2021–2028 [26]	2021	Specific	Qualitative and quantitative	Yes (mid-term)	Yes	Under available national budget, EU funds, private
National Industrial Policy Guidelines for 2021–2027 [41]	2021	Specific	Qualitative and quantitative	Yes (mid-term)	Yes	Under available national budget, EU funds, private
Action Plan for the Transition to a Circular Economy 2020–2027 [36]	2020	Specific	Qualitative and quantitative	No	Not specified	Under available national budget, EU funds, private
Environmental Policy Guidelines 2021–2027 [39]	2021	Specific	Qualitative and quantitative	No	Not specified	Under available national budget, EU funds, private

References

1. 10th anniversary of the EU Bioeconomy Strategy | European Commission, 11 February 2022. Available online: https://ec.europa.eu/info/news/10th-anniversary-eu-bioeconomy-strategy-2022-feb-11_en (accessed on 5 April 2022).
2. European Commission. Bioeconomy Strategy. Available online: https://ec.europa.eu/info/research-and-innovation/research-area/environment/bioeconomy/bioeconomy-strategy_en (accessed on 18 February 2022).
3. Wreford, A.; Bayne, K.; Edwards, P.; Renwick, A. Enabling a transformation to a bioeconomy in New Zealand. *Environ. Innov. Soc. Transit.* **2019**, *31*, 184–199. [CrossRef]
4. Muscat, A.; de Olde, E.M.; Kovacic, Z.; de Boer, I.J.M.; Ripoll-Bosch, R. Food, energy or biomaterials? Policy coherence across agro-food and bioeconomy policy domains in the EU. *Environ. Sci. Policy* **2021**, *123*, 21–30. [CrossRef]
5. Purkus, A.; Hagemann, N.; Bedtke, N.; Gawel, E. Towards a sustainable innovation system for the German wood-based bioeconomy: Implications for policy design. *J. Clean. Prod.* **2018**, *172*, 3955–3968. [CrossRef]
6. European Commission. A Bioeconomy for Europe. 2012, pp. 9–11. Available online: https://op.europa.eu/en/publication-detail/-/publication/1f0d8515-8dc0-4435-ba53-9570e47dbd51 (accessed on 28 December 2022).
7. Solbu, G. Frictions in the bioeconomy? A case study of policy translations and innovation practices. *Sci. Public Policy* **2021**, *48*, 911–920. [CrossRef]

8. Nilsson, M.; Zamparutti, T.; Petersen, J.E.; Nykvist, B.; Rudberg, P.; Mcguinn, J. Understanding Policy Coherence: Analytical Framework and Examples of Sector-Environment Policy Interactions in the EU. *Environ. Policy Gov.* **2012**, *22*, 395–423. [CrossRef]
9. Purwestri, R.C.; Hájek, M.; Hochmalová, M.; Palátová, P.; Huertas-Bernal, D.C.; García-Jácome, S.P.; Jarský, V.; Kašpar, J.; Riedl, M.; Marušák, R. The role of Bioeconomy in the Czech national forest strategy: A comparison with Sweden. *Int. For. Rev.* **2021**, *23*, 492–510. [CrossRef]
10. Kelleher, L.; Henchion, M.; O'Neill, E. Policy coherence and the transition to a bioeconomy: The case of Ireland. *Sustainability* **2019**, *11*, 7247. [CrossRef]
11. Singh, A.; Christensen, T.; Panoutsou, C. Policy review for biomass value chains in the European bioeconomy. *Glob. Transit.* **2021**, *3*, 13–42. [CrossRef]
12. OECD. *The Bioeconomy to 2030*; OECD: Paris, France, 2009; pp. 287–293. ISBN 9789264038530.
13. European Commission. *A Sustainable Bioeconomy for Europe: Strengthening the Connection between Economy, Society and the Environment*; European Commission: Brussels, Belgium, 2018; pp. 10–22. ISBN 9789279941450.
14. United Nations. THE 17 GOALS | Sustainable Development. Available online: https://sdgs.un.org/goals (accessed on 1 March 2022).
15. Fischer, K.; Stenius, T.; Holmgren, S. Swedish Forests in the Bioeconomy: Stories from the National Forest Program. *Soc. Nat. Resour.* **2020**, *33*, 896–913. [CrossRef]
16. Maier, D. The use of wood waste from construction and demolition to produce sustainable bioenergy—A bibliometric review of the literature. *Int. J. Energy Res.* **2022**, *46*, 11640–11658. [CrossRef]
17. Purwestri, R.C.; Hájek, M.; Šodková, M.; Sane, M.; Kašpar, J. Bioeconomy in the National Forest Strategy: A Comparison Study in Germany and the Czech Republic. *Forests* **2020**, *11*, 608. [CrossRef]
18. Wohlin, C.; Kalinowski, M.; Romero Felizardo, K.; Mendes, E. Successful combination of database search and snowballing for identification of primary studies in systematic literature studies. *Inf. Softw. Technol.* **2022**, *147*, 106908. [CrossRef]
19. Siddiqi, S.; Sharan, A. Keyword and Keyphrase Extraction Techniques: A Literature Review. *Int. J. Comput. Appl.* **2015**, *109*, 18–23. [CrossRef]
20. Onan, A.; Korukoğlu, S.; Bulut, H. Ensemble of keyword extraction methods and classifiers in text classification. *Expert Syst. Appl.* **2016**, *57*, 232–247. [CrossRef]
21. Henning, J. The Likert Scale. Available online: http://thefutureplace.typepad.com/the_future_place/2010/09/the-likert-scale-tarsk-14-things-all-researchers-should-know.html (accessed on 11 August 2022).
22. Sorrell, S. Improving the evidence base for energy policy: The role of systematic reviews. *Energy Policy* **2007**, *35*, 1858–1871. [CrossRef]
23. Miljand, M. Using systematic review methods to evaluate environmental public policy: Methodological challenges and potential usefulness. *Environ. Sci. Policy* **2020**, *105*, 47–55. [CrossRef]
24. Snyder, H. Literature review as a research methodology: An overview and guidelines. *J. Bus. Res.* **2019**, *104*, 333–339. [CrossRef]
25. Latvian Ministry of Agriculture. *Latvian Bioeconomy Strategy 2030*; Latvian Ministry of Agriculture: Riga, Latvia, 2018; pp. 5–22.
26. Ministry of Environmental Protection and Regional Development. National Waste Management Plan for 2021–2028. 2021. Available online: https://www.varam.gov.lv/lv/atkritumu-apsaimniekosanas-valsts-plans-2021-2028gadam-0 (accessed on 7 April 2022).
27. Zlaugotne, B.; Zihare, L.; Balode, L.; Kalnbalkite, A.; Khabdullin, A.; Blumberga, D. Multi-Criteria Decision Analysis Methods Comparison. *Environ. Clim. Technol.* **2020**, *24*, 454–471. [CrossRef]
28. Chakraborty, S. TOPSIS and Modified TOPSIS: A comparative analysis. *Decis. Anal. J.* **2022**, *2*, 100021. [CrossRef]
29. Ture, H.; Dogan, S.; Kocak, D. Assessing Euro 2020 Strategy Using Multi-criteria Decision Making Methods: VIKOR and TOPSIS. *Soc. Indic. Res.* **2019**, *142*, 645–665. [CrossRef]
30. Pachemska, T.A.; Lapevski, M.; Timovski, R. Analytical Hierarchical Process (AHP) method application in the process of selection and evaluation. In Proceedings of the UNITECH—International Scientific Conference, Online, 21–22 November 2014; pp. 373–380. Available online: https://www.researchgate.net/publication/276985609_ANALYTICAL_HIERARCHICAL_PROCESS_AHP_METHOD_APPLICATION_IN_THE_PROCESS_OF_SELECTION_AND_EVALUATION (accessed on 28 December 2022).
31. Krohling, R.A.; Pacheco, A.G.C. A-TOPSIS—An approach based on TOPSIS for ranking evolutionary algorithms. *Procedia Comput. Sci.* **2015**, *55*, 308–317. [CrossRef]
32. Balioti, V.; Tzimopoulos, C.; Evangelides, C. Multi-Criteria Decision Making Using TOPSIS Method Under Fuzzy Environment. Application in Spillway Selection. *Proceedings* **2018**, *2*, 637. [CrossRef]
33. Dolge, K.; Balode, L.; Laktuka, K.; Kirsanovs, V.; Barisa, A.; Kubule, A. *A Comparative Analysis of Bioeconomy Development in European Union Countries*; Springer Nature: Cham, Switzerland, 2022. [CrossRef]
34. Ministry of Regional Development and Local Government of Latvia. Latvian Sustainable Development Strategy 2030, Riga. 2010. Available online: http://polsis.mk.gov.lv/documents/3323 (accessed on 7 April 2022).
35. The Cabinet of Ministers. *Strategy of Latvia for the Achievement of Climate Neutrality by 2050*; The Cabinet of Ministers: Riga, Latvia, 2019; p. 50.
36. The Cabinet of Ministers. *Action Plan for the Transition to a Circular Economy 2020–2027*; The Cabinet of Ministers: Riga, Latvia, 2020.

37. Cross-Sectoral Coordination Center. *Latvian National Development Plan for 2021–2027*; Cross-Sectoral Coordination Center: Riga, Latvia, 2020.
38. Ministry of Economics. *Latvian National Energy and Climate Plan for 2021–2030*; Ministry of Economics: Riga, Latvia, 2020.
39. Ministry of Environmental Protection and Regional Development of the Republic of Latvia. Environmental Policy Guidelines 2021–2027. 2021. Available online: https://www.varam.gov.lv/lv/media/25691/download (accessed on 7 April 2022).
40. The Cabinet of Ministers. *Latvia's Adaptation to Climate Change Plan for the Period Until 2030*; The Cabinet of Ministers: Riga, Latvia, 2019.
41. The Cabinet of Ministers. *National Industrial Policy Guidelines for 2021–2027*; The Cabinet of Ministers: Riga, Latvia, 2021.
42. Ministry of Agriculture of Latvia. Latvian Food Producers Are Fully Capable of Meeting the Latvian Population's Demand for Food. Available online: https://www.zm.gov.lv/presei/latvijas-partikas-razotaji-pilniba-spej-nodrosinat-latvijas-iedzivotaj?id=12879 (accessed on 15 August 2022).
43. Korņilovs, G. Fisheries in Latvia. Available online: https://enciklopedija.lv/skirklis/31606 (accessed on 16 November 2022).
44. Ministry of Agriculture of Republic of Latvia. Characteristics of Fisheries Sector in Latvia. Available online: https://www.zm.gov.lv/en/zivsaimnieciba/#jump (accessed on 16 November 2022).
45. Benga, E. Development of fisheries in the coastal zone of the Baltic Sea and the Gulf of Riga (coastal fisheries). 2015. Available online: https://www.arei.lv/sites/arei/files/files/lapas/Zvejniecbas%20attstba%20Baltijas%20jras%20un%20Rgas%20jras%20la%20piekrastes%20josl%20piekrastes%20zveja.pdf (accessed on 28 December 2022).
46. The Ministry of Agriculture of Republic of Latvia. Agriculture in Latvia 2021. 2022. Available online: https://www.zm.gov.lv/en/lauksaimnieciba/#jump (accessed on 6 December 2022).
47. Ministry of Agriculture of Republic of Latvia. *Latvian Aquaculture Development Plan 2021–2027 (Project)*; Ministry of Agriculture of Republic of Latvia: Riga, Latvia, 2021.
48. Ministry of Agriculture of Republic of Latvia. *Fisheries Development Action Programme 2021–2027*; Ministry of Agriculture of Republic of Latvia: Riga, Latvia, 2021. Available online: https://www.zm.gov.lv/zivsaimnieciba/statiskas-lapas/ricibas-programma-zivsaimniecibas-attistibai-2021-2027-gadam?id=23594#jump (accessed on 6 December 2022).

MDPI

St. Alban-Anlage 66

4052 Basel

Switzerland

www.mdpi.com

Sustainability Editorial Office

E-mail: sustainability@mdpi.com

www.mdpi.com/journal/sustainability